AF326816

Advances in Industrial Control

Series editors

Michael J. Grimble, Glasgow, UK
Michael A. Johnson, Kidlington, UK

For further volumes:
http://www.springer.com/series/1412

Yiannis Boutalis · Dimitrios Theodoridis
Theodore Kottas · Manolis A. Christodoulou

System Identification and Adaptive Control

Theory and Applications of the Neurofuzzy and Fuzzy Cognitive Network Models

 Springer

Yiannis Boutalis
Dimitrios Theodoridis
Theodore Kottas
Department of Electrical
  and Computer Engineering
Democritus University of Thrace
Xanthi
Greece

Manolis A. Christodoulou
Kifisia
Greece

ISSN 1430-9491   ISSN 2193-1577   (electronic)
ISBN 978-3-319-06363-8   ISBN 978-3-319-06364-5   (eBook)
DOI 10.1007/978-3-319-06364-5
Springer Cham Heidelberg New York Dordrecht London

Library of Congress Control Number: 2014936844

Series Editors' Foreword

The series *Advances in Industrial Control* aims to report and encourage technology transfer in control engineering. The rapid development of control technology has an impact on all areas of control discipline. New theory, new controllers, actuators, sensors, new industrial processes, computer methods, new applications, new philosophies…, and new challenges. Much of this development work resides in industrial reports, feasibility study papers, and the reports of advanced collaborative projects. The series offers an opportunity for researchers to present an extended exposition of such new work in all aspects of industrial control for wider and rapid dissemination.

The so-called "intelligent control" movement is motivated by the idea that the types of systems that can be represented and analyzed by traditional mathematical approaches are limited in scope. Two particular system characteristics that engineers wish to accommodate, namely unknown (nonlinear) systems and "soft" information (expert knowledge, linguistic knowledge) led to the introduction of neural networks and fuzzy logic into the control engineer's toolkit.

These additions to the tools and techniques available to the control engineer can be followed in the publication lists of the *Advances in Industrial Control* monograph series and its sister series *Advanced Textbooks in Control and Signal Processing*. For neural network approaches, we can cite:

- *Neuro-control and Its Applications* by Sigeru Omatu, Marzuki Khalid and Rubiyah Yusof (ISBN 978-3-540-19965-6, 1995);
- *Neural Network Engineering in Dynamic Control Systems* edited by Kenneth J. Hunt, George R. Irwin and Kevin Warwick (ISBN 978-3-540-19973-1, 1995);
- *Adaptive Control with Recurrent High-order Neural Networks* by George A. Rovithakis and Manolis A. Christodoulou (ISBN 978-1-85233-623-3, 2000);
- *Nonlinear Identification and Control: A Neural Network Approach* by Guoping Liu (ISBN 978-1-85233-342-3, 2001);

and the widely used textbook:

- *Neural Networks for Modelling and Control of Dynamic Systems* by Magnus Norgaard, Ole Ravn, Niels K. Poulsen and Lars K. Hansen (ISBN 978-1-85233-227-3, 2000).

For the fuzzy-logic approaches, there are fewer entries, but we can cite:

- *Expert Aided Control System Design* by Colin Tebbutt (ISBN 978-3-540-19894-9, 1994);
- *Fuzzy Logic, Identification and Predictive Control* by Jairo Espinosa, Joos Vandewalle and Vincent Wertz (ISBN 978-1-85233-828-2, 2005); and
- *Advanced Fuzzy Logic Technologies in Industrial Applications* edited by Ying Bai, Hanqi Zhuang and Dali Wang (ISBN 978-1-84628-468-6, 2006).

In an attempt to gain even more flexibility for system representation, it is not surprising to find researchers combining neural networks with fuzzy-logic approaches to create a neuro-fuzzy modeling approach. Such a synthesis is developed, analyzed, and applied in the first part of this *Advances in Industrial Control* monograph, *System Identification and Adaptive Control: Theory and Applications of the Neurofuzzy and Fuzzy Cognitive Network Models* by Yiannis Boutalis, Dimitrios Theodoridis, Theodore Kottas, and Manolis Christodoulou. As can be seen from the above list of monographs, Manolis Christodoulou has previously written on *Recurrent High-order Neural Networks* with George A. Rovithakis. However, this newer monograph involves different authors and presents exciting and innovative developments for system identification and adaptive control using neuro-fuzzy models. Part II of the monograph travels in a different direction and reports how cognitive maps and fuzzy-logic concepts can be combined as fuzzy cognitive network models. Such models can be used in various ways to solve process control problems or control decision-making tasks. One strength of the monograph is the description of a number of simulation studies using benchmark nonlinear systems and case-study systems. The case studies include one laboratory motor experiment and some industrial power system problems. The examples are developed and presented to show the different features and characteristics of the neuro-fuzzy and fuzzy cognitive network identification and control methods.

In conclusion, the monograph reports and demonstrates new extensions and syntheses of the concepts of neural networks, fuzzy logic, and cognitive maps for the control of a wide range of plants and industrial processes. Control researchers and industrial engineers will find new concepts in this monograph and will undoubtedly appreciate the introductory explanatory sections and the many simulated examples as a route to comprehending these new developments in the "intelligent control" paradigm.

Glasgow, Scotland, UK   M. J. Grimble
                                                                               M. A. Johnson

Preface

Contemporary man-made engineering systems or systems associated with socioeconomical or biological processes can be particularly complex, characterized by possibly unknown nonlinearities, operating in uncertain environments. The complexity of these systems hinders the design of suitable control techniques, because the dynamical mathematical model required by "conventional" control approaches is unknown most of the times. Even in the case that the mathematical description is possible, there exist difficulties in the adaptation of the feedback controllers when the system is time varying with an unknown to the designer way. These drawbacks have led the recent research effort to "intelligent" techniques, seeking the development of new approximation models and control techniques that have the ability to learn and adapt to varying environmental conditions or internal dynamical behavior of the system.

Artificial neural networks and adaptive fuzzy systems constitute a reliable choice for modeling unknown systems, since they can be considered as universal approximators. In this sense, they can approximate any smooth nonlinear function to any prescribed accuracy in a convex compact region, provided that sufficient hidden neurons and training data or fuzzy rules are available. Recently, the combination of artificial neural networks and adaptive fuzzy systems has led to the creation of new approaches, fuzzy-neural, or *neuro-fuzzy* approaches that capture the advantages of both fuzzy logic and neural networks and intend to approach systems in a more successful way. Another modeling approach, that stems from fuzzy cognitive maps, is the creation of a *cognitive graph* capturing the causal relationships between the crucial variables of the system associated with the node values of the graph. Numerous applications and recent theoretical developments have shown that, under proper training, this model is capable of approximating the behavior of complex nonlinear systems in the engineering field and beyond.

This book is based on recent developments of the theory of Neuro-Fuzzy and the Fuzzy Cognitive Network (FCN) models and their potential applications. Its primary purpose is to present a set of alternative approaches, which would allow the design of:

- potent neuro-fuzzy system approximators and controllers able to guarantee stability, convergence, and robustness for dynamical systems with unknown nonlinearities
- appropriate cognitive graph modeling with guaranteed operational convergence and parameter identification algorithms.

The book is the outcome of the recent research efforts of its authors. Its is divided into two parts, each one being associated with each of these models. Part I is devoted to the Neuro-Fuzzy approach. It is based on the development of a new adaptive recurrent neuro-fuzzy approximation scheme, which is used for system identification and the construction of a number of controllers with guaranteed stability and robustness. The central idea in the development of the new approximation scheme is an alternative description of a classical dynamical fuzzy system, which allows its approximation by high-order neural networks (HONNs), a point that constitutes an innovative element of the presented scheme. The capabilities of the developed approximators and controllers are tested on a number of simulated benchmark problems and a real DC motor system.

Part II of the book is devoted to the FCN model. Fuzzy Cognitive Networks stem from fuzzy cognitive maps (FCM), initially introduced by Bart Kosko in 1986. An FCN is actually an operational extension of FCM which assumes, first, that it always reaches equilibrium points during its operation and second, it is in continuous interaction with the system it describes and may be used to control it. This way, the FCN is capable of capturing steady-state operational conditions of the system it describes and associates them with input values and appropriate weight sets. In the sequence, it stores the acquired knowledge in fuzzy rule-based databases, which can be used in determining subsequent control actions. Part II presents basic theoretical results related to the existence and uniqueness of equilibrium points in FCN, the adaptive weight estimation based on system operation data, the fuzzy rule storage mechanism, and the use of the entire framework to control unknown plants. The operation of the FCN framework is simulated and tested on a number of selected applications, ranging from a well-known benchmark problem to real-life potential projects, like hydroelectric power plant coordination and a novel scheme for optimal operation of a small-scale smart electric grid of renewable power plants based on real meteorological data. Through these examples, various aspects of the application of the FCN framework of operation are revealed.

Xanthi, Greece, March 2014

Yiannis Boutalis
Dimitrios Theodoridis
Theodore Kottas
Manolis A. Christodoulou

Contents

Part I
The Recurrent Neurofuzzy Model

Chapter 1
Introduction and Scope of Part I

1.1 Introduction

In our world, there are two principal objectives in the scientific study of the environment: we want to understand (identification) and to control. These two goals are in continuous interaction with each other, since deeper understanding allows firmer control, while, on the other hand, systematic application of scientific theories inevitably generates new problems which require further investigation, and so on.

In systems that change with time, the derivation of adaptive estimation and control methodologies is essential. It was the design of autopilots for high-performance aircraft which primarily motivated the research in adaptive control in the early 1950s (Ioannou and Sun 1996), while the words *adaptive systems* and *adaptive control* have been used as early as 1950 (Aseltine et al. 1958). A generic definition of adaptive systems, being systems that change to conform to new or changed circumstances, has been used to label approaches and techniques in a variety of areas. In this book, the following specific definition of adaptive control is used (Ioannou and Fidan 2006):

Adaptive control is the combination of a parameter estimator, which generates parameter estimates online, with a control law in order to control classes of plants whose parameters are completely unknown and/or could change with time in an unpredictable manner.

The choice of the parameter estimator, the choice of the control law, and the way they are combined leads to different classes of adaptive control schemes which are covered in this book.

For most engineering systems, there are two important information sources. First, we have sensors which provide numerical measurements of variables and second, we have human experts who provide linguistic instructions and descriptions concerning the system and its operation. Conventional engineering approaches can only make use of numerical information and have difficulty in incorporating linguistic information. On the other hand neural networks may exploit and learn from numerical data, while fuzzy logic systems are able to incorporate and handle linguistic information. Therefore, in the neurofuzzy approach presented in this book, we

Y. Boutalis et al., *System Identification and Adaptive Control*,
Advances in Industrial Control, DOI: 10.1007/978-3-319-06364-5_1,

try to relax this difficulty by using an underlying fuzzy description but keeping the least information coming from experts and replace it by neural networks learning from data.

System identification and control using *neurofuzzy* (NF) or *fuzzy-neural* network approaches has become a popular research topic in the past decades (Lin 1994; Li and Mukaidono 1995; Lin and Cunningham 1995; Kosmatopoulos and Christodoulou 1996; Lin and Lee 1996; Spooner and Passino 1996; Juang and Lin 1998; Jou et al. 1999; Zhang and Morris 1999; Lee and Teng 2000; Mitra and Hayashi 2000; Wu and Er 2000; Wang et al. 2001, 2002; Diao and Passino 2002; Mastorocostas and Theocharis 2002; Wang and Lee 2002; Zhou et al. 2002; Gao and Joo 2003; Lina and Wai 2003; Subudhi and Morris 2003; Kukolj and Levi 2004; Nounou and Passino 2004; Vieira et al. 2004; Leu et al. 2005; Yu and Zhang 2005; Adeli and Jiang 2006; Liu and Li 2006; Theocharis 2006; Chemachema and Belarbi 2007; Baruch et al. 2008; Sheikhzadeh et al. 2008; Leu et al. 2009). Many characteristics of the NF network contribute to this phenomenon. Some of them are, as compared to the general neural networks, faster convergence speed, and the combination of the adaptive learning capabilities from neural networks with the generality of representation from fuzzy logic (Zadeh 1965). Moreover, the NF network approach automates the design of fuzzy rules and makes the combinational learning of numerical data as well as expert knowledge expressed as fuzzy *if-then* rules possible. In contrast to the pure neural network or fuzzy system, the NF method possesses both of their advantages; it brings the low-level learning and computational power of neural networks into fuzzy systems, and provides the high-level human-like thinking and reasoning of fuzzy systems into neural networks (Kosko 1992; Lin 1994; Lin and Lee 1996).

Among the neural network structures, the recurrent ones (Rubio and Yu 2007; Freitag et al. 2011) and especially the Recurrent High-Order Neural Networks (RHONN) (Kosmatopoulos et al. 1995; Rovithakis and Christodoulou 2000) have been proven very successful because their recurrent nature makes them able to approximate dynamic systems, while the inclusion of high-order input terms allows a linear in respect to the parameters modeling, leading to identification schemes with guaranteed global minimum of the error estimation function. Recently, discrete-time RHONN new schemes of nonlinear systems identification, control, and state estimation, including stability and trajectory tracking analysis and real-time implementation have been proposed by Alanis, , and co-workers in (Alanis et al. 2010, 2011), where the NN weights learning is performed using Kalman filtering discrete-time schemes. Since the NF networks have many advantages over simple feedforward neural networks, and the recurrent neural networks are very potent approximators, it seems worth constructing a recurrent network based on a NF approach.

In this part of the book, we consider the adaptive control problem for nonlinear systems having the following form:

$$\dot{x} = f(x) + G(x) \cdot u \tag{1.1}$$

where the state $x \in R^n$ is assumed to be completely measured, the control u is in R^q, f is an unknown smooth vector field called the drift term and G is a matrix with rows containing the unknown smooth controlled vector fields with elements g_{ij}.

In the controller design based on the feedback linearization technique, the most commonly used control structure is $u = [G(x)]^{-1} \cdot [-f(x) + \upsilon]$ (for square systems, number of inputs equals with number of states), with υ being a new control variable. When the nonlinearities $f(x)$ and $G(x)$ are unknown, many adaptive control schemes have been developed (Sastry and Bodson 1989; Chen and Liu 1994; Wang 1994; Yesildirek and Lewis 1995; Spooner and Passino 1996; Ge et al. 1998), in which the unknown function $G(x)$ is usually approximated by a function approximator $\hat{g}_{ij}\left(x, W_g\right)$ (where W_g is an estimated weight or parameter matrix). Consequently, the estimate W_g must be such that $G(x)$ is nonsingular. Several attempts have been made to deal with such a problem, as follows:

1. choosing the initial parameter W_g (0) sufficiently close to the ideal value by offline training before the controller is put into operation (Chen and Liu 1994);
2. using projection algorithms to guarantee the estimate W_g inside a feasible set, in which $\hat{g}_{ij}\left(x, W_g\right) \neq 0$ [some a priori knowledge for the studied systems is required for constructing the projection algorithms, Sastry and Isidori (1989), Wang (1994), Spooner and Passino (1996), Ioannou and Fidan (2006)];
3. modifying the adaptive controller by introducing a sliding mode control portion to keep the control magnitude bounded (Yesildirek and Lewis 1995; Ge et al. 1998);
4. applying neural networks or fuzzy systems to approximate the inverse of $G(x)$ in Spooner and Passino (1996) and Sanner and Slotine (1992), which requires the upper bound of the first time derivative of $G(x)$ being known a priori.

By introducing a new method termed *parameter hopping*, a novel NF adaptive controller will be presented later in this book. The singularity issue mentioned above is completely avoided, and at the same time, the stability and control performance of the closed-loop system are guaranteed.

For unknown systems, in order to control the systems following the above scheme, the system has first to be identified by appropriate approximation model. We construct a neuro fuzzy approximation model by using recurrent high-order neural networks (RHONNs) and appropriately defined fuzzy dynamical systems. In the sequel, we present a brief introduction of RHONN and the adaptive fuzzy systems focusing mainly in the Mamdani centroid of area defuzzification procedure.

1.2 Recurrent High-Order Neural Networks

In spite of reported successful neural control applications, in the initial era of neural networks development there were not so many stability analysis in neurocontrol. To the best of our knowledge, there were a few published results, such as Poznyak et al. (1999) regarding the analysis of nonlinear systems controlled by dynamic

NNs. Among them, the works in (Rovithakis and Christodoulou 1994, 1995) utilize a particular version of recurrent HONNs to identify the nonlinear system and based on the NN model to calculate the control law.

In this section, we briefly introduce dynamic NNs in the form of the recurrent high-order neural networks (RHONNs) as introduced in Kosmatopoulos et al. (1995). Recurrent neural network (RNN) models (Kosmatopoulos and Christodoulou 1994, 1996; Kosmatopoulos et al. 1995; Rovithakis and Christodoulou 2000) are characterized by a two-way connectivity between units. This distinguishes them from feedforward neural networks, where the output of one unit is connected only to units in the next layer. In the simple case, the state history of each unit or neuron is determined by a differential equation of the form:

$$\dot{x}_i = a_i x_i + b_i \sum_j w_{ij} s_{f_j}(x) \tag{1.2}$$

where x_i is the state of the i-th neuron, a_i, b_i are real constants, w_{ij} is the synaptic weight connecting the j-th input to the i-th neuron, and s_{f_j} is the j-th input to the above neuron. Each s_{f_j} is either an external input or the state of a neuron passed through a sigmoid nonlinearity.

In a recurrent second-order neural network the total input to the neuron is not only a linear combination of the components s_{f_j}, but also of their products $s_{f_j} \cdot s_{f_k}$. Moreover, one can pursue along this line and include higher order interactions represented by triplets $s_{f_j} \cdot s_{f_k} \cdot s_{f_l}$, quadruplets, etc. This class of neural networks form is called recurrent high-order neural networks (RHONNs).

Consider now a RHONN consisting of n neurons and m inputs. The state of each neuron is governed by a differential equation of the form:

$$\dot{x}_i = a_i x_i + b_i \left[\sum_{p=1}^{L} w_{ip} \prod_{j \in I_p} s_j^{d_j(p)} \right] \tag{1.3}$$

where $[I_1, I_2, \ldots, I_L]$ is a collection of L nonordered subsets of $[1, 2, \ldots, m+n]$, w_{ip} are the (adjustable) synaptic weights of the neural network, and $d_j(p)$ are nonnegative integers. The state of the ith neuron is again represented by x_i, and $s = [s_1, s_2, \ldots, s_{n+m}]^T$ is the vector consisting of inputs to each neuron, defined by:

$$s = \begin{bmatrix} s_1 \\ \vdots \\ s_n \\ s_{n+1} \\ \vdots \\ s_{m+n} \end{bmatrix} = \begin{bmatrix} s(x_1) \\ \vdots \\ s(x_n) \\ s(u_1) \\ \vdots \\ s(u_m) \end{bmatrix} \tag{1.4}$$

where $u = [u_1, u_2, \ldots, u_m]^T$ is the external input vector to the network. The function $s(\cdot)$ is a monotone increasing, differentiable sigmoid function of the form:

$$s(x) = \alpha \frac{1}{1 + e^{-\beta x}} - \gamma \tag{1.5}$$

where α, β are positive real numbers and γ is a real number. In the special case that $\alpha = \beta = 1$, $\gamma = 0$, we obtain the logistic function, and by setting $\alpha = \beta = 2$, $\gamma = 1$, we obtain the hyperbolic tangent function; these are the sigmoid activation functions most commonly used in neural network applications. We now introduce the L-dimensional vector s_f, which is as:

$$s_f = \begin{bmatrix} s_{f_1} \\ s_{f_2} \\ \vdots \\ s_{f_L} \end{bmatrix} = \begin{bmatrix} \prod_{j \in I_1} s_j^{d_j(1)} \\ \prod_{j \in I_2} s_j^{d_j(2)} \\ \vdots \\ \prod_{j \in I_L} s_j^{d_j(L)} \end{bmatrix} \tag{1.6}$$

and hence the RHONN model in Eq. (1.3) is rewritten as:

$$\dot{x}_i = a_i x_i + b_i \left[\sum_{p=1}^{L} w_{ip} s_{f_p}(x) \right] \tag{1.7}$$

Moreover, if we define the adjustable parameter vector as:

$$w_{f_i} = b_i [w_{i1} \, w_{i2} \cdots w_{iL}]^T$$

then Eq. (1.3) and hence Eq. (1.7) becomes:

$$\dot{x}_i = a_i x_i + w_{f_i}^T s_f(x) \tag{1.8}$$

The vectors w_{f_i} represent the adjustable weights of the network, while the coefficients a_i, $i = 1, 2, \ldots, n$ are part of the underlying network architecture and are fixed during training. In order to guarantee that each neuron x_i is bounded input bounded state (BIBS) stable, we will assume that each $a_i < 0$.

The dynamic behavior of the overall network is described by expressing Eq. (1.8) in vector notation as:

$$\dot{x} = Ax + W_f s_f(x) \tag{1.9}$$

where $x = [x_1, x_2, \ldots, x_n]^T$ is the state vector, $W_f = \left[w_{f_1}, w_{f_2}, \ldots, w_{f_n} \right] \in R^{L+n}$ is the weight matrix and $A = diag\,[a_1, a_2, \ldots, a_n]$ is an $n \times n$ diagonal matrix. Since each $a_i < 0$, A is a stability matrix. Although it is not explicitly written, the vector s_f is a function of both the network state x_i and the external input u.

1.3 Functional Representation of Adaptive Fuzzy Systems

Fuzzy sets and systems have gone through substantial development since the introduction of fuzzy set theory by Zadeh (1965, 1968, 1971, 1975) about four decades ago. They have found a great variety of applications ranging from control engineering, qualitative modeling, signal processing, machine intelligence, decision making, motor industry, robotics, and so on (Sugeno 1985; Zimmermann 1991; Sugeno and Yasukawa 1993; Seker et al. 2003; Yu and Zhang 2005).

Following a similar approach with the proof of universal approximation ability of neural networks (Sanner and Slotine 1992), it has been shown (Wang and Mendel 1992) that a fuzzy system is capable of approximating any smooth nonlinear function over a convex compact region. Fuzzy basis function-based fuzzy systems are used to represent those unknown nonlinear functions. The parameters of the fuzzy systems, including membership functions that characterize linguistic values in fuzzy rules, are updated according to some adaptive laws which are derived based on Lyapunov stability analysis (Jang 1993; Wang 1994; Chen et al. 1996; Cho and Wang 1996; Ordonez et al. 1997; Hojati and Gazor 2002; Feng and Chen 2005; Labiod et al. 2005; Zhoua et al. 2005; EL-Hawwary and Elshafei 2006).

The performance, complexity, and adaptive law of an adaptive fuzzy system representation can be quite different depending upon the type of the fuzzy system (Mamdani 1976 or Takagi and Sugeno 1985). It also depends upon whether the representation is linear or nonlinear in its adjustable parameters. Adaptive fuzzy controllers depend also on the type of the adaptive fuzzy subsystems they use. Suppose that the adaptive fuzzy system is intended to approximate the nonlinear function $f(x)$. In the Mamdani type form, the following linear in the parameters fuzzy logic model is used (Wang 1994, Passino and Yurkovich 1998):

$$f(x) = \sum_{l=1}^{M} \theta_l \xi_l(x) = \theta^T \xi(x) \tag{1.10}$$

where M is the number of fuzzy rules, $\theta = (\theta_1, ..., \theta_M)^T$, $\xi(x) = (\xi_1(x), ..., \xi_M(x))^T$ and $\xi_l(x)$ is the fuzzy basis function defined by

$$\xi_l(x) = \frac{\prod_{i=1}^{n} \mu_{F_i^l}(x_i)}{\sum_{l=1}^{M} \prod_{i=1}^{n} \mu_{F_i^l}(x_i)} \tag{1.11}$$

θ_l are adjustable parameters, and $\mu_{F_i^l}$ are given membership functions of the input variables (can be Gaussian, triangular, or any other type of membership functions).

In Tagaki-Sugeno formulation, $f(x)$ is given by

$$f(x) = \sum_{l=1}^{M} g_l(x)\xi_l(x) \tag{1.12}$$

where $g_l(x) = a_{l,0} + a_{l,1}x_1 + \ldots + a_{l,n}x_n$, with $x_i, i = 1 \ldots n$ being the elements of vector x and $\xi_l(x)$ being defined in (1.11). According to Passino and Yurkovich (1998), (1.12) can also be written in the linear to the parameters form, where the adjustable parameters are all $a_{l,i}, l = 1 \ldots M, i = 1 \ldots n$.

From the above definitions it is apparent that in both, Mamdani and Tagaki-Sugeno, forms the success of the adaptive fuzzy system model in approximating the nonlinear function $f(x)$ depend on the careful selection of the fuzzy partitions of input and output variables. Also, the selected type of the membership functions and the proper number of fuzzy rules contribute to the success of the adaptive fuzzy system. This way, any adaptive fuzzy or NF approach, following a linear in the adjustable parameters formulation becomes vulnerable to initial design assumptions related to the fuzzy partitions and the membership functions chosen. In the following chapters, this drawback is largely overcome by using the concept of weighted indicator functions, which are in the sequel approximated by high-order neural network approximators (HONNs). In this way, there is not any need for initial design assumptions related to the membership values and the fuzzy partitions of the **if** part.

More precisely, the underlying functional representation, that is used in this book, is connected with the fuzzy model and the centroid of area defuzzification procedure of Mamdani type which is briefly explained in the following subsection.

1.3.1 COA Defuzzification and Indicator Functions

The centroid of area (COA) method is the most prevalent and physically appealing of all the defuzzification methods. Final crisp output when using centroid defuzzification is equal to a weighted average of centroid of consequents MFs and is given as (Chai et al. 2009):

$$f(x) = \frac{\sum_{i=1}^{r} \omega_i(x) \, v_i \bar{x}_{f_i}}{\sum_{i=1}^{r} \omega_i(x) \, v_i} \tag{1.13}$$

where ω_i is the firing strength of the i-th rule, v_i is the area of the consequent MFs of i-th rule and $\bar{x}_{f_i}$ is the i-th fuzzy center of the consequent MFs.

In a more detailed form, we can rewrite Eq. (1.13) as:

$$f(x) = \frac{\omega_1(x) \, v_1}{\sum_{i=1}^{r} \omega_i(x) \, v_i} \bar{x}_{f_1} + \cdots + \frac{\omega_r(x) \, v_r}{\sum_{i=1}^{r} \omega_i(x) \, v_i} \bar{x}_{f_r}. \tag{1.14}$$

Now, if we choose an indicator function as: $I_i(x) = \omega_i(x) v_i$ then Eq. (1.14) can be written as:

$$f(x) = \frac{I_1(x)}{\sum\limits_{i=1}^{r} I_i(x)} \bar{x}_{f_1} + \cdots + \frac{I_r(x)}{\sum\limits_{i=1}^{r} I_i(x)} \bar{x}_{f_r}. \qquad (1.15)$$

Hence, if we define $\left(I'\right)_i(x) = \dfrac{I_i(x)}{\sum\limits_{i=1}^{r} I_i(x)}$ as the weighted indicator function (WIF), the system output finally is given by the following form:

$$f(x) = \sum_{i=1}^{r} \left(I'\right)_i \cdot \bar{x}_{f_i}. \qquad (1.16)$$

Equation (1.16) provides the functional representation of the fuzzy system in terms of WIF. This representation will be used in the next chapter for deriving the proposed NF approach.

1.4 Outline of Adaptive Dynamic Identification and Control Based on the Neurofuzzy Mode

System modeling has played an important role in many engineering fields such as control, pattern recognition, communications, and so on. The main idea in conventional approaches is to find a global function of systems based on mathematical tools. However, it is well known that these methods have been found to be unsatisfactory in coping with ill-defined and uncertain systems. In order to circumvent these problems, model-free approaches using either fuzzy logic or neural networks have been proposed. Functionally, a fuzzy system or a neural network can be described as a function approximator. The capability of sufficiently complex multilayered feedforward networks to approximate an unknown mapping $f : \Re^r \to \Re^s$ arbitrarily well first has been investigated by Cybenko (1989), Funahashi (1989), Hornik et al. (1989) (all for sigmoid hidden layer activation functions).

The problem of identifying complex nonlinear systems when we have less information for the model or when we consider the controlled plant as a blackbox, something common in many physical systems, may be resolved with the help of neural network or fuzzy inference systems or a combination of them, leading to NF approaches. This is due to the universal approximation abilities of both neural networks and fuzzy inference systems (Hornik et al. 1989; Passino and Yurkovich 1998; Wang 1994). For the engineers, the stability issues are very important to be ensured before they move further and apply their NF networks to real systems. Therefore, first of all the derived identification algorithms have to be proven to be Lyapunov stable. Under these specifications many researchers such as, Ioannou and Fidan (2006), Kosko (1997), Lewis et al. (1999), Wang (1994), gave the very basics and most useful information to system identification, which is the first step in the subsequent control of unknown nonlinear dynamical systems with various nonlinearities.

In existing NF networks, in early 1990s, almost all of these systems are trained by the backpropagation (BP) algorithm (Jang 1993; Li and Mukaidono 1995; Lin 1995; Lin and Cunningham 1995; Cho and Wang 1996; Juang and Lin 1998). The major drawback of the BP algorithm is that it arises from a nonconvex optimization problem, and therefore it presents slow speed of learning and entrapment in local minima. Therefore, the optimal solution is not guaranteed and BP-powered neural approximators can not be implemented in real-time processes. Nowadays, although some special fuzzy-neural networks (fuzzy neurons and fuzzy weights) have been presented, the typical approach is to build standard neural networks which are designed to approximate a fuzzy system through the structure of neural networks, a methodology that was chosen in developing the material of part I of this book.

Suppose that a nonlinear dynamical system assumes the general form:

$$\dot{x} = f(x, u), \tag{1.17}$$

where f is unknown. The research by Kosmatopoulos et al. (1995); Kosmatopoulos and Christodoulou (1996), Rovithakis and Christodoulou (2000) introduced RHONNs to approximate f, where RHONNs ensure exponential error convergence to zero. In the next chapters, we enhance this approximator to include fuzzy logic and interweave it with RHONN to generate a new NF approximator (Boutalis et al. 2009; Theodoridis et al. 2009a, 2012). In this approximator, every HONN approximates a group of fuzzy rules associated with a center that has resulted from the fuzzy partitioning of the system output variables, or alternatively, approximates a WIF (mentioned in Sect. 1.3.1).

Figure 1.1 shows the overall scheme of the proposed NF algorithm which approximates the plant model (1.17) based on state variables measurement and input signals. The arrow that passes through the NF model is indicative of the fact that the output error is used to train the NF network. The NF network's input is the plant's output (through the sigmoidal terms) and the desired NF network output is the plant's output. The error difference between the actual input of the plant and the output of the NF network is to be minimized and can be used to train the NF network. Once an identification NF model of the plant is available, this model can be used for the design of the controller, as pointed out below.

The identification procedure can be generally divided into two phases. First the training phase, where the whole procedure is repeated consequently for several epochs using training inputs usually random values in a certain interval. Second the testing phase, where a certain input signal (depending on the bounds of the training input signal) is applied in conjunction with a fixed weight matrix (optimal weight values extracted from the previous phase) in order to test the accuracy of the approximator in reproducing the behavior of the plant, in probably unknown data.

In conventional control approaches, most of the control schemes are usually devised assuming exact knowledge of (1.17). When f is unknown, appropriate adaptive identification schemes have first to be applied. Adaptive control has a rich

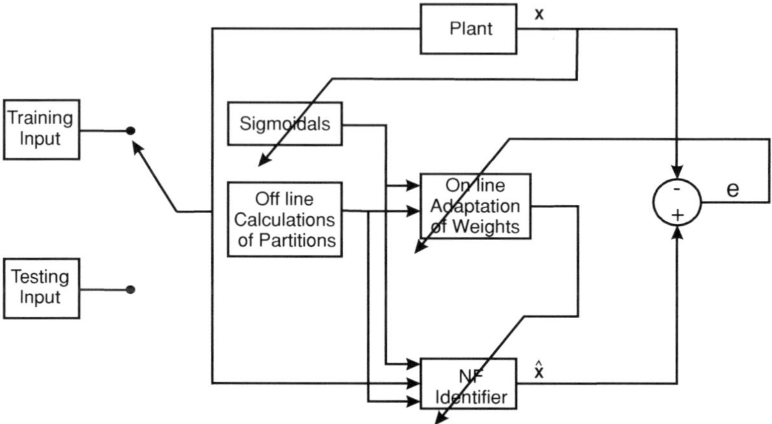

Fig. 1.1 Overall scheme of the proposed identification scheme.

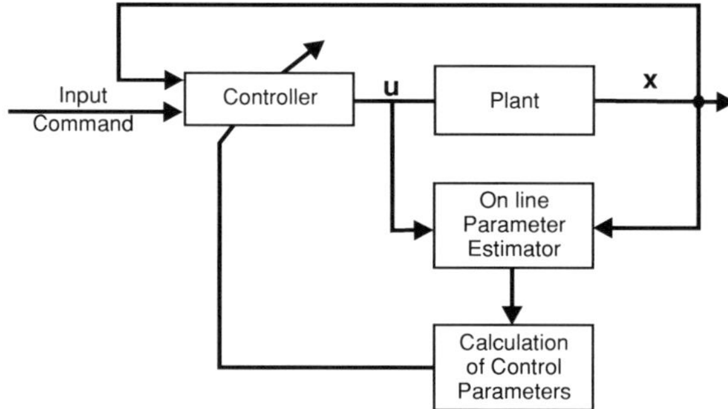

Fig. 1.2 Indirect adaptive control structure

literature full of different techniques for design, analysis, performance, and applications. Indicative lists of papers can be found in several pioneering and survey papers (Naredra and Valavani 1979; Astrom 1983; Ortega and Yu 1989; Kokotovic and Arcak 2001) and books (Krstic et al. 1995; Lin 1994; Rovithakis and Christodoulou 2000; Ioannou and Sun 1996), with the research going on until nowdays. In adaptive control literature, the indirect and alternatively the direct control approaches are used. In the indirect adaptive control schemes the dynamics of the system are first identified and then a control input is generated according to the certainty equivalence principle. The basic structure of indirect adaptive control as presented by Ioannou and Fidan (2006), is shown in Fig. 1.2.

Within the work of this book, we present a new NF approach for the indirect adaptive control of square (number of inputs = number of states) unknown nonlinear systems assuming only parameter uncertainty, (Theodoridis et al. 2008; Boutalis et al. 2009). Moreover, according to a posterior work by Theodoridis et al. (2010a) the NF indirect control of general affine in the control nonlinear dynamical systems,

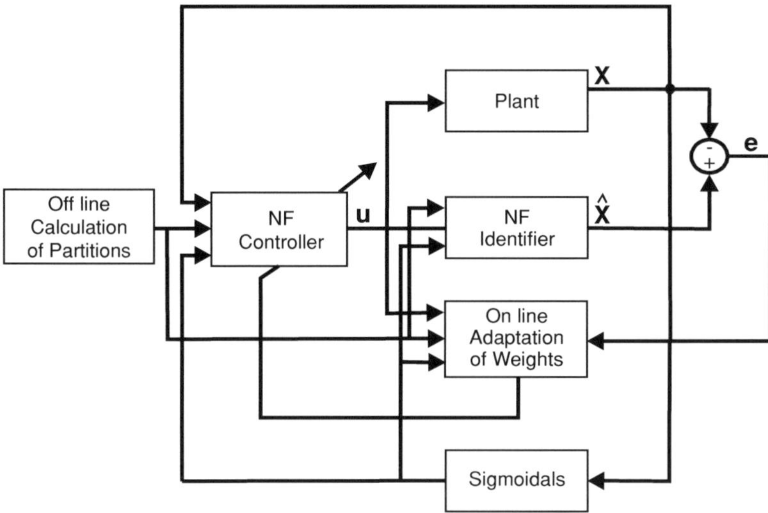

Fig. 1.3 Overall scheme of the proposed indirect adaptive NF control system

assuming both parameter and dynamic uncertainties is also treated. The proposed indirect scheme comprises two interrelated phases: first, the identification of the model and second the suitable control of the plant in order to cope with uncertainties.

In this approach, referred to as indirect adaptive F-RHONN control and is further developed in Chap. 3, the plant parameters are estimated online except from the elements of the matrices of centers, which are considered constant and, as it is shown later in Sect. 2.1, contain centers of output membership functions. The basic structure of the indirect adaptive recurrent NF controller is described in Fig. 1.3.

Since we do not have a desired output for the NF controller, the error e between the plant and the NF model is taken into account in order to adapt the weights of the NF approximator and, through it, the controller. The fuzzy centers of the plant output are determined offline before the NF controller starts to work. After that, a NF network is first trained to provide a model of the nonlinear plant. This can be used in parallel with the plant, with errors at the plant output taken into account from its NF model. Our NF network is considered as a dynamic one, by simply connecting the past NF outputs as inputs to the NF network, thus making the NF network a very complicated and highly nonlinear dynamical system. The connections do not need to be only serial, parallel, or feedback. Combinations of the three types are also permitted.

Similar to the static multilayer networks, the synaptic weights are adjusted according to a gradient descent like rule, arising after a Lyapunov stability analysis. The performance of indirect adaptive control schemes depends on the accuracy of the plant model estimation. Therefore, an initial offline training is often required before applying the adaptive control procedure, something that may be considered as a drawback in some applications.

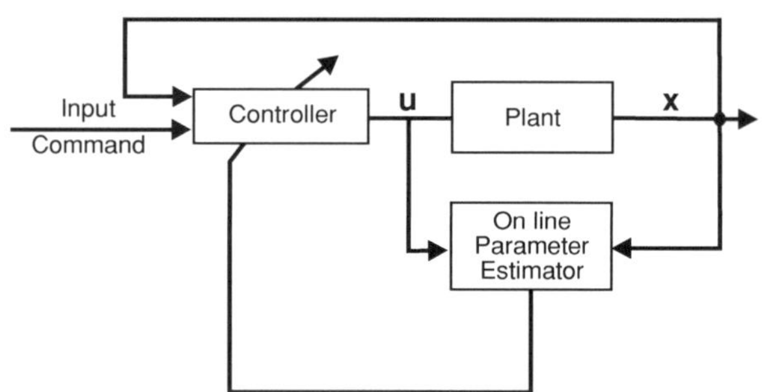

Fig. 1.4 Direct adaptive control structure

In direct adaptive control approaches the requirement of initial offline training is no longer so necessary. Figure 1.4 shows the structure of a direct adaptive control scheme as presented by Ioannou and Fidan (2006). In this case, the controller parameters are generated directly by the online estimator. In general, the ability to parameterize the plant model with respect to the desired controller parameters is what gives us the choice to use the direct adaptive control approach.

In the last decades, several NN, fuzzy, NF, or FN control approaches have been proposed based on Lyapunovs stability theory (Lewis et al. 1996; Polycarpou 1996; Wang 1994; Ge et al. 1998, 1999, 2001; Sanner and Slotine 1992; Rovithakis and Christodoulou 1995; Ciliz 2007; McLain et al. 1999; Ioannou and Tsakalis 1986; Spooner and Passino 1996; Yang 2004; Belarbi and Chemachema 2007; Zhang and Morris 1999; Zhou et al. 2002; Vieira et al. 2004; Wang et al. 2002, 2007; Subudhi and Morris 2003; Leu et al. 2005; Chemachema and Belarbi 2007). One main advantage of these schemes is that the adaptive laws were derived based on Lyapunov synthesis and, therefore, guarantee the stability of continuous time systems and the robustness of the control schemes, without the requirement for an initial offline training.

In this book, as concerning the direct adaptive control schemes and motivated by Rovithakis and Christodoulou (2000) a HONN-based NF controller is proposed (Theodoridis et al. 2009b) and used for the direct regulation of nonlinear dynamical systems under the presence of modeling errors (Theodoridis et al. 2010b, 2011b) or model order problems (Theodoridis et al. 2011a). First, the existence of disturbance is assumed and expressed as modeling error terms depending on both input and system states as well as to a not necessarily known constant value (Theodoridis et al. 2011b). Then, a robustifying analysis of the method is presented when a NF direct control scheme is applied. The proposed NF approximator may also assume a smaller number of states than the original unknown model as presented in Theodoridis et al. (2011a). Also, a direct adaptive NF controller is proposed, in order to address the adaptive tracking problems for a class of affine nonlinear systems which are exactly input–output linearizable by nonlinear state feedback.

In these approaches, referred to as *direct adaptive F-RHONN control schemes*, the control law parameters are estimated online except from the centers of the output membership function partitions of vector fields f and g_i, which are initially determined offline. Moreover, it is assumed that the actual system is parameterized using these parameters and this is an essential part of direct control schemes. The

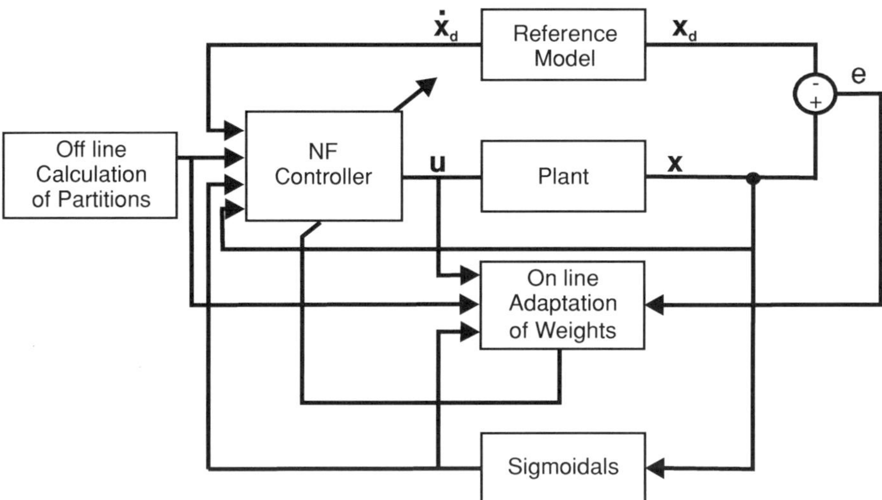

Fig. 1.5 Overall scheme of the proposed direct adaptive NF control system

basic structure of the proposed *direct adaptive F-RHONN controller* is described in
Fig. 1.5.

As we can see in Fig. 1.5, the inputs to the NF controller include the previous
reference inputs, the constant matrix of fuzzy partition centers, the weight updating
matrix, the sigmoidal vector or matrix terms, and the error between the actual plant
and the reference model output. The online parameter estimator is designed based on
the tracking error to provide the direct online weights estimate of the optimal weights
at each time t, by processing the plant input u and the output x. The weights' estimates are then used in the control law without intermediate calculations. In general,
the ability to parameterize the plant model with respect to the desired controller
parameters is what gives us the choice to use the direct adaptive control approach.

1.5 Goals and Outline of Part I

The main goals of Part I of this book are the following:

1. The presentation of a new NF approximator scheme for the identification of
   dynamical unknown nonlinear systems which presents the following distinct
   advantages in comparison with other approximation models of the literature.

   a. The union between artificial neural networks, in the form of high-order neural
      networks, and fuzzy logic in order to generate a more powerful and general
      approximator.

b. From the fuzzy logic aspect, the reduction of strong requirements concerning the careful selection of the fuzzy partitions of input and output variables, the selected type of the membership functions and the proper number of fuzzy rules which contribute to the success of the adaptive fuzzy system.

c. From the neural network aspect, the alternative approximation of weighted indicator functions with the help of multiple high-order neural networks.

2. Based on the proposed NF approximator to address several problems of direct and indirect control as well as to theoretically solidify their behavior under different uncertainties and errors during the modeling procedure.
3. The alternative use of a novel method in order to assure the existence of the control signal and the robustness of the closed-loop system.
4. The application of the derived NF controllers on a number of selected well-known and benchmark systems demonstrating various aspects of their functionality.

The material of Part I is deployed in several chapters as follows: In Chap. 2, a detailed description of the *neurofuzzy* algorithm is presented and its approximation capabilities for continuous time general form of unknown nonlinear systems are discussed. In Chap. 3, we concentrate on indirect control of multivariable systems with parametric and dynamic uncertainties. Also, in Chaps. 4 and 5, direct control schemes are presented for multi variable multi-input systems as well as for SISO systems in Brunovsky form, where several modeling variations are examined. Finally, in Chap. 6, we investigate how our approach of fuzzy-recurrent high-order neural networks performs on selected applications.

More specifically, in Chap. 2 the identification problem is discussed. This is resolved by choosing an appropriate identification model and adjusting its parameters according to some adaptive law so that the response of the model to an input signal (or a class of input signals) approximates the response of the real system for the same input. A fuzzy-recurrent high-order neural network (F-RHONN) is proposed as an approximation model. High-order neural networks (HONN) are expansions of the first-order Hopfield and Cohen-Grossberg that allow high-order connections between neurons. The underlying fuzzy model is of Mamdani type assuming a standard defuzzification procedure such as centroid of the area or weighted average. In the sequence, learning laws are employed such as gradient descent, pure least squares and robust algorithms, which ensure that the identification error converges to zero exponentially fast or to a residual set when a modeling error exists.

There are two core ideas in the proposed method, which differentiate it from the already existing ones in the international literature: (i) A number of high-order neural networks are specialized to work around some fuzzy output membership function centers, which leads to the separation of the system to neurofuzzy subsystems and (ii) the introduction of a novel method called *parameter hopping*, which replaces the known from the literature *projection method*. It is used in order to restrict the weights and to avoid the drifting of their values to infinity. Moreover, compared with other known methods such as RHONNs, it is shown to present better and faster approximation of the system.

In Chap. 3, the indirect adaptive regulation of states to zero of affine in the control unknown nonlinear dynamical systems with multiple input and states (MIMS) under the presence of dynamical and parametric uncertainties, is discussed. More specifically, initially it is assumed that the unknown system is precisely approximated by the mathematical structure of the NF controller (complete model matching) where the only unknowns are the network weights, which are estimated online by the adaptive approximator, or the generated adaptive controller. This assumption is, however, not sufficient for the modeling and control of any unknown nonlinear system. The presence of other errors and uncertainties, as regarding the modeling of the unknown system, influence inevitably the robustness of the proposed controllers and require the theoretical study and proof of the operation limits and the best parameter values of the proposed controllers. The existence of the control signal is always assured by imposing weight restriction conditions in the weight updating laws. This is is done in a unique way (as it was described in Chap. 2) that replaces the already known method of projection with the proposed hopping of the parameters in a linear direction perpendicular to some hyperplane in the subspaces of the estimated weights. Another important step of the presented approach is the reinforcement of the control signal, which is achieved after the addition of an auxiliary signal. This auxiliary signal includes as many frequencies as the parameters of the proposed model. This way, any possibility of infinite oscillation due to hopping is removed.

In Chap. 4, the direct adaptive regulation and tracking problems of unknown SISO nonlinear dynamical Brunovsky-type systems are considered, taking into account the modeling error effects. Since the plant (under study) is considered unknown, a special form of a Brunovky-type NF dynamical systems (NFDS) is proposed for its approximation. The existence of disturbance expressed as modeling error terms depending on input and system states plus a not necessarily known constant value (nonzero constant signal), is assumed. The theoretical development of this chapter is combined with a sensitivity analysis of the closed-loop, which provides a comprehensive and rigorous analysis of the stability properties. Once again it is proved, by the help of the Lyapunov stability approach, that under certain assumptions, which should apply, the system states are driven to zero or at least to a small region around zero. The existence and the boundedness of the control signal is always assured by employing the method of parameter hopping and incorporating it in the weight updating laws. In the trajectory tracking problem, a hybrid control scheme that combines an adaptive linear recurrent term with a sliding mode term, improves the system performance by compressing the external disturbances effect and the approximation errors. Similar to the regulation problem, when the NF model is not ideal and modeling errors appear, the proposed adaptive controller can at least guarantee the uniform ultimate boundedness of the tracking error to a small set of values around zero.

In Chap. 5, we extend the analysis made for SISO to MIMO systems and present the direct adaptive control of unknown multivariable multi-input affine in the control nonlinear systems and analyze robustness issues. More specifically, the unknown plant is modeled by the Neurofuzzy (NF) approximator, which is of known structure considering the neglected nonlinearities. The development is combined with a sensitivity analysis of the closed-loop in the presence of modeling imperfections and provides a comprehensive and rigorous analysis showing that the adaptive regulator can guarantee the convergence of states to zero or at least uniform ultimate boundedness of all signals in the closed-loop when a not-necessarily known modeling error is applied. A special case of modeling imperfection, which is treated separately, is the model order problem, where the oder of the system is considered unknown and the NF model assumes lower than the system order and is constructed by omitting some states. Finally, the state trajectory tracking problem is considered and analyzed, assuming always a NF approximator with various modeling imperfections. In the controllers presented, the existence and boundedness of the control signal is always assured by employing the method of "modified parameter hopping," which appears in the weight updating laws and transforms the controllers to hysteresis switching ones, avoiding theoretical issues concerning the existence of solutions of the dynamic equations due to hopping.

Finally, in Chap. 6, we present selected applications of adaptive control methods based on the NF model. Namely, we present NF-based control for trajectory tracking of robotic manipulators, simulations on chemical bioreactor based on the NF modeling and we report on the implementation performed in ACS lab for the regulation of a DC motor using the proposed NF indirect control algorithm of Chap. 3.

In the robotic example, the proposed control scheme uses a NF model to estimate system uncertainties. The function of robot system dynamics is first modeled by a fuzzy system, which in the sequel is approximated by a combination of HONNs. The overall representation is linear in respect to the unknown NN weights leading to weight adaptation laws that ensure stability and convergence to unique global minimum of the error functional. Due to the adaptive NF modeling, the proposed controller is independent of robot dynamics, since the free parameters of the NF controller are adaptively updated to cope with changes in the system and the environment. Adaptation laws for the network parameters are derived, which ensure network convergence and stable control. A weight hopping technique is also introduced to ensure that the estimated weights stay within prespecified bounds.

In the DC motor implementation, the validity and potency of the proposed NF scheme were tested in the regulation of a real separately excited DC motor. The control algorithm used, is the indirect one, presented in Chap. 3. The hardware testbed was constructed in the ACSL lab and some of its results are reported here. The experimental results indicate that the developed controller work well. Finally, in this chapter we present the chemical bioreactor application in order to illustrate in one more example the potency of our method.

References

Adeli, H., & Jiang, X. (2006). Dynamic fuzzy wavelet neural network model for structural system identification. *Journal of Structural Engineering, 132*, 102–111.

Alanis, A. Y., Sanchez, E. N., Loukianov, A. G., & Perez-Cisneros, M. A. (2010). Real-time discrete neural block control using sliding modes for electric induction motors. *IEEE Transactions on Control Systems Technology, 18*, 11–21.

Alanis, A. Y., Sanchez, E. N., Loukianov, A. G., & Perez-Cisneros, M. A. (2011). Real-time neural-state estimation. *IEEE Transactions on Neural Networks, 22*, 497–505.

Aseltine, J. A., Mancini, A. R., & Sartune, C. W. (1958). A survey of adaptive control systems. *IEEE Transactions on Automatic Control, 3*, 102–108.

Astrom, K. J. (1983). Theory and applications of adaptive control: a survey. *Automatica, 19*, 471–486.

Baruch, I. S., Lopez, R. B., Guzman, J. L. O. & Flores, J. M. (2008). A fuzzy-neural multi-model for nonlinear systems identification and control. *Fuzzy Sets and Systems, 159*(20), 2650–2667.

Belarbi, K., & Chemachema, M. (2007). Stable direct adaptive neural network controller with a fuzzy estimator of the control error for a class of perturbed nonlinear systems. *IET Control Theory and Applications, 1*, 1398–1404.

Boutalis, Y. S., Theodoridis, D. C., & Christodoulou, M. A. (2009). A new neuro fds definition for indirect adaptive control of unknown nonlinear systems using a method of parameter hopping. *IEEE Transactions on Neural Networks, 20*, 609–625.

Chai, Y., Jia, L., & Zhang, Z. (2009). Mamdani model based adaptive neural fuzzy inference system and its application. *International Journal of Information and Mathematical Sciences, 5*, 22–29.

Chemachema, M., & Belarbi, K. (2007). Robust direct adaptive controller for a class of nonlinear systems based on neural networks and fuzzy logic system. *International Journal on Artificial Intelligence Tools, 16*, 553–560.

Chen, B. S., Lee, C. H., & Chang, Y. C. (1996). Tracking design of uncertain nonlinear siso systems: adaptive fuzzy approach. *IEEE Transactions on Fuzzy Systems, 4*, 32–43.

Chen, F. C., & Liu, C. C. (1994). Adaptively controlling nonlinear continuous-time systems using multilayer neural networks. *IEEE Transactions on Automatic Control, 39*, 1306–1310.

Cho, K. B., & Wang, B. H. (1996). Radial basis function based adaptive fuzzy systems and their applications to system identification and prediction. *Fuzzy Sets and Systems, 83*, 325–339.

Ciliz, M. (2007). Adaptive backstepping control using combined direct and indirect adaptation. *Circuits, Systems and Signal Processing, 26*, 911939.

Cybenko, G. (1989). Approximation by superpositions of a sigmodial function. *Mathematics of Control, Signals and Systems, 2*, 303–314.

Diao, Y., & Passino, K. M. (2002). Adaptive neural/fuzzy control for interpolated nonlinear systems. *IEEE Transactions on Fuzzy Systems, 10*, 583–595.

EL-Hawwary, M. & Elshafei, A. (2006). Robust adaptive fuzzy control of a two-link robot arm. *International Journal of Robotics and Automation,21*, 266–272.

Feng, G., & Chen, G. (2005). Adaptive control of discrete-time chaotic systems: a fuzzy control approach. *Chaos, Solitons and Fractals, 23*, 459467.

Freitag, S., Graf, W., & Kaliske, M. (2011). Recurrent neural networks for fuzzy data. *Integrated Computer-Aided Engineering, 18*, 265–280.

Funahashi, K. (1989). On the approximate realization of continuous mappings by neural networks. *Neural Networks, 2*, 183–192.

Gao, Y., & Joo, M. (2003). Online adaptive fuzzy neural identification and control of a class of mimo nonlinear systems. *IEEE Transactions on Fuzzy Systems, 11*, 462–477.

Ge, S. S., Lee, T. H., & Harris, C. J. (1998). *Adaptive neural network control of robotic manipulators.* London: World Scientific.

Ge, S. S., Hang, C. C., & Zhang, T. (1999). Direct method for robust adaptive nonlinear control with guaranteed transient performance. *Systems and Control Letters, 37*, 275–284.

Ge, S. S., Hang, C. C., Lee, T. H., & Zhang, T. (2001). *Stable adaptive neural network control.* Norwell, MA: Kluwer.

Hojati, M., & Gazor, S. (2002). Hybrid adaptive fuzzy identification and control of nonlinear systems. *IEEE Transactions on Fuzzy Systems, 10*, 198–210.

Hornik, K., Stinchcombe, M., & White, H. (1989). Multilayer feedforward networks are universal approximators. *Neural Networks, 2*, 359–366.

Ioannou, P., & Fidan, B. (2006). *Adaptive control tutorial.* SIAM, Advances in Design and Control Series.

Ioannou, P. A., & Sun, J. (1996). *Robust adaptive control*. Englewood Cliffs, NJ: Prentice-Hall.

Ioannou, P. A. & Tsakalis, K. S. (1986). A robust direct adaptive controller. *IEEE Transactions on Automatic Control, AC-31*(11), 10331043.

Jang, J. S. R. (1993). Anfis: adaptive-network-based fuzzy inference system. *IEEE Transactions on Systems, Man and Cybernetics, 23*, 665–684.

Jou, I. C., Chang, C. J., & Chen, H. K. (1999). A hybrid neuro-fuzzy system for adaptive vehicle separation control. *Journal of VLSI Signal Processing,21*, 15–29.

Juang, C. F., & Lin, C. T. (1998). An on-line self-constructing neural fuzzy inference network and its applications. *IEEE Transactions on Fuzzy Systems, 6*, 12–32.

Kokotovic, P. V., & Arcak, M. (2001). Constructive nonlinear control: a historical perspective. *Automatica, 37*, 637–662.

Kosko, B. (1992). *Neural networks and fuzzy systems*. Englewood Cliffs, NJ: Prentice-Hall.

Kosko, B. (1997). *Fuzzy engineering*. Upper Saddle River, NJ: Prentice Hall.

Kosmatopoulos, E. B., & Christodoulou, M. A. (1994). Filtering, prediction, and learning properties of ece neural networks. *IEEE Transactions on Systems, Man, and Cybernetics, 24*, 971–981.

Kosmatopoulos, E. B., & Christodoulou, M. A. (1996). Recurrent neural networks for approximation of fuzzy dynamical systems. *International Journal of Intelligent Control and Systems, 1*, 223–233.

Kosmatopoulos, E. B., Polycarpou, M. M., Christodoulou, M. A., & Ioannou, P. A. (1995). High-order neural network structures for identification of dynamical systems. *IEEE Transactions on Neural Networks, 6*, 422–431.

Krstic, M., Kanellakopoulos, I., & Kokotovic, P. (1995). *Nonlinear and adaptive control design*. New York, NY: Wiley.

Kukolj, D., & Levi, E. (2004). Identification of complex systems based on neural and takagi-sugeno fuzzy model. *IEEE Transactions on Systems, Man, and Cybernetics, Part B, 34*, 272–282.

Labiod, S., Boucherit, M. S., & Guerra, T. M. (2005). Adaptive fuzzy control of a class of mimo nonlinear systems. *Fuzzy Sets and Systems, 151*, 59–77.

Lee, C., & Teng, C. (2000). Identification and control of dynamic systems using recurrent fuzzy neural networks. *IEEE Transactions on Fuzzy Systems, 8*, 349–366.

Leu, Y., Wang, W., & Lee, T. (2005). Observer-based direct adaptive fuzzy-neural control for nonaffine nonlinear systems. *IEEE Transactions on Neural Networks, 16*, 853–861.

Leu, Y. G., Wang, W. Y., & Li, I. H. (2009). Rga-based on-line tuning of bmf fuzzy-neural networks for adaptive control of uncertain nonlinear systems. *Neurocomputing, 72*, 2636–2642.

Lewis, F., Jaganathan, S. & Yesildirek, A. (1999). Neural network control of robot manipulators and nonlinear systems. London: Taylor and Francis.

Lewis, F. L., Yesildirek, A., & Liu, K. (1996). Multilayer neural net robot controller: Structure and stability proofs. *IEEE Transactions on Neural Networks, 7*, 1–12.

Li, R. P. & Mukaidono, M. (1995). A new approach to rule learning based on fusion of fuzzy logic and neural networks. *IEICE Transactions on Fuzzy Systems, E78-d*, 1509–1514.

Lin, C. (1995). A neural fuzzy control system with structure and parameter learning. *Fuzzy Sets and Systems, 70*, 183–212.

Lin, C. T. (1994). *Neural fuzzy control systems with structure and parameter learning*. Singapore: World Scientific.

Lin, C. T., & Lee, C. S. G. (1996). *Neural fuzzy systems: A neuro-fuzzy synergism to intelligent systems*. Englewood Cliffs, NJ: Prentice Hall.

Lin, Y. H., & Cunningham, G. A. (1995). A new approach to fuzzy-neural system modelling. *IEEE Transactions on Fuzzy Systems, 3*, 190–197.

Lina, F. J., & Wai, R. J. (2003). Robust recurrent fuzzy neural network control for linear synchronous motor drive system. *Neurocomputing, 50*, 365–390.

Liu, Y., & Li, Y. (2006). Dynamic modeling and adaptive neural-fuzzy control for nonholonomic mobile manipulators moving on a slope. *International Journal of Control, Automation, and Systems, 4*, 1–7.

Mamdani, E. (1976). Advances in the linguistic synthesis of fuzzy controllers. *International Journal of Man-Machine Studies, 8*, 669678.

Mastorocostas, P. A., & Theocharis, J. B. (2002). A recurrent fuzzy-neural model for dynamic system identification. *IEEE Transactioba on SMC-Part B, 32,* 176–190.

McLain, R. B., Henson, M. A., & Pottmann, M. (1999). Direct adaptive control of partially known nonlinear systems. *IEEE Transactions on Neural Networks, 10,* 714–721.

Mitra, S., & Hayashi, Y. (2000). Neuro-fuzzy rule generation: survey in soft computing framework. *IEEE Transactions on Neural Networks, 11,* 748–768.

Naredra, K. S., & Valavani, L. S. (1979). Direct and indirect model reference adaptive control. *Automatica, 15,* 653–664.

Nounou, H. N., & Passino, K. M. (2004). Stable auto-tuning of adaptive fuzzy/neural controllers for nonlinear discrete-time systems. *IEEE Transactions on Fuzzy Systems, 12,* 70–83.

Ordonez, R., Zumberge, J., Spooner, J. T., & Passino, K. M. (1997). Adaptive fuzzy control: experiments and comparative analyses. *IEEE Transactions on Fuzzy Systems, 5,* 167–188.

Ortega, R., & Yu, T. (1989). Robustness of adaptive controllers: a survey. *Automatica, 25,* 651–678.

Passino, K. & Yurkovich, S. (1998). *Fuzzy control.* Menlo Park: Addison Wesley Longman.

Polycarpou, M. M. (1996). Stable adaptive neural control scheme for nonlinear systems. *IEEE Transactions on Automatic Control, 41,* 447–450.

Poznyak, A. S., Yu, W., Sanchez, E. N., & Perez, J. P. (1999). Nonlinear adaptive trajectory tracking using dynamic neural networks. *IEEE Transactions on Neural Networks, 10,* 1402–1411.

Rovithakis, G. & Christodoulou, M. A. (2000). Adaptive control with recurrent high order neural networks (theory and industrial applications). Advances in Industrial Control. London: Springer London Limited.

Rovithakis, G. A., & Christodoulou, M. A. (1994). Adaptive control of unknown plants using dynamical neural networks. *IEEE Transactions on Systems, Man and Cybernetics, 24,* 400–412.

Rovithakis, G. A., & Christodoulou, M. A. (1995). Direct adaptive regulation of unknown nonlinear dynamical systems via dynamic neural networks. *IEEE Transactions on Systems, Man and Cybernetics, 25,* 1578–1594.

Rubio, J. J., & Yu, W. (2007). Nonlinear system identification with recurrent neural networks and dead-zone kalman filter algorithm. *Neurocomputing, 70,* 24602466.

Sanner, R. M., & Slotine, J. E. (1992). Gaussian networks for direct adaptive control. *IEEE Transactions on Neural Networks, 3,* 837–863.

Sastry, S. S., & Bodson, M. (1989). *Adaptive control: Stability, convergence, and robustness.* Englewood Cliffs, NJ: Prentice-Hall.

Sastry, S. S., & Isidori, A. (1989). Adaptive control of linearizable systems. *IEEE Transactions on Automatic Control, 34,* 1123–1131.

Seker, H., Odetayo, M. O., Petrovic, D., & Naguib, R. N. G. (2003). A fuzzy logic based-method for prognostic decision making in breast and prostate cancers. *IEEE Transactions on Information Technology in Biomedicine, 7,* 114–122.

Sheikhzadeh, M., Trifkovic, M., & Rohani, S. (2008). Adaptive mimo neuro-fuzzy logic control of a seeded and an unseeded anti-solvent semi-batch crystallizer. *Chemical Engineering Science, 63*(1261), 1272.

Spooner, J. T., & Passino, K. M. (1996). Stable adaptive control using fuzzy systems and neural networks. *IEEE Transactions on Fuzzy Systems, 4,* 339–359.

Subudhi, B., & Morris, A. S. (2003). Fuzzy and neuro-fuzzy approaches to control a flexible single-link manipulator. *Systems and Control Engineering, 217,* 387–399.

Sugeno, M. (1985). *Industrial applications of fuzzy control.* Amsterdam: Elsevier.

Sugeno, M., & Yasukawa, T. (1993). A fuzzy-logic-based approach to qualitative modeling. *IEEE Transactions on Fuzzy Systems, 1,* 7–31.

Takagi, T., & Sugeno, M. (1985). Fuzzy identification of systems and its applications to modeling and control. *IEEE Transactions on Systems, Man, and Cybernetics, 15,* 116–132.

Theocharis, J. B. (2006). A high-order recurrent neuro-fuzzy system with internal dynamics: application to the adaptive noise cancellation. *Fuzzy Sets and Systems, 157,* 471–500.

Theodoridis, D., Boutalis, Y. & Christodoulou, M. (2009a). A new neuro-fuzzy dynamical system definition based on high order neural network function approximators. In: *European Control Conference ECC-09, Budapest, Hungary.*

Theodoridis, D., Boutalis, Y., & Christodoulou, M. (2010a). Indirect adaptive control of unknown multi variable nonlinear systems with parametric and dynamic uncertainties using a new neuro-fuzzy system description. *International Journal of Neural Systems, 20*, 129–148.

Theodoridis, D., Boutalis, Y., & Christodoulou, M. (2010b). A new neuro-fuzzy approach with robustness analysis for direct adaptive regulation of systems in brunovsky form. *International Journal of Neural Systems, 20*, 319–339.

Theodoridis, D., Boutalis, Y., & Christodoulou, M. (2011a). Neuro-fuzzy direct adaptive control of unknown nonlinear systems with analysis on the model order problem. *Journal of Zhejiang University-SCIENCE C (Computers & Electronics), 12*, 1–16.

Theodoridis, D., Boutalis, Y., & Christodoulou, M. (2011b). Robustifying analysis of the direct adaptive control of unknown multivariable nonlinear systems based on a new neuro-fuzzy method. *Journal of Artificial Intelligence and Soft Computing Research, 1*, 59–79.

Theodoridis, D., Boutalis, Y. & Christodoulou, M. (2012). Dynamical recurrent neuro-fuzzy identification schemes employing switching parameter hopping. *International Journal of Neural Systems, 22*, 16.

Theodoridis, D. C., Christodoulou, M. A., & Boutalis, Y. S. (2008). Indirect adaptive neuro—fuzzy control based on high order neural network function approximators. *Proccedings of the 16th Mediterranean Conference on Control and Automation—MED08* (pp. 386–393). Corsica, France: Ajaccio.

Theodoridis, D. C., Boutalis, Y. S. & Christodoulou, M. A. (2009b). Direct adaptive control of unknown nonlinear systems using a new neuro-fuzzy method together with a novel approach of parameter hopping. *Kybernetica, 45*, 349–386.

Vieira, J., Dias, F. M., & Mota, A. (2004). Artificial neural networks and neuro-fuzzy systems for modelling and controlling real systems: a comparative study. *Engineering Applications of Artificial Intelligence, 17*, 265–273.

Wang, C., Liu, H., & Lin, T. (2002). Direct adaptive fuzzy-neural control with state observer and supervisory controller for unknown nonlinear dynamical systems. *IEEE Transactions on Fuzzy Systems, 10*, 39–49.

Wang, J. S., & Lee, C. S. G. (2002). Self-adaptive neuro-fuzzy inference systems for classification applications. *IEEE Transactions on Fuzzy Systems, 10*, 790–802.

Wang, L. (1994). *Adaptive fuzzy systems and control: Design and stability analysis.* Englewood Cliffs, NJ: Prentice Hall.

Wang, L. X., & Mendel, J. M. (1992). Fuzzy basis functions, universal approximation, and orthogonal least squares learning. *IEEE Transactions on Neural Networks, 3*, 807–814.

Wang, M., Chen, B., & Dai, S. L. (2007). Direct adaptive fuzzy tracking control for a class of perturbed strict-feedback nonlinear systems. *Fuzzy Sets and Systems, 158*, 2655–2670.

Wang, W., Leu, Y., & Hsu, C. (2001). Robust adaptive fuzzy-neural control of nonlinear dynamical systems using generalized projection update law and variable structure controller. *IEEE Transactions on SMC-Part B, 31*, 140–147.

Wu, S., & Er, M. J. (2000). Dynamic fuzzy neural networks a novel approach to function approximation. *IEEE Transactions on SMC-Part B, 30*, 358–364.

Yang, Y. (2004). Direct robust adaptive fuzzy control (drafc) for uncertain nonlinear systems using small gain theorem. *Fuzzy Sets and Systems, 151*, 79–97.

Yesildirek, A., & Lewis, F. L. (1995). Feedback linearization using neural networks. *Automatica, 31*, 1659–1664.

Yu, L. X., & Zhang, Y. Q. (2005). Evolutionary fuzzy neural networks for hybrid financial prediction. *IEEE Transactions on Systems, Man, and Cybernetics Part C, 35*, 244–249.

Zadeh, L. A. (1965). Fuzzy sets. *Information Control, 8*, 338–353.

Zadeh, L. A. (1968). Fuzzy algorithm. *Information Control, 12*, 94–102.

Zadeh, L. A. (1971). Similarity relations and fuzzy orderings. *Information Science, 3*, 177–200.

Zadeh, L. A. (1975). The consept of a linguistic variable and its application to approximate reasoning: i, ii, iii. *Information Sciences, 8,* 199–251.

Zhang, J., & Morris, A. J. (1999). Recurrent neuro-fuzzy networks for nonlinear process modeling. *IEEE Transactions on Neural Networks, 10,* 313–326.

Zhou, Y., Li, S., & Jin, R. (2002). A new fuzzy neural network with fast learning algorithm and guaranteed stability for manufacturing process control. *Fuzzy Sets and Systems, 132,* 201–216.

Zhoua, S., Feng, G., & Feng, C. B. (2005). Robust control for a class of uncertain nonlinear systems: adaptive fuzzy approach based on backstepping. *Fuzzy Sets and Systems, 151,* 1–20.

Zimmermann, H. J. (1991). *Fuzzy set theory and its application* (2nd ed.). Boston, MA: Kluwer.

Chapter 2
Identification of Dynamical Systems Using Recurrent Neurofuzzy Modeling

2.1 The Recurrent Neurofuzzy Model

Let us consider a nonlinear function $f(x, u)$, where $f : R^{n+m} \rightarrow R^n$ is a smooth vector field defined on a compact set $\Psi \subset R^{n+m}$, with input space $u \in U_c \subset R^m$ and state-space $x \in X \subset R^n$. Also, we assume that the dynamic equation which describes the i/o behavior of a system has the following form (Christodoulou et al. 2007; Theodoridis et al. 2009, 2012):

$$\dot{x}(t) = f(x(t), u(t)), \tag{2.1}$$

or in a per-state form:

$$\dot{x}_i(t) = f_i(x(t), u(t)), \tag{2.2}$$

where $f_i(\cdot)$, $i = 1, 2, \ldots, n$, is a continuous function and t denotes the temporal variable. In order to proceed further we have to state the following assumption:

Assumption 1 Notice that since $\Psi \subset \Re^{n+m}$ then Ψ is closed and bounded set. Also, it is noted that even if Ψ is not compact we may assume that there is a time instant T such that $(x(t), u(t))$ remain in a compact subset of Ψ for all $t < T$; i.e. if $\Psi_T := \{(x(t), u(t)) \in \Psi, t < T\}$. The interval Ψ_T represents the time period over which the approximation is to be performed.

We consider that function $f(x, u)$ is approximated by a fuzzy system using appropriate fuzzy rules. In this framework, let Ω_f be defined as the universe of discourse of $(x, u) \in X \cup U \subset R^{n+m}$ belonging to the $(j_1, j_2, \ldots, j_{n+m})$th input fuzzy patch and pointing—through the vector field $f(\cdot)$—to the subset that belongs to the $l_1, l_2, \ldots, l_n$th output fuzzy patch. Also, Ω_{f_i} is a subset of Ω_f containing input pair values associated with f_i. Furthermore, $\Omega_{f_i}^p$, with $p = 1, 2, \ldots, q$ the number of fuzzy partitions of the i-th state variable, is defined as the p-th subregion of Ω_{f_i} such that $\Omega_{f_i} = \bigcup_{p=1}^{q} \Omega_{f_i}^p$.

Y. Boutalis et al., *System Identification and Adaptive Control*,
Advances in Industrial Control, DOI: 10.1007/978-3-319-06364-5_2,
© Springer International Publishing Switzerland 2014

Definition 2.1 According to the above notation the indicator function (IF) connected to $\Omega^p_{f_i}$ is defined as follows:

$$I^p_i(x(t), u(t)) = \begin{cases} \alpha^p_i(x(t), u(t)) & \text{if } (x(t), u(t)) \in \Omega^p_{f_i} \\ 0 & \text{otherwise} \end{cases} \tag{2.3}$$

where $\alpha^p_i(x(t), u(t))$ denotes the firing strength of the rule.

Then, assuming a standard defuzzification procedure (e.g., centroid of area or weighted average, see Sect. 1.3.1), the functional representation of the fuzzy system that approximates the real one can be written as

$$\hat{f}_i(x(t), u(t)) = \frac{\sum\limits_{p=1}^{q} I^p_i \cdot \bar{x}^p_{f_i}}{\sum\limits_{p=1}^{q} I^p_i}, \tag{2.4}$$

where the summation is carried over all the available fuzzy rules.

Definition 2.2 Using the notation presented in Sect. 1.3.1, we can define as weighted IF (WIF) the following equation:

$$(I')^p_i = \frac{I^p_i}{\sum\limits_{p=1}^{q} I^p_i} \tag{2.5}$$

which is the IF defined in (2.3) divided by the sum of all IF participating in the summation of (2.4).

Thus, Eq. (2.4) can be rewritten as

$$\hat{f}_i(x(t), u(t)) = \sum\limits_{p=1}^{q} (I')^p_i \cdot \bar{x}^p_{f_i}. \tag{2.6}$$

Based on the fact that functions of high-order neurons are capable of approximating discontinuous functions (Kosmatopoulos and Christodoulou 1996; Christodoulou et al. 2007), we use high-order neural networks (HONNs) to approximate a $(I')^p_i$. Thus, we have the following definition:

Definition 2.3 A HONN is defined as:

$$N^p_i(x(t), u(t); w, k) = \sum\limits_{l=1}^{k} w^{pl}_{f_i} \prod\limits_{j \in I_l} \Phi^{d_j(l)}_j, \tag{2.7}$$

where $I_l = \{I_1, I_2, \ldots, I_k\}$ is a collection of k not-ordered subsets of $\{1, 2, \ldots, n + m\}$, $d_j(l)$ are nonnegative integers. Φ_j are the elements of the following vector:

$$\Phi = \begin{bmatrix} \Phi_1 & \cdots & \Phi_n & \Phi_{n+1} & \cdots & \Phi_{n+m} \end{bmatrix}^T = \begin{bmatrix} s(x_1) & \cdots & s(x_n) & s(u_1) & \cdots & s(u_m) \end{bmatrix}^T ,$$

where s denotes the sigmoid function defined as:

$$s(x) = \frac{\alpha}{1 + e^{-\beta x}} - \gamma, \tag{2.8}$$

with α, β being positive real numbers and γ being a real number.

Special attention has to be given in the selection of parameters α, β, γ so that $s(x)$ fulfill the persistency of excitation condition $\left(s \in [-\gamma, -\gamma + \alpha] \text{ when } \gamma < 0\right)$ required in some system identification tasks. Also, $w_{f_i}^{pl}$ is the HONN weights with $i = 1, 2, \ldots, n$, $p = 1, 2, \ldots, q$ and $l = 1, 2, \ldots, k$.

Thus, Eq. (2.7) can be written as

$$N_i^p(x(t), u(t); w, k) = \sum_{l=1}^{k} w_{f_i}^{pl} s_l(x(t), u(t)), \tag{2.9}$$

where $s_l(x(t), u(t))$ are high-order terms of sigmoid functions of the state and/or input.

The next lemma (Kosmatopoulos and Christodoulou 1996) states that a HONN of the form in Eq. (2.9) can approximate the weighted indicator function (WIF), $\left(I'\right)_i^p$.

Lemma 2.1 *Consider the WIF $\left(I'\right)_i^p$ and the family of HONN's $N_i^p(x(t), u(t); w, k)$. Then for any $\varepsilon_i^p > 0$, there is a vector of weights w and a number of k high-order connections such that:*

$$\sup_{(x(t), u(t)) \in \Psi} \left\{ \left(I'\right)_i^p (x(t), u(t)) - \sum_{l=1}^{k} w_{f_i}^{pl} s_l(x(t), u(t)) \right\} < \varepsilon_i^p$$

The magnitude of approximation error $\varepsilon_i^p > 0$ depends on the choice of the number of high-order terms.

Under the definition of WIFs and the above lemma, one could rewrite the rules of the fuzzy system as follows:

$$R_i^p : \text{IF } (x(t), u(t)) \in \Omega_{f_i}^p \text{ THEN HONN}_p \text{ is } \left(I'\right)_i^p (t) .$$

Following the above analysis and Eq. (2.6), actually we give a weighting value according to the output fuzzy partitioning, as shown in Fig. 2.1, to every HONN that participates to the estimation of $f_i(x, u)$.

As a consequence, we have the following definition:

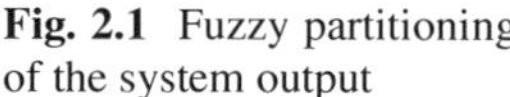

Fig. 2.1 Fuzzy partitioning of the system output

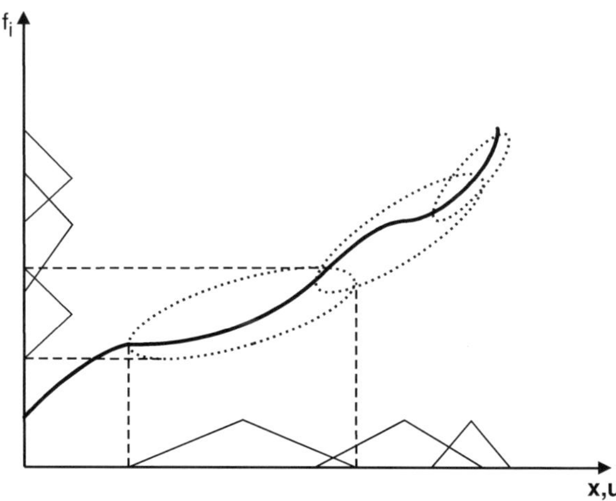

Definition 2.4 The center weighting value (CWV) $\bar{x}^p_{f_i}$ which is the p-th fuzzy center of the i-th state variable (or equivalently f_i) influences a HONN by a degree of implementation $\bar{x}^p_{f_i}$.

Therefore, rule R^p_i can be equivalently expressed as

R^p_i: IF $(x(t), u(t)) \in \Omega^p_{f_i}$ THEN HONN$_p$ is $\left(I'\right)^p_i (t)$ with CWV $\bar{x}^p_{f_i}$.

Now, we can group the rules that participate in the construction of the i-th state variable output according to the following form:

R_i: IF $(x(t), u(t)) \in \Omega_{f_i}$ THEN HONN$_1$ is $\left(I'\right)^1_i (t)$ with CWV $\bar{x}^1_{f_i}$ and HONN$_2$ is $\left(I'\right)^2_i (t)$ with CWV $\bar{x}^2_{f_i}$ and $\cdots$ and HONN$_q$ is $\left(I'\right)^q_i (t)$ with CWV $\bar{x}^q_{f_i}$.

It then follows easily that, the i-th state variable of the system output is determined as follows:

$$R_i : \text{IF } (x(t), u(t)) \in \Omega_{f_i} \text{ THEN}$$

$$f_i (x, u) = \left(I'\right)^1_i (t) \cdot \bar{x}^1_{f_i} + \left(I'\right)^2_i (t) \cdot \bar{x}^2_{f_i} + \cdots + \left(I'\right)^q_i (t) \cdot \bar{x}^q_{f_i},$$

where each $\left(I'\right)^l_i, l = 1, \ldots, q$ is replaced by the respective HONN. It is clear that the information about the antecedent partitioning of the rules as well as the number of rules is not necessary to be determined here. Therefore, the rules are not treated here in the classical way of Mamdani or Takagi-Sugeno definition but their consequent parts are determined directly from F-HONNs.

Following the above notation, Eq. (2.6) in conjunction with Eq. (2.9) can be rewritten as

$$\hat{f}_i(x(t), u(t)) = \sum_{p=1}^{q} \bar{x}^p_{f_i} \cdot \left(\sum_{l=1}^{k} w^{pl}_{f_i} \cdot s_l(x(t), u(t)) \right), \qquad (2.10)$$

or in a more compact form:

$$\dot{\hat{f}} = X_f W_f s_f(x, u). \tag{2.11}$$

An alternative, recurrent NF form of Eq. (2.1) which will be used in the subsequent analysis of this thesis is:

$$\dot{\hat{x}} = A\hat{x} + \hat{f}. \tag{2.12}$$

Considering that f is approximated by the NF model described above, Eq. (2.12) can be rewritten as

$$\dot{\hat{x}} = A\hat{x} + X_f W_f s_f(x, u), \tag{2.13}$$

where A is a $n \times n$ stable matrix, which for simplicity can be taken to be diagonal as $A = \mathrm{diag}[-a_1, -a_2, \ldots, -a_n]$, with $a_i > 0$. Also, X_f is a matrix containing the centers of the partitions of every fuzzy output variable of $f(x, u)$, $s_f(x, u)$ is a vector containing high-order combinations of sigmoid functions of the state x and control input u. Also, W_f is a matrix containing respective neural weights according to (2.9) and (2.10). For notational simplicity we assume that all output fuzzy variables are partitioned to the same number, q, of partitions. Under these specifications X_f is a $n \times n \cdot q$ block diagonal matrix of the form $X_f = \mathrm{diag}(\bar{x}_{f_1}, \bar{x}_{f_2}, \ldots, \bar{x}_{f_n})$, with $\bar{x}_{f_i}$ being a q-dimensional row vector of the form:

$$\bar{x}_{f_i} = \left[\bar{x}_{f_i}^1 \ \bar{x}_{f_i}^2 \ \cdots \ \bar{x}_{f_i}^q \right],$$

or in a more detailed form:

$$X_f = \begin{bmatrix} \bar{x}_{f_1}^1 & \cdots & \bar{x}_{f_1}^q & 0 & \cdots & 0 & 0 & \cdots & 0 \\ 0 & \cdots & 0 & \bar{x}_{f_2}^1 & \cdots & \bar{x}_{f_2}^q & 0 & \cdots & 0 \\ \cdots & \cdots & \cdots & \cdots & \cdots & \cdots & \cdots & \cdots & \cdots \\ 0 & \cdots & 0 & 0 & \cdots & 0 & \bar{x}_{f_n}^1 & \cdots & \bar{x}_{f_n}^q \end{bmatrix}. \tag{2.14}$$

Also, $s_f(x) = \left[s_1(x) \ \ldots \ s_k(x) \right]^T$, where each $s_l(x)$ with $l = 1, 2, \ldots, k$, is a high-order combination of sigmoid functions of the state variables and input signals. Finally, W_f is a $n \cdot q \times k$ matrix with neural weights. W_f assumes the form $W_f = \left[W_{f_1} \ \cdots \ W_{f_n} \right]^T$, where each W_{f_i} is a matrix $\left[w_{f_i}^{pl} \right]_{q \times k}$ and is given as:

$$W_{f_i} = \begin{bmatrix} w_{f_i}^{11} & w_{f_i}^{12} & \cdots & w_{f_i}^{1k} \\ w_{f_i}^{21} & w_{f_i}^{22} & \cdots & w_{f_i}^{2k} \\ \vdots & \vdots & \cdots & \vdots \\ w_{f_i}^{q1} & w_{f_i}^{q2} & \cdots & w_{f_i}^{qk} \end{bmatrix},$$

or in a more detailed form:

$$
W_f = \begin{bmatrix}
w_{f_1}^{11} & w_{f_1}^{12} & \cdots & w_{f_1}^{1k} \\
w_{f_1}^{21} & w_{f_1}^{22} & \cdots & w_{f_1}^{2k} \\
\vdots & \vdots & \ddots & \vdots \\
w_{f_1}^{q1} & w_{f_1}^{q2} & \cdots & w_{f_1}^{qk} \\
\vdots & \vdots & \ddots & \vdots \\
w_{f_n}^{11} & w_{f_n}^{12} & \cdots & w_{f_n}^{1k} \\
w_{f_n}^{21} & w_{f_n}^{22} & \cdots & w_{f_n}^{2k} \\
\vdots & \vdots & \ddots & \vdots \\
w_{f_n}^{q1} & w_{f_n}^{q2} & \cdots & w_{f_n}^{qk}
\end{bmatrix}.
$$

From the above definitions and Eq. (2.10), it is obvious that the accuracy of the approximation of $f_i(x, u)$ depends on the approximation abilities of HONNs and on an initial estimate of the centers of the output membership functions. These centers can be obtained by experts or by offline techniques based on gathered data. Any other information related to the input membership functions is not necessary because it is replaced by the HONNs.

Figure 2.2 shows the overall scheme of the proposed NF modeling that approximates function $f_i(x, u)$ depending only on measurements of x, u. When these measurements are given as inputs to the NF network (input layer) that includes high-order sigmoidal terms, the output of indicator layer gives the weighted IF outputs that influence the corresponding rules according to output fuzzy center (rule layer). The appropriate summation of all rules at each sampling time instant gives the overall output of the function $f_i(x, u)$ (output layer).

2.2 Approximation Capabilities of the Neurofuzzy Model

The approximation problem consists of determining whether by allowing enough high-order connections and fuzzy centers, there exist weights W_f, such that the F-RHONNs model could approximate the input–output behavior of a complex dynamical system of the form (2.1). In this equation the input u belongs to a class U_c of (piecewise continuous) admissible inputs.

By adding and subtracting Ax, where A is a Hurwitz matrix, (2.2) is rewritten as

$$
\dot{x} = Ax + g(x, u) \tag{2.15}
$$

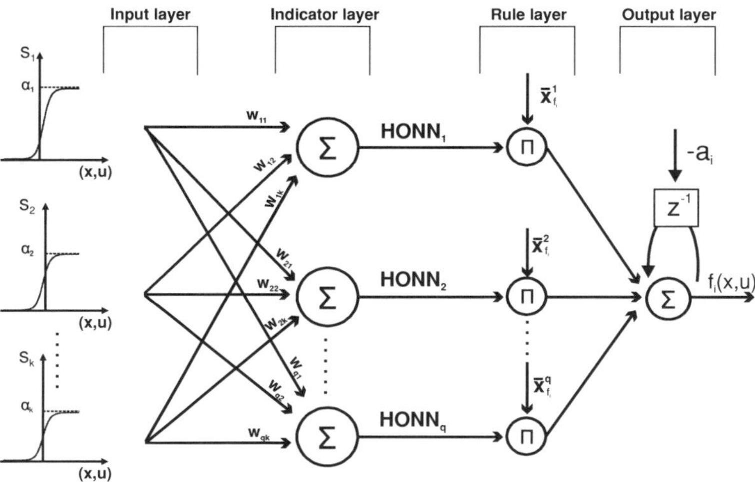

Fig. 2.2 Overall scheme of the proposed NF model that approximates function $f_i(x, u)$ using state measurements x and input signals u

where $g(x, u) := f(x, u) - Ax$.

In order to have a well-posed problem, we will impose the following mild assumptions on the system to be approximated:

Assumption 2 Given a class $U_c \subset R^q$ of admissible inputs, for any $u \in U_c$ and any finite initial condition $x(0)$, the state trajectories are uniformly bounded for any finite $T > 0$. Meaning that we do not allow systems processing trajectories that escape at infinite, in finite time T, T being arbitrarily small. Hence, $|x(T)| < \infty$.

Assumption 3 Functions f_i are continuous with respect to their arguments and satisfy a local Lipschitz condition so that (2.2) has a unique solution for any finite initial condition $x(0)$ and $u \in U_c$, in the sense of Caratheodory (Hale 1969).

Based on the above assumptions, we obtain the following theorem:

Theorem 2.1 *Suppose that the system* (2.1) *and the model* (2.10) *are initially at the same state* $\hat{x}(0) = x(0)$, *then for any* $\varepsilon > 0$ *and any finite* $T > 0$, *there exist integers* k, q, *a matrix* $W_f^\star \in R^{k \times q \times n}$ *and appropriately selected fuzzy output centers* $\bar{x}_{f_i}^p$ *such that the state* $\hat{x}(t)$ *of the F-RHONNs model* (2.10) *with k high-order connections, q fuzzy centers and weight values* $W_f = W_f^\star$ *which satisfies:*

$$\sup_{0 \le t \le T} \left|\hat{x}(t) - x(t)\right| \le \varepsilon.$$

Proof Following a procedure similar to the work of Kosmatopoulos et al. (1995), we proceed as follows: By assumption, $(x(t), u(t)) \in \Psi$ for all $t \in [0, T]$, where Ψ is a compact subset of R^{n+m}.

Let $\Psi_e = \{(x, u) \in R^{n+m} : |(x, u) - (x_y, u_y)| \leq \varepsilon, (x_y, u_y) \in \Psi\}$. It can be readily seen that Ψ_e is also a compact subset of R^{n+m} and $\Psi \subset \Psi_e$. That is, Ψ_e is ε larger than Ψ, where ε is the required degree of approximation. Since s_f is a continuous function, it satisfies a Lipschitz condition in Ψ_e, i.e., there is a constant l such that for all $(\hat{x}_1, u), (\hat{x}_2, u) \in \Psi_e$:

$$|s_f(\hat{x}_1, u) - s_f(\hat{x}_2, u)| \leq l |x_1 - x_2|. \tag{2.16}$$

In what follows, we show that the function $X_f W_f^\star s_f$ satisfies the conditions of Stone–Weirstrass Theorem (Stone 1948; Bishop 1961) and can approximate any continuous function over a compact domain.

The dynamic behavior of F-RHONNs model is described by (2.13). Since $\hat{x}(0) = x(0)$, the state error $e = \hat{x} - x$ satisfies the differential equation

$$\dot{e} = Ae + X_f W_f s_f - g(x, u), \tag{2.17}$$

where $e(0) = 0$.

Therefore, it can be readily shown that if k, q are sufficiently large, then there exist weight values $W_f = W_f^\star$ such that $X_f W_f^\star s_f(x, u)$ can approximate $g(x, u)$ to any degree of accuracy, for all (x, u) in a compact domain. Hence, there exists $W_f = W_f^\star$ such that

$$\sup_{(x,u)\in\Psi_e} \left| X_f W_f^* s_f(x, u) - g(x, u) \right| \leq \delta, \tag{2.18}$$

where δ is a constant to be designed in the sequel. The solution of (2.17) is

$$e(t) = \int_0^t e^{A(t-\tau)} X_f W_f^* s_f(\hat{x}(\tau), u(\tau)) \, d\tau - \int_0^t e^{A(t-\tau)} g(x(\tau), u(\tau)) \, d\tau$$

$$= \int_0^t e^{A(t-\tau)} X_f W_f^* s_f(\hat{x}(\tau), u(\tau)) \, d\tau - \int_0^t e^{A(t-\tau)} X_f W_f^* s_f(x(\tau), u(\tau)) \, d\tau$$

$$+ \int_0^t e^{A(t-\tau)} X_f W_f^* s_f(x(\tau), u(\tau)) \, d\tau - \int_0^t e^{A(t-\tau)} g(x(\tau), u(\tau)) \, d\tau. \tag{2.19}$$

Since A is a Hurwitz matrix, there exist positive constants c, α such that $\|e^{At}\| \leq ce^{-\alpha t}$ for all $t \geq 0$. Also, let $k = cl \|X_f W_f^\star\|$. Based on the aforementioned definitions of the constants $c, \alpha, k, \varepsilon$, let δ in (2.18) be chosen as

$$\delta = \frac{\varepsilon\alpha}{2c}e^{-\frac{k}{\alpha}} > 0. \tag{2.20}$$

First consider the case where $(x(t), u(t)) \in \Psi_e$ for all $t \in [0, T]$. Starting from (2.19), taking norms on both sides and using (2.16), (2.18) and (2.20), the following inequalities hold for all $t \in [0, T]$:

$$|e(t)| \leq \int_0^t \left\| e^{A(t-\tau)} \right\| \left\| X_f W_f^* \right\| \left| s_f(\hat{x}, u) - s_f(x, u) \right| d\tau$$

$$+ \int_0^t \left\| e^{A(t-\tau)} \right\| \left| X_f W_f^* s_f(x, u) - g(x, u) \right| d\tau$$

$$\leq \int_0^t e^{-\alpha(t-\tau)} k \left| e(\tau) \right| d\tau + \int_0^t \delta c e^{-\alpha(t-\tau)} d\tau$$

$$\leq k \int_0^t e^{-\alpha(t-\tau)} \left| e(\tau) \right| d\tau + \frac{\varepsilon}{2} e^{-\frac{k}{\alpha}}. \tag{2.21}$$

Then, using the Bellman–Gronwall Lemma (Hale 1969), we obtain:

$$|e(t)| \leq \frac{\varepsilon}{2} e^{-\frac{k}{\alpha}} \cdot e^{\int_0^t k e^{-\alpha(t-\tau)} d\tau} \leq \frac{\varepsilon}{2} e^{-\frac{k}{\alpha}} \cdot e^{\frac{k}{\alpha}} \leq \frac{\varepsilon}{2}. \tag{2.22}$$

It should be noted here that the assumption of $\hat{x}(0) = x(0)$ can be easily relaxed without affecting the conclusion of the theorem. In this case, one should consider that an exponentially fast decaying error term is added in ϵ.

The above theorem proves that if sufficiently large number of connections are allowed in F-RHONNs model then it is possible to approximate any dynamical system to any degree of accuracy. This result does not provide us with any constructive method for obtaining the optimal weights $W_f^\star$. In what follows, we consider the learning problem of adjusting the weights adaptively, such that the Neurofuzzy model identifies general dynamic systems.

2.3 Learning Algorithms for Parameter Identification

We proceed now to develop weight updating laws assuming that the unknown system is modeled exactly by an F-RHONNs architecture of the form (2.13). In the next section, we extend this analysis to cover the case where there exists a nonzero mismatch between the system and F-RHONNs model with optimal weight values, that is, we assume the existence of modeling errors.

Following the standard practice in system identification algorithms, we will assume that the input $u(t)$ and the state $x(t)$ remain bounded for all $t \geq 0$. Based on the definition of $s_f(x, u)$, X_f as given by (2.8), (2.14) this implies that $s_f(x, u)$, X_f are also bounded. In the sections that follow, we present different approaches for estimating the unknown parameters $\left(w_{f_i}^{pl}\right)^{\star}$ of F-RHONNs model.

2.3.1 Simple Gradient Descent

In developing this identification scheme, we start again from the differential equation that describes the unknown system with no modeling error which is given by

$$\dot{x}_i = -a_i x_i + \bar{x}_{f_i} W_{f_i}^{\star} s_f(x, u). \tag{2.23}$$

Based on (2.23), the identifier is now chosen as

$$\dot{\hat{x}}_i = -a_i \hat{x}_i + \bar{x}_{f_i} W_{f_i} s_f(x, u), \tag{2.24}$$

where W_{f_i} is again the estimate of the unknown optimal weight matrix $W_{f_i}^{\star}$. In this case, the state error $e_i = \hat{x}_i - x_i$ satisfies

$$\dot{e}_i = -a_i e_i + \bar{x}_{f_i} \tilde{W}_{f_i} s_f(x, u), \tag{2.25}$$

where $\tilde{W}_{f_i} = W_{f_i} - W_{f_i}^{\star}$.

The next theorem gives the error F-RHONNs model with the gradient method for adjusting the weights.

Theorem 2.2 *Consider the error F-RHONNs model given by (2.25) whose weights are adjusted according to equation*

$$\dot{W}_{f_i} = -\bar{x}_{f_i}^T e_i s_f^T P_i. \tag{2.26}$$

Then for $i = 1, 2, \ldots, n$, the following properties are guaranteed:

1. e_i, $\tilde{W}_{f_i} \in L_{\infty}$, $e_i \in L_2$,
2. $\lim_{t \to \infty} e_i(t) = 0$,
3. $\lim_{t \to \infty} \dot{W}_{f_i}(t) = 0$.

Proof (1) Consider the Lyapunov candidate function:

$$V = \frac{1}{2} \sum_{i=1}^{n} e_i^2 + \frac{1}{2} \sum_{i=1}^{n} \text{tr} \left\{ \tilde{W}_{f_i} P_i^{-1} \tilde{W}_{f_i}^T \right\}. \tag{2.27}$$

Taking the time derivatives of the Lyapunov function candidate (2.27) and after substituting Eq. (2.25) we obtain

$$\dot{V} = \sum_{i=1}^{n} e_i \dot{e}_i + \sum_{i=1}^{n} \text{tr}\left\{\dot{W}_{f_i} P_i^{-1} \tilde{W}_{f_i}^T\right\}$$

$$= -\sum_{i=1}^{n} a_i \,|e_i|^2 + \sum_{i=1}^{n} \left(e_i \bar{x}_{f_i} \tilde{W}_{f_i} s_f + \text{tr}\left\{\dot{W}_{f_i} P_i^{-1} \tilde{W}_{f_i}^T\right\}\right)$$

$$= -\sum_{i=1}^{n} a_i \,|e_i|^2. \tag{2.28}$$

Considering that in deriving (2.28) we assumed that

$$\text{tr}\left\{\dot{W}_{f_i} P_i^{-1} \tilde{W}_{f_i}^T\right\} = -e_i \bar{x}_{f_i} \tilde{W}_{f_i} s_f,$$

and using matrix trace properties we result in Eq. (2.26).

Thus, $\dot{V}$ is negative semidefinite. Since $\dot{V} \leq 0$, we conclude that $V \in L_\infty$, which implies that e_i, $\tilde{W}_{f_i} \in L_\infty$. Furthermore, $W_{f_i} = \tilde{W}_{f_i} + W_{f_i}^*$ is also bounded. Since V is a nonincreasing function of time and bounded from below, the $\lim_{t\to\infty} V = V_\infty$ exists; therefore, by integrating $\dot{V}$ from 0 to ∞ we have

$$\int_0^\infty \sum_{i=1}^{n} a_i \,|e_i|^2 dt \leq [V(0) - V_\infty] < \infty,$$

which implies that $e_i \in L_2$.

(2) Since $e_i \in L_2 \cap L_\infty$, using Barbalat's Lemma we conclude that $\lim_{t\to\infty} e_i(t) = 0$.

(3) Finally, using the boundedness of $\bar{x}_{f_i}$, $s_f(x, u)$ and the convergence of $e_i(t)$ to zero, we have that $\dot{W}_{f_i}$ also converges to zero (Ioannou and Fidan 2006).

Remark 2.1 The above theorem does not imply that the weight estimation error $\tilde{W}_{f_i} = W_{f_i} - W_{f_i}^*$ converges to zero. In order to achieve convergence of the weights to their correct value, the additional assumption of persistent excitation needs to be imposed on the vector $s_f(x, u)$ because $\bar{x}_{f_i}$ satisfies this condition by definition. In particular, $s_f \in R^k$ is said to be *persistently exciting* if there exist positive scalars β_1, β_2 and T such that for all $t \geq 0$:

$$\beta_1 I \leq \int_t^{t+T} s_f(\tau) s_f^T(\tau) \, d\tau \leq \beta_2 I, \tag{2.29}$$

where I is the $k \times k$ identity matrix.

This can be achieved if the constant γ in Eq. (2.8) is selected such that: $s_f \cdot s_f^T > 0$.

2.3.2 Pure Least Squares

The basic idea behind least squares (LS) method is to fit a mathematical model to a sequence of observed data by minimizing the sum of the squares of the difference between the observed and computed data. This way, any noise or inaccuracies in the observed data are expected to have less effect on the accuracy of the mathematical model (Ioannou and Fidan 2006).

The method is simple to apply and analyze in the case where the unknown parameters appear in a linear form, such as in Eq. (2.23). The pure LS algorithm can be thought as a gradient algorithm with a time-varying learning rate and could be written as follows:

$$\dot{W}_{f_i} = -\frac{\bar{x}_{f_i}^T e_i z_i^T P_i}{\left|\bar{x}_{f_i}\right|^2}, \quad W_{f_i}(0) = W_{f0}, \tag{2.30}$$

$$\dot{P}_i = -\frac{P_i z_i z_i^T P_i}{n_s^2}, \quad P_i(0) = P_0,$$

where P_i is the gain matrix which is positive definite and $n_s^2 \geq 1$ is a normalization signal designed to guarantee that $\frac{z_i}{n_s}$ is bounded, with z_i defined in the following lemma 2.2. The property of n_s is used to establish the boundedness of the estimated parameters even when z_i is not guaranteed to be bounded. A straightforward choice for n_s is $n_s^2 = 1 + \alpha z_i^T z_i$, $\alpha > 0$. If z_i is bounded, we can take $\alpha = 0$. The following lemma is useful in the development of the adaptive identification algorithm, which is presented in this section.

Lemma 2.2 *The system described by* Eq. (2.23) *can be expressed as*

$$\dot{z}_i = -a_i z_i + s_f, \ z_i(0) = 0, \tag{2.31}$$

$$x_i = \bar{x}_{f_i} W_{f_i}^\star z_i + e^{-a_i t} x_i(0). \tag{2.32}$$

Proof From (2.31) after integrating we have

$$z_i(t) = \int_0^t e^{-a_i(t-\tau)} s_f(x(\tau), u(\tau))\, d\tau,$$

therefore,

$$\bar{x}_{f_i} W_{f_i}^\star z_i + e^{-a_i t} x_i(0) = e^{-a_i t} x_i(0) + \int_0^t e^{-a_i(t-\tau)} \bar{x}_{f_i} W_{f_i}^\star s_f(x(\tau), u(\tau))\, d\tau.$$

$$\tag{2.33}$$

Using (2.32), the right-hand side of (2.33) is equal to $x_i(t)$ and this concludes the proof.

Using the above lemma the dynamical system is described by the following equation:

$$x_i = \bar{x}_{f_i} W^{\star}_{f_i} z_i + \varepsilon_i, \tag{2.34}$$

where $\varepsilon_i = e^{-a_i t} x_i(0)$ is an exponentially decaying term that appears when a nonzero initial state is applied. After ignoring the exponentially decaying term ε_i (Rovithakis and Christodoulou 2000), the F-RHONNs model can be written as

$$\hat{x}_i = \bar{x}_{f_i} W_{f_i} z_i. \tag{2.35}$$

The state error equation $e_i = \hat{x}_i - x_i$, after substituting (2.34), (2.35) becomes

$$e_i = \bar{x}_{f_i} \tilde{W}_{f_i} z_i - \varepsilon_i. \tag{2.36}$$

The cost function $J(W_{f_i})$ is chosen as

$$J(W_{f_i}) = \frac{\sum\limits_{i=1}^{n} e_i^2}{2} = \frac{\sum\limits_{i=1}^{n} \left[\left(\bar{x}_{f_i} W_{f_i} z_i - \bar{x}_{f_i} W^{\star}_{f_i} z_i \right) - \varepsilon_i \right]^2}{2}. \tag{2.37}$$

If we use the LS method described by (2.30) and (2.31), a problem that may be encountered in the application of the LS's algorithm is that P_i may become arbitrarily small and thus slow down adaptation in some directions. Therefore, we can use one of various modifications that prevent $P_i(t)$ from going to zero as follows: if the smallest eigenvalue of $P_i(t)$ becomes smaller than ρ_1 then $P_i(t)$ is reset to $P_i(t) = \rho_0 I$, where $\rho_0 \geq \rho_1 > 0$ are some design constants (Rovithakis and Christodoulou 2000).

Theorem 2.3 *The pure LS algorithm given by (2.30), (2.31) guarantees the following properties:*

1. $e_i, \dot{W}_{f_i} \in L_2 \cap L_\infty, \quad W_{f_i}, P_i \in L_\infty.$
2. $\lim_{t \to \infty} e_i(t) = 0,$
3. $\lim_{t \to \infty} \dot{W}_{f_i}(t) = 0.$

Proof (1) From (2.31) we have that $\dot{P}_i \leq 0$, i.e., $P_i(t) \leq P_0$. Because $P_i(t)$ is nonincreasing and bounded from below (i.e., $P_i(t) = P_i^T(t) \geq 0$) it has a limit, i.e.,

$$\lim_{t \to \infty} P_i(t) = \bar{P}_i,$$

where $\bar{P}_i = \bar{P}_i^T \geq 0$ is a constant positive definite diagonal matrix and thus $P_i \in L_\infty$.

Let us now consider the Lyapunov candidate function

$$V = \frac{1}{2} \sum_{i=1}^{n} \left[\left(\bar{x}_{f_i} \tilde{W}_{f_i} \right) P_i^{-1} \left(\bar{x}_{f_i} \tilde{W}_{f_i} \right)^T + \int_t^{\infty} \varepsilon_i^2 (\tau) \, d\tau \right]. \tag{2.38}$$

Taking the time derivatives of the Lyapunov function candidate (2.38) and considering the equations, $\frac{d}{dt} \left(P^{-1} \right) = -P^{-1} \dot{P} P^{-1}$, (2.30), (2.31), (2.36) we obtain

$$\begin{aligned}
\dot{V} &= \sum_{i=1}^{n} \left(\bar{x}_{f_i} \dot{W}_{f_i} P_i^{-1} \tilde{W}_{f_i}^T \bar{x}_{f_i}^T \right) + \frac{1}{2} \sum_{i=1}^{n} \left[\left(\bar{x}_{f_i} \tilde{W}_{f_i} \right) \dot{P}_i^{-1} \left(\bar{x}_{f_i} \tilde{W}_{f_i} \right)^T - \varepsilon_i^2 \right] \\
&= -\sum_{i=1}^{n} \left(\bar{x}_{f_i} \tilde{W}_{f_i} z_i e_i \right) + \frac{1}{2} \sum_{i=1}^{n} \left[\frac{\left| \bar{x}_{f_i} \tilde{W}_{f_i} z_i \right|^2}{n_s^2} - \varepsilon_i^2 \right] \\
&= -\sum_{i=1}^{n} \left[e_i \left(e_i + \varepsilon_i \right) \right] + \frac{1}{2} \sum_{i=1}^{n} \left(\frac{(e_i + \varepsilon_i)^2}{n_s^2} - \varepsilon_i^2 \right) \\
&= -\frac{1}{2} \sum_{i=1}^{n} \left[e_i^2 + (e_i + \varepsilon_i)^2 \right] + \frac{1}{2} \sum_{i=1}^{n} \frac{(e_i + \varepsilon_i)^2}{n_s^2} \\
&\leq -\frac{1}{2} \sum_{i=1}^{n} e_i^{\,2} \leq 0.
\end{aligned} \tag{2.39}$$

Equation (2.39) implies that $V \in L_\infty$, and therefore $\tilde{W}_{f_i} \in L_\infty$. Then, Eq. (2.36) in conjunction with the boundedness of z_i, gives $e_i \in L_\infty$. Furthermore, $W_{f_i} = \tilde{W}_{f_i} + W_{f_i}^*$ is also bounded. Since V is a nonincreasing function of time and bounded from below, the $\lim_{t \to \infty} V = V_\infty$ exists; therefore, by integrating $\dot{V}$ from 0 to ∞ we have

$$\frac{1}{2} \int_0^{\infty} \sum_{i=1}^{n} e_i^{\,2} \leq [V(0) - V_\infty] < \infty,$$

which implies that $e_i \in L_2$. From (2.30) we have

$$\left\| \dot{W}_{f_i} \right\| \leq \frac{|e_i|\,|z_i|\,\|P_i\|}{\left| \bar{x}_{f_i} \right|}. \tag{2.40}$$

Since $\bar{x}_{f_i}, P_i, z_i, e_i \in L_\infty$, and $e_i \in L_2$, we have $\dot{W}_{f_i} \in L_2 \cap L_\infty$.

(2) Since $e_i \in L_2 \cap L_\infty$, using Barbalat's Lemma we conclude that $\lim_{t \to \infty} e_i(t) = 0$.

(3) Finally, using the boundness of $\bar{x}_{f_i}$, $s_f(x, u)$ and the convergence of $e_i(t)$ to zero, we have that $\dot{W}_{f_i}$ also converges to zero (Ioannou and Fidan 2006).

Also, if a persistency of excitation condition such as Remark 2.1 is valid then $W_{f_i}(t) \to W_{f_i}^\star$ as $t \to \infty$.

2.4 Robust Learning Algorithms

Due to an insufficient number of high-order terms or fuzzy output centers in the F-RHONNs model, we have to deal with unmodeled dynamics, noises, disturbances, and other frequently encountered uncertainties. In such cases, if standard adaptive laws are used for updating the weights, then the presence of modeling error in problems related to learning in dynamic environments may cause the adjusted weight values (and consequently the estimation error e_i) to drift to infinity. Examples of such behavior can be found in the adaptive control literature of linear systems (Ioannou and Fidan 2006).

In this section, we modify the weight updating laws to avoid the parameter drift phenomenon. To formulate the problem we note that by adding and subtracting $-a_i x_i + \sum_{l=1}^{k} \bar{x}_{f_i} W_{f_i}^{l\star} s_l(x, u)$, the dynamic behavior of each state of the system (2.2) can be expressed by the following differential equation:

$$\dot{x}_i = -a_i x_i + \sum_{l=1}^{k} \bar{x}_{f_i} W_{f_i}^{l\star} s_l(x, u) + \mu_i(t), \qquad (2.41)$$

where the modeling error $\mu_i(t)$ is given by

$$\mu_i(t) = f_i(x(t), u(t)) + a_i x_i(t) - \sum_{l=1}^{k} \bar{x}_{f_i} W_{f_i}^{l\star} s_l(x(t), u(t)). \qquad (2.42)$$

The unknown optimal weight matrix $W_{f_i}^{l\star}$ is defined as the value of the weight vector $W_{f_i}^{l}$ that minimizes the L_∞-norm difference between $f_i(x, u) + a_i x_i$ and $\sum_{l=1}^{k} \bar{x}_{f_i} W_{f_i}^{l} s_l(x, u)$ for all $(x, u) \in \Psi \subset R^{n+m}$, subject to the constraint that $\left| x_{f_i} \cdot W_{f_i}^{l} \right| \leq \rho_l$, where ρ_l is a large design constant.

The region Ψ denotes the smallest compact subset of R^{n+m} that includes all the values that (x, u) can take, i.e., $(x(t), u(t)) \in \Psi$ for all $t \geq 0$. Since by assumption $u(t)$ is uniformly bounded and the dynamical system to be identified is bounded input bounded output (BIBO) stable, the existence of such Ψ is ensured.

In particular, for $i = 1, 2, \ldots, n$, the optimal weight vector $W_{f_i}^{l\star}$ is defined as

$$W_{f_i}^{l\star} := \arg \min_{\left| \bar{x}_{f_i} W_{f_i}^{l} \right| \leq \rho_l} \left\{ \sup_{(x,u) \in \Psi} \left| f_i(x, u) + a_i x_i - \sum_{l=1}^{k} \bar{x}_{f_i} W_{f_i}^{l} s_l(x, u) \right| \right\}.$$

The formulation developed above follows the methodology of Kosmatopoulos et al. (1995) closely. Using this formulation, we now have a system of the form (2.41)

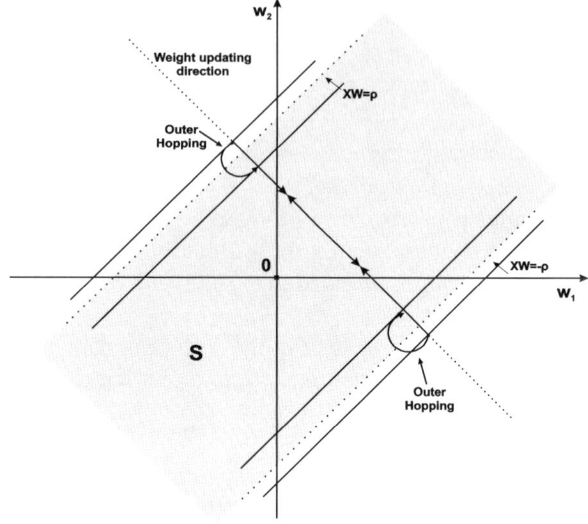

Fig. 2.3 Pictorial representation of outer parameter hopping during the identification procedure

instead of (2.23). It is noted that since $x(t)$ and $u(t)$ are bounded, the modeling error $\mu_i(t)$ is also bounded, i.e., $\sup_{t\geq 0} |\mu_i(t)| \leq \bar{\mu}_i$ for some finite constant $\bar{\mu}_i$.

Now, it is also of practical use to ensure that $\sum_{l=1}^{k} \bar{x}_{f_i} W_{f_i}^l s_l(x, u)$ does not approach even temporarily infinity because in this case the method may become algorithmically unstable. To avoid this situation we have to ensure that $\left| \bar{x}_{f_i} \cdot W_{f_i}^l \right| < \rho_l$, with ρ_l being a design parameter determining an external limit for $\bar{x}_{f_i} \cdot W_{f_i}^l$. We note that, since $\bar{x}_{f_i}$ and $W_{f_i}^l$ are row and column vectors, respectively, and since $\bar{x}_{f_i}$ has constant values, their product is linear in respect to the elements of $W_{f_i}^l$ and $\bar{x}_{f_i} \cdot W_{f_i}^l$ can describe a hyperplane. In the sequel, we consider the forbidden hyperplanes being defined by the equation $\left| \bar{x}_{f_i} \cdot W_{f_i}^l \right| = \rho_l$. When the weight vector reaches one of the forbidden hyperplanes $\left| \bar{x}_{f_i} \cdot W_{f_i}^l \right| = \rho_l$ and the direction of updating is toward the forbidden hyperplane, a *parameter hopping* is introduced that moves the weights inside the restricted area. A more analytical and general description of the novel method of *parameter hopping* is presented in Chap. 3 Sect. 3.2.2.

The above procedure is depicted in Fig. 2.3, in a simplified two-dimensional representation. The magnitude of hopping is $-\dfrac{\kappa_l P_i^l \left(\bar{x}_{f_i} W_{f_i}^l (\bar{x}_{f_i})^T \right)}{\operatorname{tr}\{(\bar{x}_{f_i})^T \bar{x}_{f_i}\}}$ being determined by following the vectorial proof given in Chap. 3 (where $b = W_{f_i}^l$ and our plane is described by equation $\bar{x}_{f_i} \cdot W_{f_i}^l = \rho_l$, with $\bar{x}_{f_i}$ the normal to it), with κ_l a positive constant (such as, $0 < \kappa_l P_i^l < 1$) decided appropriately from the designer and P_i^l is the l-th element of the gain matrix P_i.

In what follows, we develop a robust learning algorithm based on the F-RHONNs identifier employing the parameter hopping. Hence, the identifier is chosen as in

(2.24) where $W^l_{f_i}$ is the estimate of the unknown optimal weight matrix $W^{l\star}_{f_i}$. Using (2.24), (2.41), the state error $e_i = \hat{x}_i - x_i$ satisfies:

$$\dot{e}_i = -a_i e_i + \bar{x}_{f_i} \tilde{W}_{f_i} s_f(x, u) - \mu_i(t), \tag{2.43}$$

or in a more detailed form:

$$\dot{e}_i = -a_i e_i + \sum_{l=1}^{k} \bar{x}_{f_i} \tilde{W}^l_{f_i} s_l - \mu_i(t). \tag{2.44}$$

Owing to the presence of the modeling error $\mu_i(t)$, the learning law given by (2.26) is modified by performing *parameter hopping*, when $\bar{x}_{f_i} \cdot W^l_{f_i}$ reaches the outer forbidden planes as depicted in Fig. 2.3. $\bar{x}_{f_i} \cdot W^l_{f_i}$ is confined in space $S = \left\{ \bar{x}_{f_i} \cdot W^l_{f_i} : \left| \bar{x}_{f_i} \cdot W^l_{f_i} \right| \leq \rho_l \right\}$, lying between these hyperplanes. The weight updating law for $W^l_{f_i}$ can now be expressed as

$$\dot{W}^l_{f_i} = -\left(\bar{x}_{f_i} \right)^T e_i s_l P^l_i - \frac{\sigma_l \kappa_l P^l_i \left(\bar{x}_{f_i} W^l_{f_i} \left(\bar{x}_{f_i} \right)^T \right)}{\text{tr}\left\{ \left(\bar{x}_{f_i} \right)^T \bar{x}_{f_i} \right\}}, \tag{2.45}$$

with

$$\sigma_l = \begin{cases} 0 & \text{if } \bar{x}_{f_i} W^l_{f_i} = \pm\rho_l \\ & \text{and } \bar{x}_{f_i} \dot{W}^l_{f_i} <> 0 \; . \\ 1 & \text{otherwise} \end{cases} \tag{2.46}$$

In the current notation, the "$\pm$" symbol has a one to one correspondence with the "$<>$" one, meaning that "$+$" case corresponds to "$<$" case and "$-$" case corresponds to "$>$" case.

The above weight adjustment law is the same as (2.26) if $\bar{x}_{f_i} W^l_{f_i}$ belongs to a hypersphere of radius ρ_l. If initially $\bar{x}_{f_i} W^l_{f_i}(0)$ belongs to this hypersphere, one strategy that can be followed is to apply a "hopping" to the weight updating equation whenever a vector is approaching the forbidden outer hyperplane and is directed toward it. The "hopping" could send the weight back to the desired hyperspace allowing thus the algorithm to search the entire space for a better weight solution. Thus, in the case that the weights leave this hypersphere, the weight adjustment law is modified by the addition of a *hopping* term $-\frac{\kappa_l P^l_i \left(\bar{x}_{f_i} W^l_{f_i} \left(\bar{x}_{f_i} \right)^T \right)}{\text{tr}\left\{ \left(\bar{x}_{f_i} \right)^T \bar{x}_{f_i} \right\}}$, whose objective is to prevent the weight values from drifting to infinity. This modification appeared first in Boutalis et al. (2009). As it is explained in Chap. 3 (Remark 3.2) the weight hopping does not affect the existence of solutions of the dynamic equations of the model, so that Lyapunov stability arguments can be safely applied.

Now, we are ready to state the following theorem:

Theorem 2.4 *Consider the F-RHONNs model given by* (2.24) *whose weights are adjusted according to* (2.45), (2.46). *Then for* $i = 1, \ldots, n$ *and* $l = 1, \ldots, k$ *the following properties are guaranteed:*

1. $e_i, W_{f_i}^l, \dot{W}_{f_i}^l \in L_\infty$,
2. there exist constants r, s such that :
   $\int_0^t |e_i(t)|^2 \, d\tau \leq r + s \int_0^t |\mu_i(t)|^2 \, d\tau$.

Proof (1) Consider the Lyapunov candidate function:

$$V = \frac{1}{2} \sum_{i=1}^n \left(|\bar{x}_{f_i}|^2 e_i^2 \right) + \frac{1}{2} \sum_{i=1}^n \sum_{l=1}^k \left[\left(\bar{x}_{f_i} \tilde{W}_{f_i}^l \right)^T \left(P_i^l \right)^{-1} \left(\bar{x}_{f_i} \tilde{W}_{f_i}^l \right) \right]. \tag{2.47}$$

Taking the time derivatives of the Lyapunov function candidate (2.47) and taking into account (2.44), (2.45) we obtain

$$\dot{V} = \sum_{i=1}^n \left(|\bar{x}_{f_i}|^2 e_i \dot{e}_i \right) + \sum_{i=1}^n \sum_{l=1}^k \left[\left(\bar{x}_{f_i} \dot{W}_{f_i}^l \right)^T \left(P_i^l \right)^{-1} \left(\bar{x}_{f_i} \tilde{W}_{f_i}^l \right) \right]$$

$$= -\sum_{i=1}^n \left(a_i |\bar{x}_{f_i}|^2 e_i^2 + |\bar{x}_{f_i}|^2 e_i \mu_i \right)$$

$$+ |\bar{x}_{f_i}|^2 \sum_{i=1}^n \left(\sum_{l=1}^k \bar{x}_{f_i} e_i \tilde{W}_{f_i}^l s_l - \sum_{l=1}^k \bar{x}_{f_i} e_i \tilde{W}_{f_i}^l s_l \right)$$

$$+ \sum_{i=1}^n \frac{|\bar{x}_{f_i}|^2 \sum_{l=1}^k \sigma_l \kappa_l \left(\bar{x}_{f_i} W_{f_i}^l \right) \left(\bar{x}_{f_i} \tilde{W}_{f_i}^l \right)}{|\bar{x}_{f_i}|^2}$$

$$= -|\bar{x}_{f_i}|^2 \sum_{i=1}^n \left(a_i |e_i|^2 + e_i \mu_i \right) - \sum_{i=1}^n \sum_{l=1}^k \left[\sigma_l \kappa_l \left(\bar{x}_{f_i} W_{f_i}^l \right) \left(\bar{x}_{f_i} \tilde{W}_{f_i}^l \right) \right].$$

$$\tag{2.48}$$

Since $\tilde{W}_f = W_f - W_f^*$, we have that

$$\left(\bar{x}_{f_i} W_{f_i}^l \right) \left(\bar{x}_{f_i} \tilde{W}_{f_i}^l \right) = \bar{x}_{f_i} \left(\tilde{W}_{f_i}^l + W_{f_i}^{l*} \right) \left(\bar{x}_{f_i} \tilde{W}_{f_i}^l \right)$$

$$= \left(\bar{x}_{f_i} \tilde{W}_{f_i}^l \right) \left(\bar{x}_{f_i} \tilde{W}_{f_i}^l \right) + \left(\bar{x}_{f_i} W_{f_i}^{l*} \right) \left(\bar{x}_{f_i} \tilde{W}_{f_i}^l \right)$$

$$= \left| \bar{x}_{f_i} \tilde{W}_{f_i}^l \right|^2 + \left(\bar{x}_{f_i} W_{f_i}^{l*} \right) \left(\bar{x}_{f_i} \tilde{W}_{f_i}^l \right)$$

$$= \frac{1}{2} \left| \bar{x}_{f_i} \tilde{W}_{f_i}^l \right|^2 + \frac{1}{2} \left[\left| \bar{x}_{f_i} \tilde{W}_{f_i}^l \right|^2 + 2 \left(\bar{x}_{f_i} W_{f_i}^{l*} \right) \left(\bar{x}_{f_i} \tilde{W}_{f_i}^l \right) \right]$$

$$= \frac{1}{2} \left| \bar{x}_{f_i} \tilde{W}_{f_i}^l \right|^2 + \frac{1}{2} \left| \bar{x}_{f_i} W_{f_i}^l \right|^2 - \frac{1}{2} \left| \bar{x}_{f_i} W_{f_i}^{l*} \right|^2. \tag{2.49}$$

Since, by definition, $\left|\bar{x}_{f_i} \cdot W_{f_i}^{l*}\right| \leq \rho_l$ and $\left|\bar{x}_{f_i} \cdot W_{f_i}^{l}\right| > \rho_l$ for $\sigma_l = 1$, we have that:

$$\sum_{i=1}^{n}\sum_{l=1}^{k}\left[\frac{\sigma_l \kappa_l}{2}\left(\left|\bar{x}_{f_i} W_{f_i}^{l}\right|^2 - \left|\bar{x}_{f_i} W_{f_i}^{l*}\right|^2\right)\right] \geq 0,$$

therefore (2.48) becomes:

$$\dot{V} \leq \sum_{i=1}^{n}\left[-\left|\bar{x}_{f_i}\right|^2\left(a_i\left|e_i\right|^2 + e_i \mu_i\right)\right] - \sum_{i=1}^{n}\left[\sum_{l=1}^{k}\frac{\sigma_l \kappa_l}{2}\left|\bar{x}_{f_i}\tilde{W}_{f_i}^{l}\right|^2\right] \quad (2.50)$$

$$\leq \sum_{i=1}^{n}\left[-\frac{\left|\bar{x}_{f_i}\right|^2}{2}a_i\left|e_i\right|^2 - \sum_{l=1}^{k}\frac{\kappa_l}{2}\left|\bar{x}_{f_i}\tilde{W}_{f_i}^{l}\right|^2\right]$$

$$+ \sum_{i=1}^{n}\left[(1-\sigma_l)\sum_{l=1}^{k}\frac{\kappa_l}{2}\left|\bar{x}_{f_i}\tilde{W}_{f_i}^{l}\right|^2\right]$$

$$- \sum_{i=1}^{n}\left[\frac{\left|\bar{x}_{f_i}\right|^2}{2}\left(a_i\left|e_i\right|^2 + 2\mu_i e_i\right)\right]$$

$$\leq - \sum_{i=1}^{n}\frac{\left|\bar{x}_{f_i}\right|^2}{2}a_i\left|e_i\right|^2 - \sum_{i=1}^{n}\sum_{l=1}^{k}\frac{\kappa_l}{\lambda_{\max}\left(P_i^{-1}\right)}V\left(e_i, \tilde{W}_{f_i}^{l}\right)$$

$$+ \sum_{i=1}^{n}\left[(1-\sigma_l)\sum_{l=1}^{k}\frac{\kappa_l}{2}\left|\bar{x}_{f_i}\tilde{W}_{f_i}^{l}\right|^2 + \mu_i^2\right], \quad (2.51)$$

where $\lambda_{\max}\left(P_i^{-1}\right) > 0$ denotes the maximum eigenvalue of P_i^{-1}. Since,

$$\sum_{i=1}^{n}\left[(1-\sigma_l)\sum_{l=1}^{k}\frac{\kappa_l}{2}\left|\bar{x}_{f_i}\tilde{W}_{f_i}^{l}\right|^2\right] = \begin{cases} \sum_{i=1}^{n}\sum_{l=1}^{k}\frac{\kappa_l}{2}\left|\bar{x}_{f_i}\tilde{W}_{f_i}^{l}\right|^2 & \text{if } \bar{x}_{f_i}W_{f_i}^{l} = \pm\rho_l \\ & \text{and } \bar{x}_{f_i}\dot{W}_{f_i}^{l} >< 0 \\ 0 & \text{otherwise} \end{cases},$$

we obtain $\sum_{i=1}^{n}\left[(1-\sigma_l)\sum_{l=1}^{k}\frac{\kappa_l}{2}\left|\bar{x}_{f_i}\tilde{W}_{f_i}^{l}\right|^2\right] \leq \sum_{l=1}^{k}\kappa_l \rho_l^2$. Hence (2.51) can be written in the form

$$\dot{V} \leq -d - bV + c,$$

where $d = \sum_{i=1}^{n}\frac{a_i\left|\bar{x}_{f_i}\right|^2}{2}\left|e_i\right|^2, b = \sum_{i=1}^{n}\sum_{l=1}^{k}\frac{\kappa_l}{\lambda_{\max}\left(P_i^{-1}\right)}$ and $c = \sum_{i=1}^{n}\left(\sum_{l=1}^{k}\kappa_l \rho_l^2 + \bar{\mu}_i^2\right)$

with $\bar{\mu}_i$ an upper bound for μ_i. Therefore, when $V\left(e_i, \tilde{W}_{f_i}^{l}\right) \geq V_0 = \frac{c}{b}$, we have $\dot{V} \leq 0$, which in the sequel implies that V is bounded. Hence, $\tilde{W}^l_{f_i} \in L_\infty$ and $\mu_i \in L_\infty$ thus from (2.44) we result to $e_i \in L_\infty$. Furthermore, using (2.45) and the fact that $\bar{x}_{f_i}, e_i, s_l, P^l_i, W^l_{f_i} \in L_\infty$, we obtain $\dot{W}^l_{f_i} \in L_\infty$.

(2) Continuing the analysis, we note that by deleting the second square term in (2.50) we obtain

$$
\begin{aligned}
\dot{V} &\leq -\sum_{i=1}^{n} \left|\bar{x}_{f_i}\right|^2 \left(a_i \left|e_i\right|^2 + e_i \mu_i\right) \\
&\leq \sum_{i=1}^{n} \left[-\frac{\left|\bar{x}_{f_i}\right|^2}{2} a_i \left|e_i\right|^2 + \mu_i^2\right].
\end{aligned}
\tag{2.52}
$$

Integrating both sides of (2.52) yields

$$
\begin{aligned}
V(t) - V(0) &\leq \sum_{i=1}^{n} \left(-\frac{\left|\bar{x}_{f_i}\right|^2 a_i}{2} \int_0^t e_i^2(\tau)\, d\tau\right) + \sum_{i=1}^{n} \left(\int_0^t \mu_i^2(\tau)\, d\tau\right) \\
&\leq -\frac{\left|\bar{x}_{f_i}\right|^2 a_i}{2} \int_0^t \left|e_i(\tau)\right|^2 d\tau + \int_0^t \left|\mu_i(\tau)\right|^2 d\tau.
\end{aligned}
$$

Therefore,

$$
\int_0^t \left|e_i(\tau)\right|^2 d\tau \leq \frac{2}{\left|\bar{x}_{f_i}\right|^2 a_i} [V(t) - V(0)] + \frac{1}{\left|\bar{x}_{f_i}\right|^2 a_i} \int_0^t \left|\mu_i(\tau)\right|^2 d\tau
$$

$$
\leq r + s \int_0^t \left|\mu_i(\tau)\right|^2 d\tau
$$

where $r := \left(\frac{2}{\left|\bar{x}_{f_i}\right|^2 a_i}\right) \sup_{t \geq 0} [V(t) - V(0)]$ and $s := \frac{1}{\left|\bar{x}_{f_i}\right|^2 a_i}$. This proves the second part of Theorem 2.4.

One can observe that if the modeling error is removed, i.e., $\mu_i = 0$, then the *parameter hopping* will not guarantee the ideal properties of the adaptive law since it introduces a disturbance of the order of the design constant κ_l. This is one of the main drawbacks of *parameter hopping* that is removed with the next remark. One of the advantages of *parameter hopping* is that no assumption about bounds or location of the unknown $W^{l*}_{f_i}$ is made.

Remark 2.2 The drawback of parameter hopping is eliminated using a switching term κ_s, which activates the small feedback term around the integrator when the magnitude of $\bar{x}_{f_i} W^l_{f_i}$ exceeds a certain value ρ_0. The assumptions we make in this case are that $\left| \bar{x}_{f_i} W^{l*}_{f_i} \right| \le \rho_0$ and ρ_0 is known. Since ρ_0 is arbitrary, it can be chosen to be high enough to guarantee $\left| \bar{x}_{f_i} W^{l*}_{f_i} \right| \le \rho_0$ in the case where limited or no information is available about the location of $W^{l*}_{f_i}$. The switching parameter constant is given by

$$
\kappa_s(t) = \begin{cases}
0 & \text{if } \left| \bar{x}_{f_i} W^l_{f_i} \right| \le \rho_0 \\[2mm]
\left(\dfrac{\left| \bar{x}_{f_i} W^l_{f_i} \right|}{\rho_0} - 1 \right)^{q_0} \kappa_0 & \text{if } \rho_0 < \left| \bar{x}_{f_i} W^l_{f_i} \right| \le 2\rho_0 \;, \\[2mm]
\kappa_0 & \text{if } \left| \bar{x}_{f_i} W^l_{f_i} \right| > 2\rho_0
\end{cases}
\tag{2.53}
$$

where q_0 is any finite integer and ρ_0, κ_0 are design constants satisfying $\rho_0 > \left| \bar{x}_{f_i} W^{l*}_{f_i} \right|$ and $\kappa_0 > 0$. The switching from 0 to κ_0 is continuous to guarantee the existence and uniqueness of solution of the weight updating differential equation.

The gradient algorithm with the switching parameter constant κ_s given by (2.53) is described as

$$
\dot{W}^l_{f_i} = -\left(\bar{x}_{f_i} \right)^T p_i e_i s_l - \frac{\kappa_s P^l_i \left(\bar{x}_{f_i} W^l_{f_i} \left(\bar{x}_{f_i} \right)^T \right)}{\mathrm{tr} \left\{ \left(\bar{x}_{f_i} \right)^T \bar{x}_{f_i} \right\}}.
\tag{2.54}
$$

As shown in Ioannou and Fidan (2006), the adaptive law (2.53), (2.54) retains all the properties of (2.45), (2.46) and, in addition, guarantees the existence of a unique solution, in the sense of Caratheodory (Hale 1969). The issue of existence and uniqueness of solutions in adaptive systems is treated in detail in Polycarpou and Ioannou (1993) and Ioannou and Fidan (2006).

2.5 Simulation Results

To demonstrate the performance of the proposed identification scheme, we present simulations testing its approximation abilities. First, we present the identification of a system having identical model structure with the NF model, where we investigate the weight convergence to their optimal values. Next, we compare the proposed F-RHONNs scheme against the simple RHONNs in approximating the angular positions of joints 1, 2 of a two link robotic manipulator.

Table 2.1 Parameters of F-RHONNs simulations

Parameters	F-RHONNs values
Recursion constant	$a = 0.5$
Sigmoidal	$\alpha = 4$
	$\beta = 0.3$
	$\gamma = -1$
High-order terms	First order
	$s_f = (s(x), s(u))$
Fuzzy centers	$\bar{x}_f = \begin{bmatrix} 1.5 & 3 & 4 \end{bmatrix}$
Learning rate	$P = 0.05$
Initial weights	$W_f = [0]$
Optimal weights, scenario 1	$W_f^\star = \begin{bmatrix} -0.0908 & 0.2809 \\ -0.1816 & 0.5618 \\ -0.2421 & 0.7491 \end{bmatrix}$
Optimal weights, scenario 2	$W_f^\star = \begin{bmatrix} 0.818 & 0.298 \\ 2.392 & 0.878 \\ 3.408 & 1.252 \end{bmatrix}$

2.5.1 Parameter Identification in a Known Model Structure

In order to test the ability of the presented modeling and identification approach in regard to the convergence of weights to their optimal values during the identification procedure, we create a known NF structure of the form:

$$\dot{x} = -ax + X_f W_f^\star s_f, \tag{2.55}$$

and an F-RHONNs approximator:

$$\dot{\hat{x}} = -a\hat{x} + X_f W_f s_f, \tag{2.56}$$

with the same parameter values as shown in Table 2.1, except the weights that are adjusted according to the simple gradient descent learning law (2.26).

In the sequel, we consider two different scenarios.

Scenario 1: In this scenario, the train phase consists of 2,500 epochs where each one holds for 3 s, with sampling time 10^{-2} s and control input randomly selected values in the interval $[-1, 1]$. We recall that both x, u participate in the model through s_f. Every epoch trains our algorithm with the same data coming from the known NF structure (2.55). The initial values for the NF structure as well as for the F-RHONNs identifier are $\begin{bmatrix} x(0) \end{bmatrix} = \begin{bmatrix} \hat{x}(0) \end{bmatrix} = \begin{bmatrix} 0.2 \end{bmatrix}$.

After each epoch, the weight evolution is plotted together with their optimal values, as shown in Fig. 2.4. One can observe the fine behavior of the identifier that finally converges almost perfectly to the known NF structure.

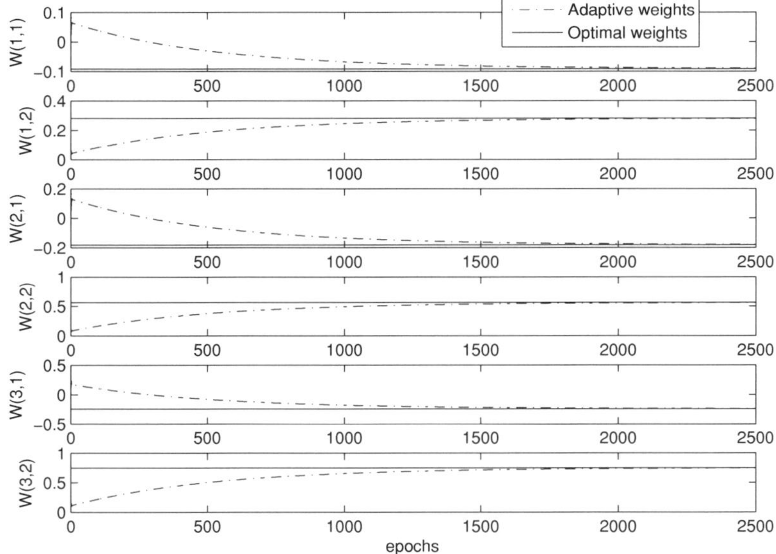

Fig. 2.4 Weight convergence to their optimal values during the training phase by using gradient descent learning law

Scenario 2: In this scenario, we repeat the same procedure with the same parameters as before except the optimal weight matrix target and the learning law, which is now the pure LS with Eqs. (2.30), (2.31). As we can see in Fig. 2.5, the weights converge faster to their optimal values when we use adaptive learning rate.

2.5.2 Two Link Robot Arm

The planar two-link revolute arm shown in Fig. 2.6 is used extensively in the literature for easy simulation of robotic controllers.

Its dynamics are given as (Lewis et al. 1993):

$$\dot{x}_1 = x_3$$
$$\dot{x}_2 = x_4$$
$$[\dot{x}_3 \ \dot{x}_4]^T = -M^{-1}(N + \tau),$$ (2.57)

where $x_1 = \theta_1$ is the angular position of joint 1, $x_2 = \theta_2$ is the angular position of joint 2, $x_3 = \dot{\theta}_1$ is the angular velocity of joint 1, and $x_4 = \dot{\theta}_2$ is the angular velocity of joint 2. Also, the matrices M, N have the following form:

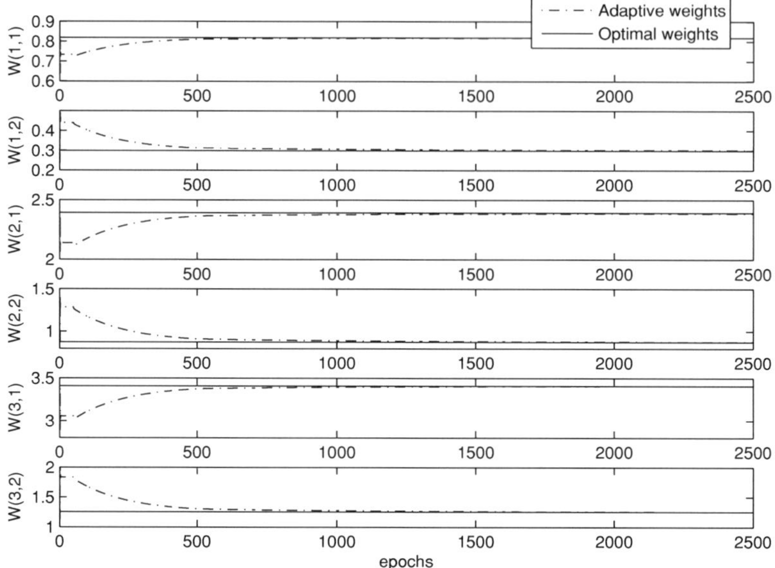

Fig. 2.5 Weight convergence to their optimal values during the training phase using pure LS algorithm

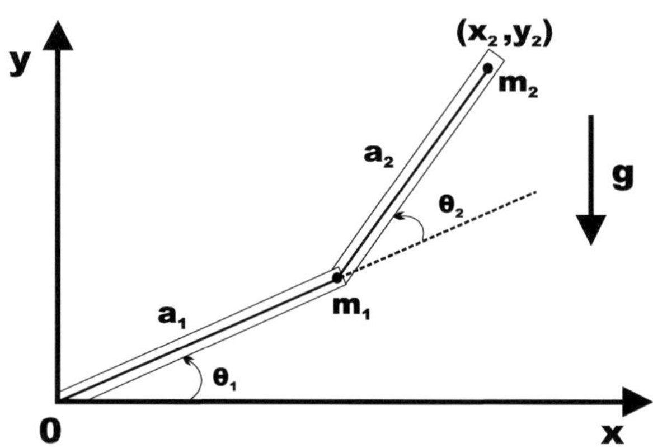

Fig. 2.6 Two-link planar elbow arm

$$M = \begin{bmatrix} M(1,1) & M(1,2) \\ M(2,1) & M(2,2) \end{bmatrix},$$

with

$$M(1,1) = (m_1 + m_2)\,a_1^2 + m_2 a_2^2 + 2m_2 a_1 a_2 \cos(x_2),$$
$$M(1,2) = m_2 a_2^2 + m_2 a_1 a_2 \cos(x_2),$$

$$M(2, 1) = m_2 a_2^2 + m_2 a_1 a_2 \cos(x_2),$$
$$M(2, 2) = m_2 a_2^2,$$

and

$$N = \begin{bmatrix} N(1, 1) \\ N(2, 1) \end{bmatrix},$$

with

$$N(1, 1) = -m_2 a_1 a_2 \left(2 x_3 x_4 + x_4^2 \right) \sin(x_2)$$
$$+ (m_1 + m_2) g a_1 \cos(x_1) + m_2 g a_2 \cos(x_1 + x_2),$$
$$N(2, 1) = m_2 a_1 a_2 x_3^2 \sin(x_2) + m_2 g a_2 \cos(x_1 + x_2).$$

We took the arm parameters as $a_1 = a_2 = 1\,\text{m}$, $m_1 = m_2 = 1\,\text{kg}$. In the simulations carried out, the aim is not the control of the system but only to test the identification performance of the proposed scheme. Therefore, we use Eqs. (2.57) as a means for deriving training data. These data help the designer to choose the appropriate fuzzy centers with the help of one of the well-known clustering methods such as fuzzy c-mean clustering. After that, the simulations take place in two different phases, which are presented below.

2.5.2.1 Training Phase

In this phase, our main purpose is to calculate the optimal weight matrix $W_f^\star$ after training our F-RHONNs model with the gradient descent algorithm given by (2.26), in conjunction with the switching parameter hopping condition given by Eqs. (2.53), (2.54). The training phase consists of 1,000 epochs where each one lasts for 2 s, with sampling time 10^{-3} s. Every epoch trains our algorithm with the same data coming from the real system with the same inputs randomly selected in the interval $[-1, 1]$. The initial values for the real system states are:

$$\begin{bmatrix} x_1(0) & x_2(0) & x_3(0) & x_4(0) \end{bmatrix} = \begin{bmatrix} -0.8 & -0.4 & 0 & 0 \end{bmatrix},$$

while for the F-RHONNs algorithm are

$$\begin{bmatrix} \hat{x}_1(0) & \hat{x}_2(0) & \hat{x}_3(0) & \hat{x}_4(0) \end{bmatrix} = \begin{bmatrix} 0 & 0 & 0 & 0 \end{bmatrix}.$$

The weights extracted from every epoch becomes the initial values for the weights during the next run (epoch).

In Tables 2.2 and 2.3 we present the parameter values that have been used for the simulations of F-RHONNs and RHONN approaches (as described in Sect. 1.2 and Rovithakis and Christodoulou (2000)), respectively.

Table 2.2 Parameters of F-RHONNs algorithm and updating law for robot two link simulations

Parameters	F-RHONNs values
Recursion constant	$a_1 = 7.7$
	$a_2 = 4.04$
Sigmoidal	$\alpha = 6.08$
	$\beta = 0.3$
	$\gamma = -7.696$
High-order terms	First order
	$s_f = (s(x_1), s(x_2), s(x_3), s(x_4), s(u_1), s(u_2))$
Fuzzy centers	$\bar{x}_{f_1} = \begin{bmatrix} -2.1 & -1.7 & -1.2 & -0.5 \end{bmatrix}$
	$\bar{x}_{f_2} = \begin{bmatrix} -1.8 & -1.4 & 1.5 & 1.8 \end{bmatrix}$
Learning rate	$P_1 = 0.00428$
	$P_2 = 0.00384$
Initial weights	$W_{f_1} = [0]$
	$W_{f_2} = [0]$
Hopping constants	$\kappa_l = 0.675$
	$\rho_l = 181$

Table 2.3 Parameters of RHONN algorithm and updating law for robot two-link simulations

Parameters	RHONN values
Recursion constant	$a_1 = 5.06$
	$a_2 = 5.19$
Sigmoidal	$\alpha = 7.93$
	$\beta = 0.25$
	$\gamma = -7.74$
High-order terms	First order
	$s_f = (s(x_1), s(x_2), s(x_3), s(x_4), s(u_1), s(u_2))$
Learning rate	$P_1 = 0.00692$
	$P_2 = 0.00766$
Initial weights	$W_{f_1} = [0]$
	$W_{f_2} = [0]$

2.5.2.2 Testing Phase

When the training stops, we proceed to test the abilities of the trained NF model using as input signal, values constraint in the interval $[-1, 1]$, which have the following form:

$$u_1(k) = \sum_{j=1}^{16} \frac{j}{196} \cdot \sin\left(\frac{\pi}{100} \cdot j \cdot k\right), \tag{2.58}$$

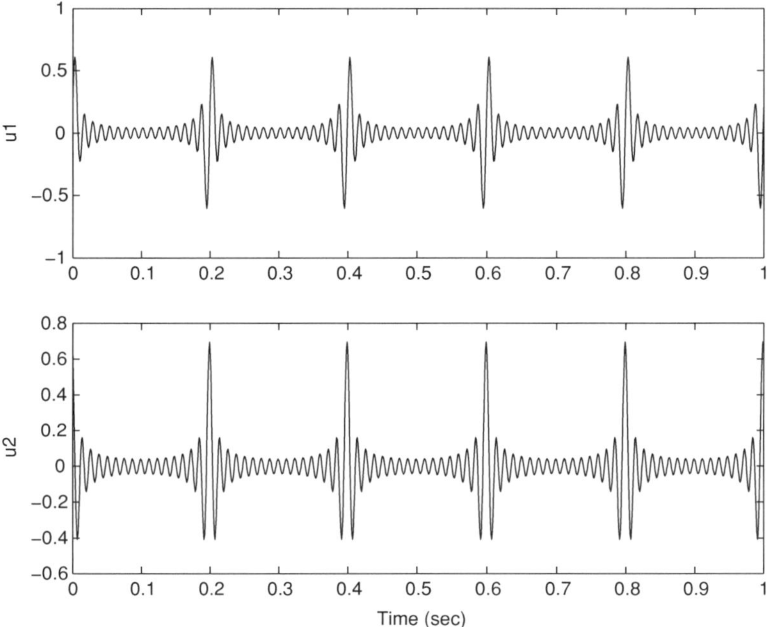

Fig. 2.7 Evolution of input signals u_1 and u_2

$$u_2(k) = \sum_{j=1}^{16} \frac{j}{196} \cdot \cos\left(\frac{\pi}{100} \cdot j \cdot k\right).\tag{2.59}$$

Figure 2.7 shows the evolution of the input signals (2.58) and (2.59), respectively.

The weight matrix derived from the training phase, which corresponds to the angular position of joint 1, 2, is given for the F-RHONNs as

$$W^{\star}_{f_{1,2}} = \begin{bmatrix} -2.0785 & -0.0284 & -0.2853 & 0.0154 & 1.2161 & 1.1566 \\ -4.9884 & -0.0680 & -0.6847 & 0.0370 & 2.9187 & 2.7759 \\ -7.0669 & -0.0964 & -0.9699 & 0.0524 & 4.1348 & 3.9325 \\ -8.7297 & -0.1190 & -1.1982 & 0.0647 & 5.1076 & 4.8578 \\ -0.1399 & -2.5484 & 0.0234 & -0.7536 & 1.5823 & 1.8583 \\ -0.1089 & -1.9821 & 0.0182 & -0.5862 & 1.2307 & 1.4453 \\ 0.1166 & 2.1236 & -0.0195 & 0.6279 & -1.3186 & -1.5485 \\ 0.1399 & 2.5484 & -0.0234 & 0.7536 & -1.5823 & -1.8583 \end{bmatrix}$$

while for the RHONN as:

$$W^{\star}_{f_{1,2}} = \begin{bmatrix} 8.0216 & -0.1592 & 1.8826 & -0.0931 & -4.7609 & -5.0891 \\ -0.7083 & 9.1152 & -0.0492 & 1.9712 & -4.9517 & -5.4681 \end{bmatrix}.$$

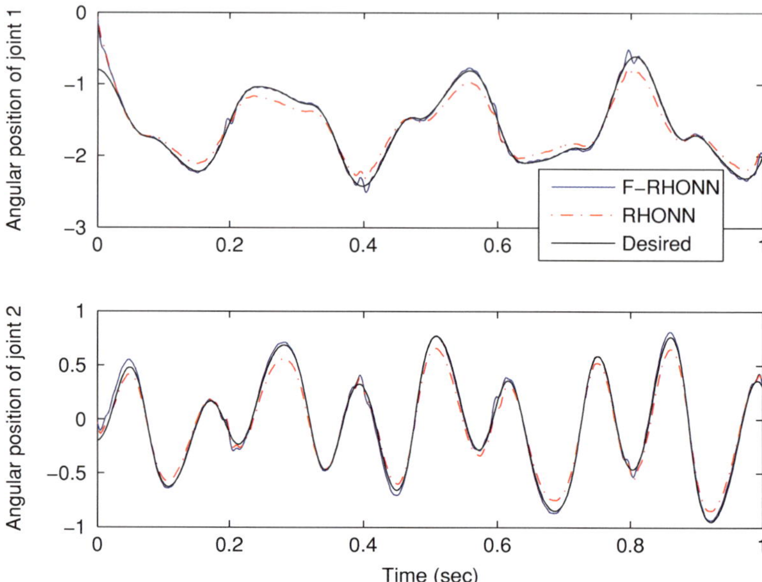

Fig. 2.8 Approximation of robot angular position 1 and 2

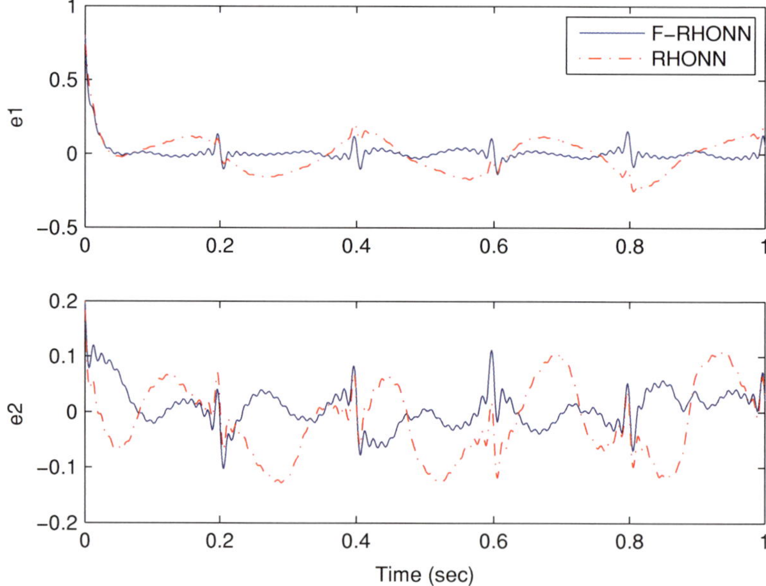

Fig. 2.9 Identification errors of robot angular positions 1 and 2 when we use different initial conditions

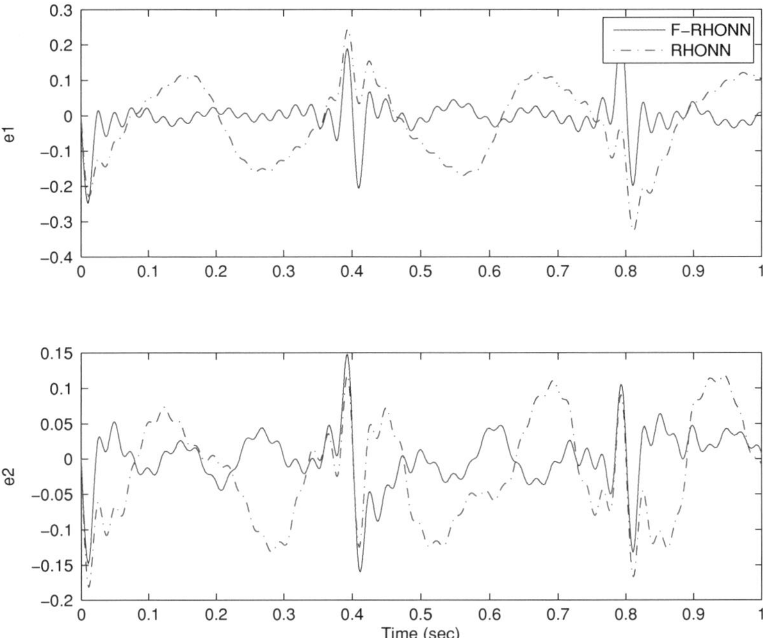

Fig. 2.10 Identification errors of robot's angular positions 1 and 2 when we use the same initial condition

Finally, we run our algorithm and RHONN approach for the above optimal weight matrices and the same control sequence with sampling time 10^{-3} s for 1 s. It is our intention to compare the approximation abilities of the proposed dynamic NF network (2.24) with RHONNs, (Rovithakis 2000) in approximating Eq. (2.57).

Figure 2.8 gives the approximation of robot states x_1, x_2 while Fig. 2.9 presents the evolution of identification errors, for RHONN and F-RHONNs models. The MSE measured as 0.0042, 0.0056 for F-RHONNs and 0.0148, 0.0194 for RHONN concerning states x_1, x_2, respectively. One can see that the dynamic NF networks are more powerful than the simple neural networks.

Once again, we test our approximator capabilities against RHONNs changing the frequency of input signals (divided by 2) and the initial values of both methods that are equal to the initial values of real system. Figure 2.10 shows that when the initial conditions of the robotic system and F-RHONNs identifier become equal the behavior of our approximator remains or becomes even better.

2.6 Summary

In this chapter, we presented a new recurrent neurofuzzy model, termed F-RHONN, for the identification of unknown nonlinear dynamical systems. It is based on multiple HONN approximators and the fuzzy partitioning of a fuzzy system output. Every HONN is specialized to work in a small region of the whole unknown system around the fuzzy output center.

In the sequel, we investigated the approximation capabilities of the proposed F-RHONNs when they are trained by different algorithms such as gradient descent and pure LS, which are proved to be Lyapunov stable. Furthermore, we examine the robustness of the training algorithms by employing a novel approach of switching parameter hopping instead of the classical σ-modification and prove once again that it is Lyapunov stable.

In the simulation results section, we demonstrated the approximation capabilities of our approach by presenting the convergence of identifier weights to their optimal values when we use a known model structure. In the sequel, the F-RHONN model was compared with simple RHONN showing the superiority of the NF model in approximating two link robot arm state variables.

References

Bishop, E. (1961). A generalization of the Stone–Weierstrass theorem. *Pacific Journal of Mathematics, 11,* 777–783.

Boutalis, Y. S., Theodoridis, D. C., & Christodoulou, M. A. (2009). A new neuro fds definition for indirect adaptive control of unknown nonlinear systems using a method of parameter hopping. *IEEE Transactions on Neural Networks, 20,* 609–625.

Christodoulou, M. A., Theodoridis, D. C., & Boutalis, Y. S. (2007). Building optimal fuzzy dynamical systems description based on recurrent neural network approximation. *Conference of Networked Distributed Systems for Intelligent Sensing and Control* (pp. 82–93). Greece: Kalamata.

Hale, J. (1969). *Ordinary differential equations.* New York: Wiley-Interscience.

Ioannou, P., & Fidan, B. (2006). Advances in design and control. In *Adaptive control tutorial.* Philadelphia: SIAM.

Kosmatopoulos, E. B., & Christodoulou, M. A. (1996). Recurrent neural networks for approximation of fuzzy dynamical systems. *International Journal of Intelligent Control and Systems, 1,* 223–233.

Kosmatopoulos, E. B., Polycarpou, M. M., Christodoulou, M. A., & Ioannou, P. A. (1995). High-order neural network structures for identification of dynamical systems. *IEEE Transactions Neural Networks, 6,* 422–431.

Lewis, F. L., Abdallah, C. T., & Dawson, D. M. (1993). *Control of robot manipulators.* New York: Macmillan.

Polycarpou, M., & Ioannou, P. (1993). On the existence and uniquness of solutions in adaptive control systems. *IEEE Transactions on Automatic Control, 38,* 474–479.

Rovithakis, G. & Christodoulou, M. A. (2000). Adaptive control with recurrent high order neural networks (theory and industrial applications). In *Advances in industrial control.* London: Springer.

Rovithakis, G. A. (2000). Performance of a neural adaptive tracking controller for multi-input nonlinear dynamical systems in the presence of additive and multiplicative external disturbances. *IEEE Transaction SMC—Part A, 30,* 720–730.

Stone, M. H. (1948). The generalized weierstrass approximation theorem. *Mathematics Magazine*, *21*, 167–184.

Theodoridis, D., Boutalis, Y., & Christodoulou, M. (2009). A new neuro-fuzzy dynamical system definition based on high order neural network function approximators. In *European Control Conference ECC-09*. Budapest, Hungary.

Theodoridis, D., Boutalis, Y., & Christodoulou, M. (2012). Dynamical recurrent neuro-fuzzy identification schemes employing switching parameter hopping. *International Journal of Neural Systems, 22*, 16 pages.

Chapter 3
Indirect Adaptive Control Based on the Recurrent Neurofuzzy Model

3.1 Neurofuzzy Identification of Affine in the Control Systems

We consider affine in the control, nonlinear dynamical systems of the form:

$$\dot{x} = f(x) + G(x) \cdot u, \tag{3.1}$$

where the state $x \in R^n$ is assumed to be completely measured, the control u is in R^q, f is an unknown smooth vector field called the drift term and G is a matrix with its rows representing the unknown smooth controlled vector fields $g_i, i = 1, 2, \ldots, n$. In order the above system to be controllable, we assume that $g_i(x) \neq 0$.

The above class of continuous-time nonlinear systems are called affine, because in (3.1) the control input appears linear with respect to g_i. The main reason for considering this class of nonlinear systems is that many of the systems encountered in engineering, are by nature or design, affine.

The following mild assumptions are also imposed on (3.1), to guarantee the existence and uniqueness of solution for any finite initial condition and $u \in U_c$.

Assumption 1 Given a class $U_c \subset R^q$ of admissible inputs, then for any $u \in U_c$ and any finite initial condition, the state trajectories are uniformly bounded for any finite $T > 0$. Meaning that we do not allow systems processing trajectories which escape at infinite, in finite time T, T being arbitrarily small. Hence, $|x(T)| < \infty$.

Assumption 2 The vector fields f, g_i are continuous with respect to their arguments and satisfy a local Lipchitz condition so that the solution $x(t)$ of (3.1) is unique for any finite initial condition and $u \in U_c$.

Y. Boutalis et al., *System Identification and Adaptive Control*,
Advances in Industrial Control, DOI: 10.1007/978-3-319-06364-5_3,
© Springer International Publishing Switzerland 2014

3.1.1 Neurofuzzy Modeling

We are using an affine in the control fuzzy dynamical system, which approximates
the system in (3.1) and uses two or more fuzzy subsystem blocks (Theodoridis et al.
2009a) for the approximation of $f_i(x)$ and $g_{ij}(x)$ according to F-RHONNs model
which has been described in Chap. 2, Sect. 2.1 and can be written as follows:

$$\hat{f}_i(x) = -a_i \hat{x}_i + \sum \bar{x}_{f_i}^p \times (I')_{f_i}^p(x), \tag{3.2}$$

$$\hat{g}_{ij}(x) = \sum \bar{x}_{g_{ij}}^p \times (I')_{g_{ij}}^p(x), \tag{3.3}$$

where $a_i > 0$, $(I')_{f_i}$, $(I')_{g_{ij}}$ are appropriately weighted fuzzy rule indicator
functions.

According to Sect. 2.1, every WIF can be approximated with the help of a suitable
HONN. Therefore, every $(I')_{f_i}$, $(I')_{g_{ij}}$ can be replaced with a corresponding HONN
as follows:

$$\hat{f}_i(x|W_f) = -a_i \hat{x}_i + \sum \bar{x}_{f_i}^p \times N_{f_i}^p(x), \tag{3.4}$$

$$\hat{g}_{ij}(x|W_g) = \sum \bar{x}_{g_{ij}}^p \times N_{g_{ij}}^p(x), \tag{3.5}$$

where N_{f_i}, $N_{g_{ij}}$ are appropriate HONNs.

So, the optimal approximation of $f(x)$ and $G(x)$ subfunctions of the dynamical
system becomes:

$$f_i(x|W_f^*) = -a_i x_i + \sum \bar{x}_{f_i}^p \times (N^*)_{f_i}^p(x), \tag{3.6}$$

$$g_{ij}(x|W_g^*) = \sum \bar{x}_{g_{ij}}^p \times (N^*)_{g_{ij}}^p(x). \tag{3.7}$$

Next, following the rationale presented in Sect. 2.1 and rearranging NN terms
so that the summations are carried over the number of the fuzzy output partitions,
the affine in the control fuzzy dynamical system in (3.2), (3.3) is replaced by the
following equivalent affine F-RHONNs, which depends on the centers of the fuzzy
output partitions $\bar{x}_{f_i}$ and $\bar{x}_{g_{ij}}$:

$$\dot{\hat{x}}_i = \hat{f}_i(x|W_f) + \hat{g}_{ij}(x|W_g)u_{ij}$$

$$= -a_i \hat{x}_i + \sum_{p=1}^{q} \bar{x}_{f_i}^p \cdot \sum_{l=1}^{k} w_{f_i}^{pl} \cdot s_l(x) + \sum_{j=1}^{m} \left(\sum_{p=1}^{q} \bar{x}_{g_{ij}}^p \cdot \sum_{l=1}^{k} w_{g_{ij}}^{pl} \cdot s_l(x) \right) \cdot u_{ij},$$

$$\tag{3.8}$$

or in a more compact form:

$$\dot{\hat{x}} = A\hat{x} + X_f W_f s_f(x) + X_g W_g S_g(x)u, \tag{3.9}$$

where A is a $n \times n$ stable matrix, which for simplicity can be taken to be diagonal as $A = \mathrm{diag}[-a_1, -a_2, \ldots, -a_n]$, with $a_i > 0$. Also, X_f, X_g are matrices containing the centers of the partitions of every fuzzy output variable $f_i(x)$ and $g_{ij}(x)$, respectively, $s_f(x)$ is a vector and $S_g(x)$ is a matrix containing high order combinations of sigmoid functions of the state x and W_f, W_g are matrices containing respective neural weights according to Sect. 2.1 and Eq. (3.9). The dimensions and the contents of all the above matrices are chosen so that $X_f W_f s_f(x)$ is a $n \times 1$ vector and $X_g W_g S_g(x)$ is a $n \times m$ matrix. For notational simplicity we assume that all output fuzzy variables are partitioned to the same number, q, of partitions. Under these specifications X_f is a $n \times n \cdot q$ block diagonal matrix of the form $X_f = \mathrm{diag}(\bar{x}_{f_1}, \bar{x}_{f_2}, \ldots, \bar{x}_{f_n})$ with $\bar{x}_{f_i}$ being a q-dimensional row vector of the form:

$$\bar{x}_{f_i} = \begin{bmatrix} \bar{x}_{f_i}^1 & \bar{x}_{f_i}^2 & \cdots & \bar{x}_{f_i}^q \end{bmatrix},$$

or in a more detailed form:

$$X_f = \begin{bmatrix} \bar{x}_{f_1}^1 & \cdots & \bar{x}_{f_1}^q & 0 & \cdots & 0 & 0 & \cdots & 0 \\ 0 & \cdots & 0 & \bar{x}_{f_2}^1 & \cdots & \bar{x}_{f_2}^q & 0 & \cdots & 0 \\ \multicolumn{9}{c}{\dotfill} \\ 0 & \cdots & 0 & 0 & \cdots & 0 & \bar{x}_{f_n}^1 & \cdots & \bar{x}_{f_n}^q \end{bmatrix},$$

where $\bar{x}_{f_i}^p$ with $p = 1, 2, \ldots, q$, denotes the center of the p-th partition of f_i. Also, $s_f(x) = \begin{bmatrix} s_1(x) & \ldots & s_k(x) \end{bmatrix}^T$, where each $s_l(x)$ with $l = 1, 2, \ldots, k$, is a high order combination of sigmoid functions of the state variables and W_f is a $n \cdot q \times k$ matrix with neural weights. W_f assumes the form $W_f = \begin{bmatrix} W_{f_1} & \cdots & W_{f_n} \end{bmatrix}^T$, where each W_{f_i} is a matrix $\begin{bmatrix} w_{f_i}^{pl} \end{bmatrix}_{m \times k}$ as described in Sect. 2.1. X_g is a $n \times n \cdot m \cdot q$ block diagonal matrix of the form $X_g = \mathrm{diag}(\bar{x}_{g_{1j}}, \bar{x}_{g_{2j}}, \ldots, \bar{x}_{g_{nj}})$ with each $\bar{x}_{g_{ij}}$ $(i = 1, 2, \ldots, n, \ j = 1, 2, \ldots, m)$ being a q-dimensional row vector of the form:

$$\bar{x}_{g_{ij}} = \begin{bmatrix} \bar{x}_{g_{ij}}^1 & \bar{x}_{g_{ij}}^2 & \cdots & \bar{x}_{g_{ij}}^q \end{bmatrix},$$

where $\bar{x}_{g_{ij}}^k$ denotes the center of the k-th fuzzy partition of g_{ij}. In a more detailed form:

$$X_g = \begin{bmatrix} \bar{x}_{g11} & \cdots & \bar{x}_{g1m} & 0 & \cdots & 0 & 0 & \cdots & 0 \\ 0 & \cdots & 0 & \bar{x}_{g21} & \cdots & \bar{x}_{g2m} & 0 & \cdots & 0 \\ \multicolumn{9}{c}{\dotfill} \\ 0 & \cdots & 0 & 0 & \cdots & 0 & \bar{x}_{gn1} & \cdots & \bar{x}_{gnm} \end{bmatrix}.$$

Also, W_g is a $n \cdot m \cdot q \times n \cdot m$ block diagonal matrix of the form:

$$W_g = \mathrm{diag}\left(\left[W_{g1}, W_{g2}, \ldots, W_{gn}\right]\right),$$

where each W_{gi} with $i = 1, 2, \ldots, n$ is also a block $m \cdot q \times m$ diagonal matrix of the form:

$$W_{gi} = \begin{bmatrix} w_{gi1}^{11} & 0 & \cdots & 0 \\ w_{gi1}^{21} & 0 & \cdots & 0 \\ \vdots & \vdots & \cdots & \vdots \\ w_{gi1}^{q1} & 0 & \cdots & 0 \\ 0 & w_{gi2}^{11} & \cdots & 0 \\ 0 & w_{gi2}^{21} & \cdots & 0 \\ \vdots & \vdots & \cdots & \vdots \\ 0 & w_{gi2}^{q1} & \cdots & 0 \\ 0 & 0 & \ddots & 0 \\ 0 & 0 & \cdots & w_{gim}^{11} \\ 0 & 0 & \cdots & w_{gim}^{21} \\ \vdots & \vdots & \cdots & \vdots \\ 0 & 0 & \cdots & w_{gim}^{q1} \end{bmatrix}.$$

Finally, S_g is a $n \cdot m \times m$ matrix of the form $S_g = [S_{g1}, S_{g2}, \ldots, S_{gn}]^T$, where each S_{gi} is a diagonal $m \times m$ matrix $S_{gi} = \mathrm{diag}(s_i, \ldots, s_i)$ with the diagonal element $s_i(x)$ being a high order combination of sigmoid functions of the state variables. Thus, matrix S_g in a detailed form can be written as:

$$S_g = \begin{bmatrix} s_1^{1,1} & \cdots & 0 \\ \vdots & \ddots & \vdots \\ 0 & \vdots & s_1^{m,m} \\ s_2^{1,1} & \cdots & 0 \\ \vdots & \ddots & \vdots \\ 0 & \cdots & s_2^{m,m} \\ \vdots & \vdots & \vdots \\ s_n^{1,1} & \cdots & 0 \\ \vdots & \ddots & \vdots \\ 0 & \cdots & s_n^{m,m} \end{bmatrix}.$$

Remark 3.1 It has to be mentioned here that the employed NF model, finally given by (3.9), offers some advantages over other fuzzy or neural adaptive models. Considering the model from the adaptive fuzzy system (AFS) point of view, the main advantage is that the proposed approach is much less vulnerable to initial design assumptions because there is no need for a-priori information related to the IF part of the rules (type and centers of membership functions, number of rules). This information is replaced by the use of HONNs. There is only the requirement for some a-priori knowledge of the centers of the output membership functions (THEN part), which can be further relaxed if one uses a bilinear parametric model according to Boutalis et al. (2013). Considering the model from the NN point of view, the final representation of the dynamic equations is actually a combination of high order neural networks, each one being specialized in approximating a function related to a corresponding center of output state membership function. This way, instead of having one large HONN trying to approximate the entire system, we have many, probably smaller, specialized HONNs. Conceptually, this strategy is expected to present better approximation results; this is also verified in the simulations section. Moreover, as it will be seen in Sect. 3.2.1, due to the particular bond of each HONN with one center of an output state membership function, the existence of the control law is ensured by introducing a novel technique of parameter "hopping" in the corresponding weight updating laws.

3.1.2 Adaptive Parameter Identification

We assume the existence of only parameter uncertainty, so, we can take into account that the actual system (3.1) can be modeled by the following NF form:

$$\dot{x} = Ax + X_f W_f^\star s_f(x) + X_g W_g^\star S_g(x)u. \tag{3.10}$$

Define now, the error between the identifier states and the real states as:

$$e = \hat{x} - x. \tag{3.11}$$

Then taking the derivative of Eq. (3.11) and after substituting Eqs. (3.9) and (3.10) we obtain the error equation:

$$\dot{e} = Ae + X_f \tilde{W}_f s_f(x) + X_g \tilde{W}_g S_g(x)u, \tag{3.12}$$

where $\tilde{W}_f = W_f - W_f^\star$ and $\tilde{W}_g = W_g - W_g^\star$. Regarding the identification of W_f and W_g in (3.9) we are now able to state the following theorem.

Theorem 3.1 *Consider the identification scheme given by* (3.12). *The learning laws:*

(a) *For the elements of* W_{f_i}:

$$\dot{w}_{f_i}^{pl}(x) = -\bar{x}_{f_i}^{p}\gamma_f p_i e_i s_l(x) \tag{3.13}$$

or equivalently in vector form $\dot{W}_{f_i}^{l}(x) = -\left(\bar{x}_{f_i}\right)^{T}\gamma_f p_i e_i s_l(x)$, *with* $i = 1,\ldots,n$, $p = 1,\ldots,q$ *and* $l = 1,\ldots,k$.

(b) *For the elements of* $W_{g_{ij}}$:

$$\dot{w}_{g_{ij}}^{p}(x) = -\bar{x}_{g_{ij}}^{p}\gamma_g p_i e_i u_j s_i(x) \tag{3.14}$$

or equivalently $\dot{W}_{g_{ij}}(x) = -\left(\bar{x}_{g_{ij}}\right)^{T}\gamma_g p_i e_i u_j s_i(x)$, *with* $i = 1,\ldots,n$ *and* $j = 1,\ldots,m$ *guarantees the following properties:*

- $e, \tilde{W}_f, \tilde{W}_g \in L_\infty, \quad e \in L_2,$
- $\lim_{t\to\infty} e(t) = 0, \ \lim_{t\to\infty} \dot{W}_f(t) = 0, \ \lim_{t\to\infty} \dot{W}_g(t) = 0.$

Proof Consider the Lyapunov function candidate:

$$V\left(e, \tilde{W}_f, \tilde{W}_g\right) = \frac{1}{2}e^T Pe + \frac{1}{2\gamma_f}\mathrm{tr}\left\{\tilde{W}_f^T \tilde{W}_f\right\} + \frac{1}{2\gamma_g}\mathrm{tr}\left\{\tilde{W}_g^T \tilde{W}_g\right\}.$$

Where $P > 0$ is chosen to satisfy the Lyapunov equation:

$$PA + A^T P = -I.$$

Taking the derivative of the Lyapunov function candidate we get:

$$\dot{V} = -\frac{1}{2}\|e\|^2 + e^T PX_f \tilde{W}_f s_f + e^T PX_g \tilde{W}_g S_g u + \frac{1}{\gamma_f}\mathrm{tr}\{\dot{W}_f^T \tilde{W}_f\} + \frac{1}{\gamma_g}\mathrm{tr}\{\dot{W}_g^T \tilde{W}_g\},$$

which results in:

$$\dot{V} = -\frac{1}{2}\|e\|^2,$$

when

$$\frac{1}{\gamma_f}\mathrm{tr}\{\dot{W}_f^T \tilde{W}_f\} = -e^T PX_f \tilde{W}_f s_f, \tag{3.15}$$

$$\frac{1}{\gamma_g}\mathrm{tr}\{\dot{W}_g^T \tilde{W}_g\} = -e^T PX_g \tilde{W}_g S_g u. \tag{3.16}$$

Then, taking into account the form of W_f and W_g and the matrix trace property that for $\mathrm{tr}\{\dot{X}^T \tilde{X}\} = A\tilde{X}B \Rightarrow \dot{X}^T = BA \Rightarrow \dot{X} = A^T B^T$, the above equations result in the following updating laws, whose element-wise learning laws are given in (3.13), (3.14).

$$\dot{W}_f = -\gamma_f X_f^T P e s_f^T, \tag{3.17}$$

$$\dot{W}_g = -\gamma_g X_g^T P e u^T S_g^T. \tag{3.18}$$

Thus, $\dot{V}$ is negative semidefinite. Since $\dot{V} \leq 0$, we conclude that $V \in L_\infty$, which implies that e, $\tilde{W}_f$, $\tilde{W}_g \in L_\infty$. Furthermore, $W_f = \tilde{W}_f + W_f^\star$ and $W_g = \tilde{W}_g + W_g^\star$ are also bounded. Since V is a nonincreasing function of time and bounded from below, the $\lim_{t \to \infty} V = V_\infty$ exists; therefore, by integrating $\dot{V}$ from 0 to ∞ we have:

$$\int_0^\infty |e|^2 \, dt \leq [V(0) - V_\infty] < \infty,$$

which implies that $e \in L_2$.

Since $e \in L_2 \cap L_\infty$, using Barbalat's Lemma (Popov 1973), we conclude that $\lim_{t \to \infty} e(t) = 0$.

Finally, using the boundedness of X_f, X_g, $s_f(x)$, $S_g(x)$, u, and the convergence of $e(t)$ to zero, we have that $\dot{W}_f$ and $\dot{W}_g$ also converges to zero (Ioannou and Fidan 2006). $\qquad\square$

3.2 The Indirect Control Scheme

The state regulation problem is known as our attempt to force the state to zero from an arbitrary initial value by applying appropriate feedback control to the plant input. However, the problem as it is stated above for the system (3.1), is very difficult or even impossible to be solved since the vector fields f, g_i are assumed to be completely unknown. To overcome this problem, following the NF representation described above, we approximate the unknown plant by the following model:

$$\dot{\hat{x}} = A\hat{x} + X_f W_f s_f(x) + X_g W_g S_g(x) u. \tag{3.19}$$

If we assume only parameter uncertainty then the unknown plant could be described by the following model arriving from the same NF representation:

$$\dot{x} = Ax + X_f W_f^\star s_f(x) + X_g W_g^\star S_g(x) u, \tag{3.20}$$

where the weight values W_f^* and W_g^* are unknown.

Since, $W_f^\star$ and $W_g^\star$ are unknown, our solution consists of designing a control law $u(x, W_f, W_g)$ and appropriate update laws for W_f and W_g to guarantee convergence of the state to zero and in some cases, which will be analyzed in the following sections, boundedness of x and of all signals in the closed-loop.

3.2.1 Parametric Uncertainties

Following the work in Theodoridis et al. (2008), Boutalis et al. (2009), we assume in this subsection the existence of only parameter uncertainty, so, we can take into account that the actual system (3.1) can be modeled by the NF form of Eq. (3.20). Furthermore, the actual system is assumed to be quadratic (number of inputs equals to number of states) and the error equation is given by Eq. (3.12).

Our objective is to find suitable control and learning laws to drive both e and x to zero, while all other signals in the closed-loop remain bounded. Taking u to be equal to:

$$u = -\left[X_g W_g S_g(x)\right]^{-1}\left[X_f W_f s_f(x)\right], \tag{3.21}$$

and substituting it into (3.19) we finally obtain:

$$\dot{\hat{x}} = A\hat{x}. \tag{3.22}$$

In the next theorem the weight updating laws are given, which can serve both the identification and the control objectives provided that the updating of the weights of matrix W_g does not compromise the existence of $\left[X_g W_g S_g(x)\right]^{-1}$.

Theorem 3.2 *Consider the identification scheme given by (3.12). Provided that* $\left[X_g W_g S_g(x)\right]^{-1}$ *exists the learning laws:*

(a) *For the elements of* W_{f_i}:

$$\dot{w}^{pl}_{f_i}(x) = -\bar{x}^p_{f_i}\gamma_f p_i e_i s_l(x) \tag{3.23}$$

or equivalently $\dot{W}^l_{f_i}(x) = -\left(\bar{x}_{f_i}\right)^T \gamma_f p_i e_i s_l(x)$, *with* $i = 1,\ldots,n$, $p = 1,\ldots,q$ *and* $l = 1,\ldots,k.$

(b) *For the elements of* W_{g_i}:

$$\dot{w}^{p}_{g_i}(x) = -\bar{x}^p_{g_i}\gamma_g p_i e_i u_i s_i(x) \tag{3.24}$$

or equivalently $\dot{W}_{g_i}(x) = -\left(\bar{x}_{g_i}\right)^T \gamma_g p_i e_i u_i s_i(x)$, *with* $i = 1,\ldots,n$ *and* $p = 1,\ldots,q$ *guarantees the following properties:*

- $e, \hat{x}, \tilde{W}_f, \tilde{W}_g \in L_\infty$, $e, \hat{x} \in L_2$,
- $\lim_{t\to\infty} e(t) = 0$, $\lim_{t\to\infty} \hat{x}(t) = 0$,
- $\lim_{t\to\infty} \dot{W}_f(t) = 0$, $\lim_{t\to\infty} \dot{W}_g(t) = 0$.

Proof Consider the Lyapunov function candidate:

$$V\left(e, \hat{x}, \tilde{W}_f, \tilde{W}_g\right) = \frac{1}{2}e^T Pe + \frac{1}{2}\hat{x}^T P\hat{x} + \frac{1}{2\gamma_f}\text{tr}\left\{\tilde{W}_f^T \tilde{W}_f\right\} + \frac{1}{2\gamma_g}\text{tr}\left\{\tilde{W}_g^T \tilde{W}_g\right\}.$$

(3.25)

Where $P > 0$ is chosen to satisfy the Lyapunov equation:

$$PA + A^T P = -I.$$

Taking the derivative of the Lyapunov function candidate (3.25) and taking into account (3.22) we get:

$$\dot{V} = -\frac{1}{2}\|e\|^2 - \frac{1}{2}\|\hat{x}\|^2 + e^T PX_f \tilde{W}_f s_f + e^T PX_g \tilde{W}_g S_g u$$

$$+ \frac{1}{\gamma_g}\text{tr}\{\dot{W}_f^T \tilde{W}_f\} + \frac{1}{\gamma_g}\text{tr}\{\dot{W}_g^T \tilde{W}_g\}$$

$$\leq -\frac{1}{2}\|e\|^2 - \frac{1}{2}\|\hat{x}\|^2,$$

when

$$\dot{W}_f = -\gamma_f X_f^T Pes_f^T, \tag{3.26}$$

$$\dot{W}_g = -\gamma_g X_g^T Peu^T S_g. \tag{3.27}$$

Thus, $\dot{V}$ is negative semi-definite. Since $\dot{V} \leq 0$ we conclude that $V \in L_\infty$, which implies that $e, \hat{x}, \tilde{W}_f, \tilde{W}_g \in L_\infty$. Furthermore, $W_f = \tilde{W}_f + W_f^\star$ and $W_g = \tilde{W}_g + W_g^\star$ are also bounded. Since V is a nonincreasing function of time and bounded from below, the $\lim_{t\to\infty} V = V_\infty$ exists; therefore, by integrating $\dot{V}$ from 0 to ∞ we have:

$$\int_0^\infty |e|^2\, dt + \int_0^\infty |\hat{x}|^2\, dt \leq [V(0) - V_\infty] < \infty,$$

which implies that $e, \hat{x} \in L_2$.

Since $e, \hat{x} \in L_2 \cap L_\infty$, using Barbalat's Lemma (Popov 1973) we conclude that $\lim_{t\to\infty} e(t) = 0$ and $\lim_{t\to\infty} \hat{x}(t) = 0$.

Finally, using the boundedness of u, $s_f(x)$, $S_g(x)$, X_f, X_g, and the convergence of $e(t)$ to zero, we have that $\dot{W}_f$ and $\dot{W}_g$ also converges to zero (Ioannou and Fidan 2006). $\square$

Lemma 3.1 *The control law (3.21) can be also extended to the following form:*

$$u = -\left[X_g W_g S_g(x)\right]^{-1} \cdot \left(X_f W_f s_f(x) + Kx\right), \tag{3.28}$$

where K is appropriate positive definite diagonal gain matrix. With this control law the negativity of the derivative of the Lyapunov function is further enhanced. Therefore, term Kx is actually acting as a robustifying term.

Proof Indeed, by using the extended control law (3.28) the state estimate dynamics become:

$$\dot{\hat{x}} = A\hat{x} - Kx. \tag{3.29}$$

Then, using the weight updating laws given in theorem 3.2 and Eq. (3.29), the derivative of the Lyapunov function becomes:

$$
\begin{aligned}
\dot{V} &= -\frac{1}{2}e^T e - \frac{1}{2}\hat{x}^T\hat{x} - xKP\hat{x} \\
&= -\frac{1}{2}\|e\|^2 - \frac{1}{2}\|\hat{x}\|^2 - \hat{x}^T KP\hat{x} + e^T KP\hat{x} \\
&\leq -\frac{1}{2}\|e\|^2 - \frac{1}{2}\|\hat{x}\|^2 - \lambda_{\min}(KP)\|\hat{x}\|^2 + \|e\|\,\|KP\|\,\|\hat{x}\| \\
&\leq -\begin{bmatrix}\|e\| & \|\hat{x}\|\end{bmatrix}\begin{bmatrix}1/2 & -\|KP\| \\ 0 & 1/2 + \lambda_{\min}(KP)\end{bmatrix}\begin{bmatrix}\|e\| \\ \|\hat{x}\|\end{bmatrix} < 0.
\end{aligned}
$$

$\square$

Because of $\frac{1}{2} > 0$, $\begin{vmatrix}1/2 & -\|KP\| \\ 0 & 1/2 + \lambda_{\min}(KP)\end{vmatrix} = \frac{1}{2}\left(\frac{1}{2} + \lambda_{\min}(KP)\right) > 0$ when $\lambda_{\min}(KP) > 0$, the above derivative of Lyapunov candidate function is negative semi-definite. The last statement is valid by definition of matrices K and P such that KP is positive definite.

In the sequel, we have to ensure the existence of the control signal. Thus, in the following subsection we present a novel method called *parameter hopping*, which comes to replace the well known from the literature *projection*.

3.2.2 The Method of Parameter Hopping

Theorem 3.2 presented previously in Sect. 3.2.1 is valid when the control law signal in (3.28) exists. Therefore, the existence of $\left[X_g W_g S_g(x)\right]^{-1}$ has to be ensured. Since the submatrices of $S_g(x)$ are diagonal with the diagonal elements $s_i(x) \neq 0$ and X_g, W_g are block diagonal, the existence of the inverse is ensured when $\bar{x}_{g_i} \cdot W_{g_i} \neq 0$, $\forall i = 1, \ldots, n$. Therefore, W_{g_i} has to be confined such that $\left|\bar{x}_{g_i} \cdot W_{g_i}\right| \geq \theta_i > 0$, with θ_i being a small positive design parameter (usually in the range of $[0.001, 0.01]$). In case the boundary defined by the above confinement is nonlinear, the updating of W_{g_i} can be modified by using a projection algorithm, (Ioannou and Fidan 2006). However, in our case the boundary surface is linear and the direction of updating is normal to it because $\nabla\left[\bar{x}_{g_i} \cdot W_{g_i}\right] = \bar{x}_{g_i}$. Therefore, the projection of the updating vector on the boundary surface is of no use. Instead, using concepts from multidimensional vector geometry we modify the updating law such that, when the weight vector approaches

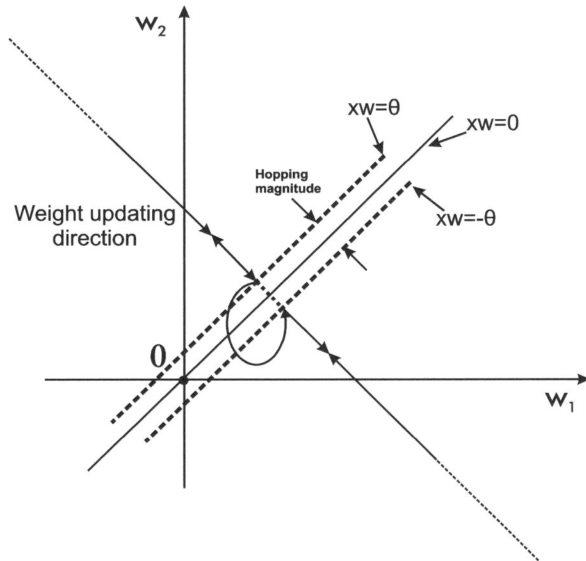

Fig. 3.1 Pictorial Representation of parameter hopping

(within a safe distance θ_i) the forbidden hyperplane $\bar{x}_{g_i} \cdot W_{g_i} = 0$ and the direction of updating is toward the forbidden hyperplane, it introduces a *hopping* which drives the weights in the direction of the updating but on the other side of the space, where here the weight space is divided into two sides by the forbidden hyperplane. For example, let the weight updating hopping occurs at the t_h time instant. Then, if the weights at t_h^- time instant lies in the space determined by $\bar{x}_{g_i} \cdot W_{g_i} < -\theta_i$ then, after performing hopping the weights move into the space determined by $\bar{x}_{g_i} \cdot W_{g_i} > \theta_i$ and from t_h^+ on they continue their updating direction. This procedure is depicted in Fig. 3.1, where a simplified two-dimensional representation is given. Before presenting the theorem that introduces the weight hopping in the weight updating laws, it is necessary to give elements from vector geometry.

3.2.2.1 Vector Geometry and Weight Hopping

In selecting the terms involved in parameter hopping, we start from the vector definition of a line of a plane and the distance of a point to a plane. The equation of a line in vector form is given by:

$$r = a + \lambda t,$$

where a is the position vector of a given point of the line, t is a vector in the direction of the line and λ is a real scalar. By giving different numbers to λ we get different points of the line each one represented by the corresponding position vector r. The vector equation of a plane can be defined by using one point of the plane and a vector

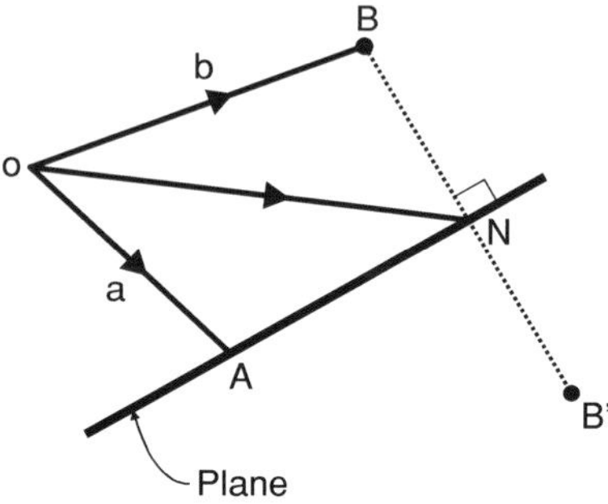

Fig. 3.2 Vector explanation of parameter hopping

normal to it. In this case:

$$r \cdot n = a \cdot n = d,$$

is the equation of the plane, where a is the position vector of a given point on the plane, n is a vector normal to the plane, and d is a scalar. When the plane passes through zero, then $d = 0$. To determine the distance of a point B with position vector b from a given plane, we consider Fig. 3.2 and combine the above definitions as follows.

Line BN is perpendicular to the plane and is described by the vector equation $r = b + \lambda n$, where n is normal to the plane vector. However, point N also lies on the plane and in case the plane passes through zero:

$$r \cdot n = 0 \Rightarrow (b + \lambda n) \cdot n = 0 \Rightarrow \lambda = \frac{-b \cdot n}{\|n\|}.$$

Apparently, if one wants to get the position vector of B' (the symmetrical of B in respect to the plane), this is given by:

$$r = b - 2\frac{b \cdot n}{\|n\|}n.$$

Quite the same, if the plane does not pass through the origin then:

$$\lambda = \frac{d - b \cdot n}{\|n\|} \quad and \quad r = b + 2\frac{d - b \cdot n}{\|n\|}n.$$

In our case $b = W_{g_i}, n = \bar{x}_{g_i}$, and the hopping magnitude becomes $2\frac{\left(\bar{x}_{g_i} W_{g_i} (\bar{x}_{g_i})^T\right)}{|\bar{x}_{g_i}|^2}$. We can equivalently write it in respect to the weight error vector if we consider that the weight estimation error is always a vector starting from the adaptive weights and directed always to the optimal weight values as can be seen in Fig. 3.3 and according to equation $\tilde{W}_{g_i} = W_{g_i} - W_{g_i}^\star$. Thus, when the adaptive weight values W_{g_i} reach the forbidden planes the hopping condition is activated and the hopping magnitude

Fig. 3.3 Weights and weights error on the adaptation direction

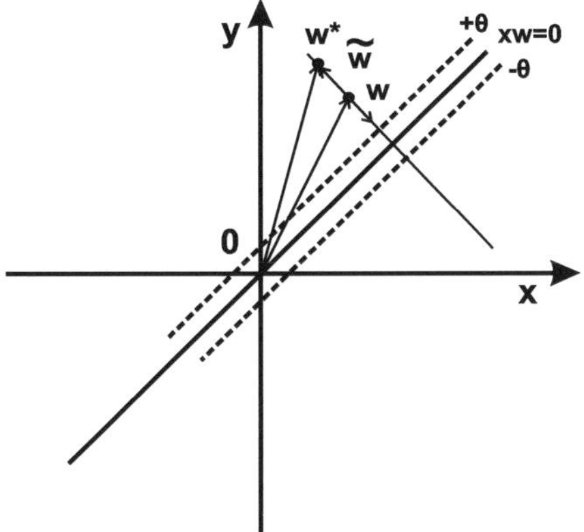

$\dfrac{\left(\bar{x}_{g_i}\, W_{g_i}\,(\bar{x}_{g_i})^T\right)}{\left|\bar{x}_{g_i}\right|^2}$ can be equivalently replaced by $\theta \cdot \dfrac{\tilde{W}_{g_i}}{\left\|\tilde{W}_{g_i}\right\|}$, where $\dfrac{\tilde{W}_{g_i}}{\left\|\tilde{W}_{g_i}\right\|}$ is the weight estimation unit error vector.

3.2.2.2 Modification of the Adaptation Laws

We are now ready to incorporate the results of previous section in our analysis. Theorem 3.3 below introduces this *hopping* in the weight updating law.

Theorem 3.3 *Consider the control scheme* (3.12), (3.28), (3.29). *The updating law:*

(a) *For the elements of W_{f_i} given by* (3.23).
(b) *For the elements of W_{g_i} given by the modified form:*

$$\dot{W}_{g_i} = \begin{cases} -\left(\bar{x}_{g_i}\right)^T \gamma_g p_i e_i u_i s_i(x) & \text{if } \left|\bar{x}_{g_i} \cdot W_{g_i}\right| > \theta_i \\ & \text{or } \bar{x}_{g_i} \cdot W_{g_i} = \pm\theta_i \\ & \text{and } \bar{x}_{g_i} \cdot \dot{W}_{g_i} <> 0 \\ -\left(\bar{x}_{g_i}\right)^T \gamma_g p_i e_i u_i s_i(x) - & \text{otherwise} \\ \dfrac{2\left(\bar{x}_{g_i}\, W_{g_i}\,(\bar{x}_{g_i})^T\right)}{tr\{(\bar{x}_{g_i})^T \bar{x}_{g_i}\}} \end{cases}$$

guarantees the properties of theorem 3.2 and ensures the existence of the control signal.

Proof The first part of the weight updating equation is used when the weights are at a certain distance (condition if $\left|\bar{x}_{g_i} \cdot W_{g_i}\right| > \theta_i$) from the forbidden plane or at the safe limit (condition $\left|\bar{x}_{g_i} \cdot W_{g_i}\right| = \pm\theta_i$) but with the direction of updating moving the weights far from the forbidden plane (condition $\bar{x}_{g_i} \cdot \dot{W}_{g_i} <> 0$).

In the second part of $\dot{W}_{g_i}$, term $-\frac{2(\bar{x}_{g_i} W_{g_i} (\bar{x}_{g_i})^T)}{tr\{(\bar{x}_{g_i})^T \bar{x}_{g_i}\}}$ determines the magnitude of weight *hopping* (Theodoridis et al. 2009b), which as explained in the previous section, has to be two times the distance of the current weight vector to the forbidden hyperplane. Therefore, the **existence** of the control signal is ensured because the weights never reach the forbidden plane.

Let that $W_{g_i}^\star$ contains the actual unknown values of W_{g_i} such that $\left|\bar{x}_{g_i} \cdot W_{g_i}^\star\right| >> \theta_i$ and that $\tilde{W}_{g_i} = W_{g_i} - W_{g_i}^\star$. Then, as explained in the previous section, the weight hopping can be equivalently written with respect to $\tilde{W}_{g_i}$ as: $-\frac{2\theta_i \tilde{W}_{g_i}}{\|\tilde{W}_{g_i}\|}$. Under this consideration the modified updating law is rewritten as:

$$\dot{W}_{g_i} = - (\bar{x}_{g_i})^T \gamma_g p_i e_i u_i s_i(x) - \frac{2\theta_i \tilde{W}_{g_i}}{\|\tilde{W}_{g_i}\|}.$$

With this updating law it can be easily verified that:

$$\dot{V} \leq - \left[\|e\| \ \|\hat{x}\|\right] \begin{bmatrix} 1/2 & -\|KP\| \\ 0 & 1/2 + \lambda_{\min}(KP) \end{bmatrix} \begin{bmatrix} \|e\| \\ \|\hat{x}\| \end{bmatrix} - \sigma_i \Theta_i, \tag{3.30}$$

with Θ_i being a positive constant expressed as $\Theta_i = \sum 2\theta_i \frac{\left((\tilde{W}_{g_i})^T \tilde{W}_{g_i}\right)}{\|\tilde{W}_{g_i}\|}$ for all time, where the summation includes all weight vectors which require hopping. Also, $\sigma_i = 1$ when the hopping condition is activated or $\sigma_i = 0$ when the hopping condition is deactivated. Therefore, the negativity of $\dot{V}$ is actually enhanced. $\square$

Remark 3.2 Similar to traditional projection methods, the hopping method actually transforms the control method to a switching one (Ioannou and Fidan 2006; Yesildirek and Lewis 1995). However, switching control methods may cause discontinuities to the right-hand side of the system's closed-loop dynamics (Kosmatopoulos and Ioannou 1999, 2002). In adaptive control literature (Polycarpou and Ioannou 1993; Kosmatopoulos and Ioannou 1999) it has been shown that, under mild assumptions, these discontinuities do not affect the existence and uniqueness of solutions of system's dynamics. Depending on the type of discontinuity and the direction of the vector field before and after the point of discontinuity the system may have a Caratheodory or a Filippov (Filippov 1998) solution (Polycarpou and Ioannou 1993; Cortes 2008).

Here, when "hopping" is performed, the direction of updating before and after "hopping" is kept intact. Similarly, any discontinuity to the right-hand side of the closed-loop dynamics implied by the hopping action will retain this characteristic, that is, the vector fields before and after "hopping" will have the same direction. Moreover, if the hopping magnitude includes a modification factor, say $\kappa > 1$, to become $-\frac{2\kappa(\bar{x}_{g_i} W_{g_i} (\bar{x}_{g_i})^T)}{tr\{(\bar{x}_{g_i})^T \bar{x}_{g_i}\}}$, the weight vector goes to a safe distance from the "hopping" condition border $\pm\theta_i$ and keeps "moving" to the same direction. This way, it cannot immediately perform hopping to the opposite direction and the danger of an infinite "hopping" is avoided. That is, with this modification term, the proposed scheme performs as a hysteresis switching control scheme (Morse et al. 1992; Weller and Goodwin 1994), where the closed-loop system possesses a unique solution and Lyapunov stability arguments can be safely applied.

3.2.3 Parametric Plus Dynamic Uncertainties

In this subsection, we examine a more general case where parametric and dynamic uncertainties are present (Boutalis et al. 2008; Theodoridis et al. 2010). When unmodeled dynamics are present, the model at (3.20) is no longer considered sufficient and has to be extended to include unmodeled dynamics. In the following, we extend this representation within the framework of singular perturbations (Kokotovic et al. 1986) to include the case where dynamic uncertainties are present. Following this framework we can assume that the unknown plant (3.1) can be completely described by:

$$\dot{x} = Ax + X_f W_f^* s_f(x) + X_g W_g^* S_g(x)u + F(x, W_f, W_g)A_0^{-1}X_0 W_0 u + F(x, W_f, W_g)z, \tag{3.31}$$

$$\mu\dot{z} = A_0 z + X_0 W_0 u, \quad z \in \Re^r, \tag{3.32}$$

where z is the state of the unmodeled dynamics and $\mu > 0$ is a small singular perturbation scalar. Defining now the identifier error as Eq. (3.11), then from Eqs. (3.9), (3.31) and (3.32) we obtain the error equation:

$$\dot{e} = Ae + X_f \tilde{W}_f s_f(x) + X_g \tilde{W}_g S_g(x)u - F(x, W_f, W_g)A_0^{-1}X_0 W_0 u - F(x, W_f, W_g)z, \tag{3.33}$$

with

$$\mu\dot{z} = A_0 z + X_0 W_0 u, \quad z \in \Re^r, \tag{3.34}$$

where $\tilde{W}_f = W_f - W_f^*$, $\tilde{W}_g = W_g - W_g^*$. Also, $F(x, W_f, W_g)$, $X_0 W_0 u$, $X_f \tilde{W}_f s_f(x)$, $X_g \tilde{W}_g S_g(x)u$ are bounded and differentiable with respect to their arguments for every $\tilde{W}_f \in B_{\tilde{W}_f}$, $\tilde{W}_g \in B_{\tilde{W}_g}$ and all $x \in B_x$, where B_x denotes a ball in an appropriate multidimensional space of real numbers. Further, we assume that the unmodeled dynamics are asymptotically stable for all $x \in B_x$. That is, we assume that there exists a constant $c > 0$ such that:

$$Re\,\lambda\{A_0\} \leq -c < 0.$$

Note that $\dot{z}$ is large since μ is small and hence, the unmodeled dynamics are fast. For a singular perturbation from $\mu > 0$ to $\mu = 0$ we obtain:

$$z = -A_0^{-1}X_0 W_0 u.$$

Since the unmodeled dynamics are asymptotically stable, the existence of A_0^{-1} is ensured. Following the singular perturbation theory (Kokotovic et al. 1986), we express the state z as:

$$z = h(x, \eta) + \eta, \tag{3.35}$$

where $h(x, \eta)$ is defined as the quasi-steady state of z and η is its fast transient. In our case,

$$h(x, \eta) = -A_0^{-1} X_0 W_0 u.$$

Substituting (3.35) into (3.33), (3.34) we obtain the singularly perturbed model as:

$$\dot{e} = Ae + X_f \tilde{W}_f s_f(x) + X_g \tilde{W}_g S_g(x)u - F(x, W_f, W_g)\eta, \tag{3.36}$$

$$\mu\dot{\eta} = A_0\eta - \mu\dot{h}\left(e, \tilde{W}_f, \tilde{W}_g, \eta, u\right). \tag{3.37}$$

Remark 3.3 $F(x, W_f, W_g)A_0^{-1} X_0 W_0 u$, $F(x, W_f, W_g)z$ in (3.31) can be viewed as correction terms in the input vector fields and in the drift term of (3.20) in the sense that the unknown system can now be described by the NF model plus correction terms.

Using concepts from singular perturbation theory and following the previous analysis, we can assume that the unknown plant (3.1) can be completely described by the NF form:

$$\dot{x} = Ax + X_f W_f^* s_f(x) + X_g W_g^* S_g(x)u + F(x, W_f, W_g)\eta, \tag{3.38}$$

in conjunction with Eq. (3.37). In (3.37),

$$\dot{h}\left(e, \tilde{W}_f, \tilde{W}_g, \eta, u\right) = \frac{\vartheta h}{\vartheta e}\dot{e} + \frac{\vartheta h}{\vartheta \tilde{W}_f}\dot{\tilde{W}}_f + \frac{\vartheta h}{\vartheta \tilde{W}_g}\dot{\tilde{W}}_g + \frac{\vartheta h}{\vartheta u}\dot{u} + \frac{\vartheta h}{\vartheta \eta}\dot{\eta}.$$

However, in the control case, u is a function of $e, \tilde{W}_f, \tilde{W}_g$. Therefore, $\dot{h}\left(e, \tilde{W}_f, \tilde{W}_g, \eta, u\right)$ is equal to:

$$\dot{h}\left(e, \tilde{W}_f, \tilde{W}_g, \eta, u\right) = \frac{\vartheta h}{\vartheta e}\dot{e} + \frac{\vartheta h}{\vartheta \tilde{W}_f}\dot{\tilde{W}}_f + \frac{\vartheta h}{\vartheta \tilde{W}_g}\dot{\tilde{W}}_g + \frac{\vartheta h}{\vartheta \eta}\dot{\eta}. \tag{3.39}$$

Before proceeding any further, we need to prove the following lemma.

Lemma 3.2 *The derivative of $h\left(e, \tilde{W}_f, \tilde{W}_g, \eta, u\right)$ is bounded according to the inequality $\left\|\dot{h}\left(e, \tilde{W}_f, \tilde{W}_g, \eta, u\right)\right\| \leq \rho_1 \|e\| + \rho_2 \|\eta\|$, where ρ_1, ρ_2 are positive constants.*

Proof The proof proceeds similar to Kokotovic et al. (1986), Rovithakis and M.A.Christodoulou (2000). Differentiating $h\left(e, \tilde{W}_f, \tilde{W}_g, \eta, u\right)$ we obtain Eq. (3.39) and after substituting Eqs. (3.36), (3.37) we have:

$$\dot{h} = h_e \left(Ae + X_f \tilde{W}_f s_f + X_g \tilde{W}_g S_g u - F\eta\right) + \frac{\vartheta h}{\vartheta \tilde{W}_f} \dot{\tilde{W}}_f + \frac{\vartheta h}{\vartheta \tilde{W}_g} \dot{\tilde{W}}_g + h_\eta \left(\frac{A_0 \eta}{\mu} - \dot{h}\right),$$
(3.40)

or equivalently,

$$\left(1 + h_\eta\right) \dot{h} = h_e Ae + h_e X_f \tilde{W}_f s_f + h_e X_g \tilde{W}_g S_g u + \frac{\vartheta h}{\vartheta \tilde{W}_f} \dot{\tilde{W}}_f + \frac{\vartheta h}{\vartheta \tilde{W}_g} \dot{\tilde{W}}_g$$

$$+ h_\eta \left(\frac{A_0 \eta}{\mu} - \eta F\right).$$
(3.41)

Therefore, we have:

$$\|\dot{h}\| \leq \frac{\|h_e Ae\|}{\|(1 + h_\eta)\|} + \frac{\left\|h_e X_f \tilde{W}_f s_f\right\|}{\|(1 + h_\eta)\|} + \frac{\left\|h_e X_g \tilde{W}_g S_g u\right\|}{\|(1 + h_\eta)\|} + \frac{\left\|h_{\tilde{W}_g} \dot{\tilde{W}}_g\right\|}{\|(1 + h_\eta)\|}$$

$$+ \frac{\left\|h_{\tilde{W}_f} \dot{\tilde{W}}_f\right\|}{\|(1 + h_\eta)\|} + \frac{\left\|h_\eta \left(\frac{A_0}{\mu} + F\right) \eta\right\|}{\|(1 + h_\eta)\|}$$

$$\leq k_4 \|e\| + k_3 \|e\| + k_2 \|e\| + k_1 \|e\| + k_0 \|e\| + \rho_2 \|\eta\|.$$
(3.42)

Hence,

$$\left\|\dot{h}\left(e, \tilde{W}_f, \tilde{W}_g, \eta, u\right)\right\| \leq \rho_1 \|e\| + \rho_2 \|\eta\|,$$

where $\rho_1 = k_0 + k_1 + k_2 + k_3 + k_4$. $\square$

In the sequel, our objective is to find suitable control and learning laws to drive both e, $\hat{x}$ and η to zero, while all other signals in the closed-loop remain bounded. Thus, we select the control signal u to be equal to:

$$u = \left[X_g W_g S_g(x)\right]^+ \left\{-X_f W_f s_f(x) + Ke + I_h u_\rho e\right\},$$
(3.43)

where $[\cdot]^+$ means pseudo-inverse in Moore-Penrose sense and K is a positive definite diagonal matrix selected by the designer. Furthermore, I_h is the identity $n \times n$ matrix and

$$u_\rho = \left[\sum_{i=1}^{m^2 n^2 q} a \cdot \sin\left(\omega_i \cdot t\right)\right],$$
(3.44)

with ω_i different choices of frequencies ($\neq 0$) which are not multiples of each other and α is a very small positive number ($0 < \alpha << 1$) selected appropriately from the designer. This signal u_ρ is also upper bounded as $\left|u_\rho\right| \leq \bar{u}_\rho$. The purpose of this auxiliary term $(I_h u_\rho e)$ in the control signal will be explained in the next paragraphs.

After, substituting (3.43) into (3.8) we obtain:

$$\dot{\hat{x}} = A\hat{x} + Ke + I_h u_\rho e. \tag{3.45}$$

Provided that $e \to 0$ and since A is a stable matrix, $\hat{x}$ and consequently x go to zero and therefore the control objective is fulfilled. As ti will be proved in theorem 3.4 below, the weight updating laws for the submatrices of W_f, W_g given by:

$$\dot{W}^l_{f_i} = -\left(\bar{x}_{f_i}\right)^T p_i e_i s_l(x) d_{f_l}, \tag{3.46}$$

and

$$\dot{W}^l_{g_{ij}} = -\left(\bar{x}_{g_{ij}}\right)^T p_i e_i u_j s_l(x) d_{g_l}, \tag{3.47}$$

can serve the purpose of driving e to zero.

There are, however, two issues that need special consideration. In many adaptive control approaches, it is possible for the identifier to become uncontrollable, although the physical system is controllable (Kosmatopoulos and Ioannou 1999, 2002). This is actually due to the fact that the control law of (3.43) may become uncomputable if $\left[X_g W_g S_g(x)\right]^+$ does not exist or $\left\|X_g W_g S_g(x)\right\|$ becomes unbounded mainly due to unbounded growth of W_g. Therefore, the existence of $\left[X_g W_g S_g(x)\right]^+$ and the boundedness of $\left\|X_g W_g S_g(x)\right\|$ has to be ensured. Since the submatrices of $S_g(x)$ are diagonal with the diagonal elements $s_i(x) \neq 0$ and X_g, W_g are block diagonal, the existence of the pseudo-inverse is ensured when $\bar{x}_{g_{ij}} \cdot W^l_{g_{ij}} \neq 0$, $\forall i = 1, \dots, n$ and $\forall j = 1, \dots, q$. Therefore, $W^l_{g_{ij}}$ has to be confined such that $\left|\bar{x}_{g_{ij}} \cdot W^l_{g_{ij}}\right| \geq \theta^l_{ij} > 0$, with θ^l_{ij} being a positive design parameter. Similarly, the boundedness of $\left\|X_g W_g S_g(x)\right\|$ is ensured when $\left|\bar{x}_{g_{ij}} \cdot W^l_{g_{ij}}\right| \leq \rho^l_{ij}$, with $\rho^l_{ij} > 0$ being a positive design parameter. As mentioned in the beginning of Sect. 3.2.2, in our case, the boundary surfaces are linear (are actually hyperplanes) and the direction of updating is normal to them, because $\nabla\left[\bar{x}_{g_{ij}} \cdot W^l_{g_{ij}}\right] = \bar{x}_{g_{ij}}$. Therefore, we use again concepts from multidimensional vector geometry to modify the updating law such that, when the weight vector approaches (within safe distances θ^l_{ij}, ρ^l_{ij}) the forbidden hyperplanes $\bar{x}_{g_{ij}} \cdot W^l_{g_{ij}} = 0, \bar{x}_{g_{ij}} \cdot W^l_{g_{ij}} = \pm w_m$ and the direction of updating is toward the forbidden hyperplane, it introduces a *hopping* or *modified hopping*, which drives the weights in the direction of the updating but on the other side of the space, where here the weight space is divided into two sides by the forbidden hyperplane.

For example, let the weight updating hopping occurs at the t_h time instant. Then, if the weights at t_h^- time instant lies in the space determined by $\bar{x}_{g_{ij}} \cdot W^l_{g_{ij}} < -\theta^l_{ij}$ then, after performing hopping the weights move into the space determined by $\bar{x}_{g_{ij}} \cdot W^l_{g_{ij}} >$

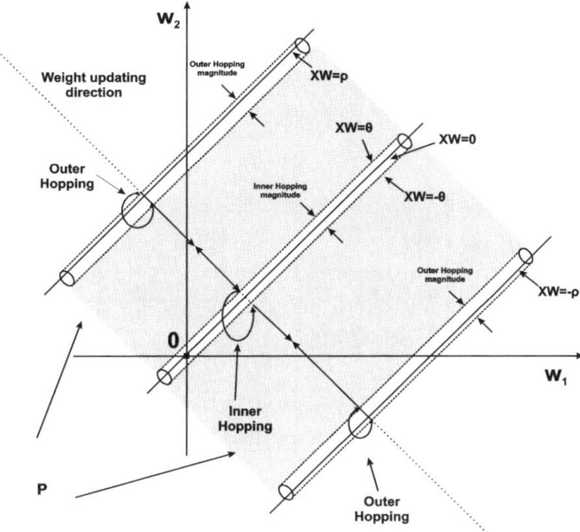

Fig. 3.4 Pictorial representation of inner and outer modified parameter hopping

θ_{ij}^{l} and from t_h^{+} on they continue their updating direction. This procedure is depicted in Fig. 3.4, in a simplified two-dimensional representation.

It is worth noting here that the weight updating law (3.47) is a gradient- based law. As shown in Sect. 2.3.2 and is supported by the adaptive estimation literature, (Ioannou and Fidan 2006), a normalized version of the weight updating law may serve the purpose of keeping the updating of the weights reasonably bounded. The normalization takes the form of a learning rate, which is normalized by the norm of the regressor vector, containing, in the general form, functions of the measurable variables. In a recent work, (Mandic 2004) the normalization of the learning rate becomes adaptive following the dynamic behavior of the measurable signal showing better performance and robustness compared to other gradient-based algorithms. In the presented approach, we follow the approach of weight hopping described above, which can guarantee that the weights remain bounded provided that their initial values lie in a bounded hypersphere.

In this case, the learning law (3.47) for the adaptation of the weights W_g is modified and becomes:

$$\dot{W}_{g_{ij}}^{l} = -\left(\bar{x}_{g_{ij}}\right)^{T} p_i e_i u_j s_l(x) d_{g_l} - \frac{2\sigma_{ij}^{l}\left(\bar{x}_{g_{ij}} W_{g_{ij}}^{l}(\bar{x}_{g_{ij}})^{T}\right)}{tr\{(\bar{x}_{g_{ij}})^{T}\bar{x}_{g_{ij}}\}}$$

$$-\frac{(1-\sigma_{ij}^{l})\kappa_{g_{ij}}^{l,\text{outer}}\left(\bar{x}_{g_{ij}} W_{g_{ij}}^{l}(\bar{x}_{g_{ij}})^{T}\right)}{\text{tr}\{(\bar{x}_{g_{ij}})^{T}\bar{x}_{g_{ij}}\}}, \tag{3.48}$$

where

$$
\sigma_{ij}^{l} = \begin{cases} 0 & \begin{aligned} &\text{if } \bar{x}_{g_{ij}}\, W_{ij}^{l} = \pm\rho_{ij}^{l} \\ &\text{and } \bar{x}_{g_{ij}}\, \dot{W}_{ij}^{l} <> 0 \end{aligned} \\[1em] 1 & \begin{aligned} &\text{if } \bar{x}_{g_{ij}}\, W_{ij}^{l} = \pm\theta_{ij}^{l} \\ &\text{and } \bar{x}_{g_{ij}}\, \dot{W}_{ij}^{l} >< 0 \end{aligned} \end{cases} ,
\tag{3.49}
$$

expresses the inner hopping condition when $\sigma_{ij}^{l} = 1$ or the outer hopping condition when $\sigma_{ij}^{l} = 0$. Also, $\kappa_{g_{ij}}^{l,\text{outer}}$ is a small positive constant value ($0 < \kappa_{g_{ij}}^{l,\text{outer}} < 1$) for the outer hopping, chosen by the designer. Due to the hopping condition, the control law (3.43) actually indirectly performs like a switching controller, (Kosmatopoulos and Ioannou 1999, 2002) where the switching here is determined by the weight hopping.

During the above operation, it is not impossible for the adaptive scheme to enter an infinite hopping loop. In principle, this might happen when, after hopping occurs, the conditions are such that the input signal is not persistently exciting. In this case, the adaptation procedure might not drive the weights towards the actual weights (presumed to be far from the forbidden area) but towards the forbidden area and then cause hopping again. Signal u_ρ introduced in the control law (3.43) is persistently exciting, (Ioannou and Fidan 2006; Vamvoudakis and Lewis 2009) and may serve the purpose of avoiding such a situation.

We are now ready to state and prove the following theorem.

Theorem 3.4 *The control scheme given by* (3.36), (3.37), (3.43), (3.45) *is asymptotically stable for all* $\mu \in (0, \mu_0)$, *where* $\mu_0 = \frac{1}{2}\left(\frac{1}{2\gamma_1\gamma_2+\gamma_3}\right)$. *Furthermore, the learning laws* (3.46), (3.48) *guarantees the existence of* $\left[X_g W_g S_g(x)\right]^{+}$ *and the following properties:*

- $e, \hat{x}, \tilde{W}_f, \tilde{W}_g \in L_\infty, \quad e, \hat{x} \in L_2,$
- $\lim_{t\to\infty} e(t) = 0, \ \lim_{t\to\infty} \hat{x}(t) = 0, \ \lim_{t\to\infty} \eta(t) = 0,$
- $\lim_{t\to\infty} \dot{W}_f(t) = 0, \quad \lim_{t\to\infty} \dot{W}_g(t) = 0.$

Proof In order to prove the above theorem, we can distinguish two possible cases.

Case 1: When the hopping condition is inactive.
We consider the Lyapunov function candidate:

$$
V = \frac{1}{2}\gamma_1 e^T P e + \frac{1}{2}\gamma_1 \hat{x}^T P \hat{x} + \frac{1}{2}\gamma_2 \eta^T P_0 \eta + \frac{1}{2}\gamma_1 \text{tr}\{\tilde{W}_f^T D_f^{-1} \tilde{W}_f\} + \frac{1}{2}\gamma_1 \text{tr}\{\tilde{W}_g^T D_g^{-1} \tilde{W}_g\}.
\tag{3.50}
$$

Where D_f, D_g are positive definite $k \times k$ gain matrices which are selected by the designer, $\gamma_1, \gamma_2, \gamma_3$ are positive constants which satisfy some relations defined in the remaining proof and $P, P_0 > 0$ are chosen to satisfy the Lyapunov equations:

$$
PA + A^T P = -I,
$$
$$
P_0 A_0 + A_0^T P_0 = -I.
$$

Taking the time derivatives of the Lyapunov function candidate (3.50) and taking into account (3.36), (3.37), (3.45) we obtain:

$$\dot{V} = -\frac{\gamma_1}{2}\,\|e\|^2 + \gamma_1 e^T P X_f \tilde{W}_f s_f + \gamma_1 e^T P X_g \tilde{W}_g S_g u + \gamma_1 e^T P F \eta - \frac{\gamma_1}{2}\,\|\hat{x}\|^2$$

$$+\, \gamma_1 \hat{x}^T P K e + \gamma_1 u_\rho \hat{x}^T P e - \frac{\gamma_2}{2\mu}\,\|\eta\|^2 - \gamma_2 \eta^T P_0 \dot{h}^T + \gamma_1 \mathrm{tr}\left\{\dot{W}_f^T D_f^{-1} \tilde{W}_f\right\}$$

$$+\, \gamma_1 \mathrm{tr}\left\{\dot{W}_g^T D_g^{-1} \tilde{W}_g\right\}.$$

Let us consider that:

$$\mathrm{tr}\{\dot{W}_f^T D_f^{-1} \tilde{W}_f\} = -e^T P X_f \tilde{W}_f s_f,$$

$$\mathrm{tr}\{\dot{W}_g^T D_g^{-1} \tilde{W}_g\} = -e^T P X_g \tilde{W}_g S_g u,$$

then using matrix trace properties the following compact form of the weight updating law arrives:

$$\dot{W}_f = -X_f^T P e s_f^T D_f, \tag{3.51}$$

$$\dot{W}_g = -X_g^T P e u^T S_g^T D_g. \tag{3.52}$$

Taking into account the form of W_f and W_g, and the involved submatrices, the above equations can also be written as (3.46), (3.47), where the learning laws are given in respect to each submatrix.

Then, $\dot{V}$ assumes the form:

$$\dot{V} \leq -\frac{\gamma_1}{2}\,\|e\|^2 - \frac{\gamma_1}{2}\,\|\hat{x}\|^2 - \frac{\gamma_2}{2\mu}\,\|\eta\|^2 + \gamma_1 \hat{x}^T \left(PK + Pu_\rho\right) e - \gamma_1 e^T P F \eta$$

$$-\, \gamma_2 \eta^T P_0 \dot{h}^T$$

$$\leq -\frac{\gamma_1}{2}\,\|e\|^2 - \frac{\gamma_1}{2}\,\|\hat{x}\|^2 - \frac{\gamma_2}{2\mu}\,\|\eta\|^2 + \gamma_1\,\|\hat{x}\|\,\|S\|\,\|e\| + \gamma_1\,\|e\|\,\|P\|\,\|F\|\,\|\eta\|$$

$$+\, \gamma_2\,\|\eta\|\,\|P_0\|\,\|\dot{h}\|\,,$$

where $S = PK + P\bar{u}_\rho$. According to Lemma 3.2 and provided that the following inequalities hold: $\|P\|\,\|F\| \leq \gamma_2$, $\rho_1\,\|P_0\| \leq \gamma_1$, $\rho_2\,\|P_0\| \leq \gamma_3$, we have:

$$\dot{V} \leq -\frac{\gamma_1}{2}\,\|e\|^2 - \frac{\gamma_1}{2}\,\|\hat{x}\|^2 + \gamma_1\,\|\hat{x}\|\,\|S\|\,\|e\| - \frac{\gamma_2}{2\mu}\,\|\eta\|^2 + \gamma_1 \gamma_2\,\|e\|\,\|\eta\|$$

$$+\, \gamma_2\,(\rho_1\,\|e\| + \rho_2\,\|\eta\|)\,\|\eta\|\,\|P_0\|$$

$$\leq -\frac{\gamma_1}{2}\,\|e\|^2 - \frac{\gamma_1}{2}\,\|\hat{x}\|^2 + \gamma_1\,\|\hat{x}\|\,\|S\|\,\|e\| - \frac{\gamma_2}{2\mu}\,\|\eta\|^2 + 2\gamma_1 \gamma_2\,\|e\|\,\|\eta\|$$

$$+\, \gamma_2 \gamma_3\,\|\eta\|^2\,,$$

which finally takes the form:

$$\dot{V} \leq -\frac{\gamma_1}{2}\|e\|^2 -\frac{\gamma_1}{2}\|\hat{x}\|^2 -\gamma_2\left(\frac{1}{2\mu}-\gamma_3\right)\|\eta\|^2 +2\gamma_1\gamma_2\|e\|\|\eta\|+\gamma_1\|\hat{x}\|\|S\|\|e\|,\tag{3.53}$$

or

$$\dot{V} \leq -\begin{bmatrix}\|e\| & \|\eta\| & \|\hat{x}\|\end{bmatrix}\begin{bmatrix}\frac{\gamma_1}{2} & -\gamma_1\gamma_2 & -\gamma_1\|S\| \\ -\gamma_1\gamma_2 & \gamma_2\left(\frac{1}{2\mu}-\gamma_3\right) & 0 \\ 0 & 0 & \frac{\gamma_1}{2}\end{bmatrix}\begin{bmatrix}\|e\| \\ \|\eta\| \\ \|\hat{x}\|\end{bmatrix}.\tag{3.54}$$

The 3×3 matrix in (3.54) is positive definite when

$$\mu \leq \mu_0 = \frac{1}{2}\left(\frac{1}{2\gamma_1\gamma_2+\gamma_3}\right).$$

Since $\dot{V}\leq 0$, we conclude that $V\in L_\infty$, which implies that $e,\hat{x},\eta,\tilde{W}_f,\tilde{W}_g\in L_\infty$. Furthermore, $W_f=\tilde{W}_f+W_f^*$, $W_g=\tilde{W}_g+W_g^*$ are also bounded. Since V is a nonincreasing function of time and bounded from below, the $\lim_{t\to\infty}V=V_\infty$ exists; therefore, by integrating $\dot{V}$ from 0 to ∞ we have:

$$\frac{\gamma_1}{2}\int_0^\infty\|e\|^2\,dt + \frac{\gamma_1}{2}\int_0^\infty\|\hat{x}\|^2\,dt + \gamma_2\left(\frac{1}{2\mu}-\gamma_3\right)\int_0^\infty\|\eta\|^2\,dt - 2\gamma_1\gamma_2\int_0^\infty\|e\|\|\eta\|\,dt -$$

$$\gamma_1\int_0^\infty\|\hat{x}\|\|S\|\|e\|\,dt \leq [V(0)-V_\infty] < \infty,$$

which implies that $e,\hat{x},\eta\in L_2$.

Since $u,A_0,\dot{h}(e,\tilde{W}_f,\tilde{W}_g),s_f,S_g$ are bounded, $e,\hat{x},\eta\in L_\infty$. Since $e,\hat{x},\eta\in L_2\cap L_\infty$, using Barbalat's Lemma we conclude that $\lim_{t\to\infty}e(t)=0$, $\lim_{t\to\infty}\hat{x}(t)=0$ and $\lim_{t\to\infty}\eta(t)=0$. Hence, we have that:

$$\lim_{t\to\infty}e(t)=\lim_{t\to\infty}\hat{x}(t)-\lim_{t\to\infty}x(t)=0.$$

Thus,

$$\lim_{t\to\infty}x(t)=0.$$

Furthermore, using the boundedness of u,s_f,S_g, and the convergence of $e(t)$ to zero, we have that $\dot{W}_f,\dot{W}_g$ also converges to zero (Ioannou and Fidan 2006).

Case 2: When the hopping condition is activated.

In order the properties of theorem 3.4 to be valid it suffices to show that by using the modified updating law for $W_{g_{ij}}^l$ the negativity of the time derivative of Lyapunov function is not compromised. Indeed the **if** part of the modified form (3.48) of $\dot{W}_{g_{ij}}^l$ is exactly the same with (3.47) and therefore according to case 1 of this theorem the negativity of $\dot{V}$ is in effect. The **if** part is used when the weights are at a certain

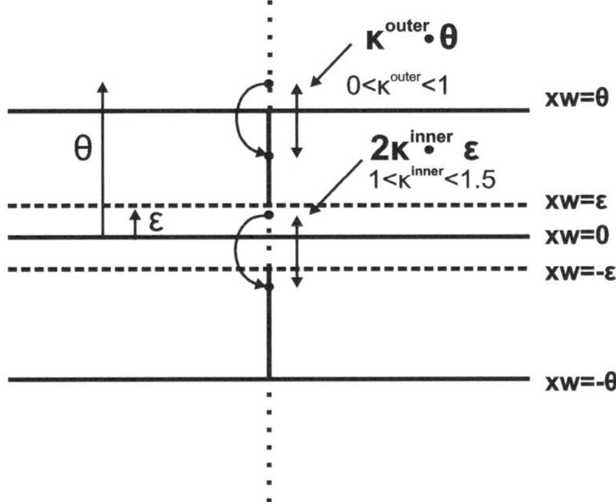

Fig. 3.5 Inner and outer hopping at a distance which depends on an appropriate selection of κ^{inner}, κ^{outer} constant values

distance (conditions if $\left| \bar{x}_{g_{ij}} \cdot W^l_{g_{ij}} \right| > \theta^l_{ij}$ and $\left| \bar{x}_{g_{ij}} \cdot W^l_{g_{ij}} \right| < \rho^l_{ij}$) from the forbidden planes or at the safe limits (conditions $\left| \bar{x}_{g_{ij}} \cdot W^l_{g_{ij}} \right| = \theta^l_{ij}$ and $\left| \bar{x}_{g_{ij}} \cdot W^l_{g_{ij}} \right| = \rho^l_{ij}$) but with the direction of updating moving the weights far from the forbidden hyperplanes (conditions $\bar{x}_{g_{ij}} \cdot \dot{W}^l_{g_{ij}} \leq 0$ and $\bar{x}_{g_{ij}} \cdot \dot{W}^l_{g_{ij}} \geq 0$ respectively).

In the **otherwise** part of $\dot{W}^l_{g_{ij}}$, terms:

$$-\frac{2 \left(\bar{x}_{g_{ij}} W^l_{g_{ij}} (\bar{x}_{g_{ij}})^T \right)}{\text{tr}\{(\bar{x}_{g_{ij}})^T \bar{x}_{g_{ij}}\}} \quad \text{and} \quad -\frac{\kappa^{l,\text{outer}}_{g_{ij}} \left(\bar{x}_{g_{ij}} W^l_{g_{ij}} (\bar{x}_{g_{ij}})^T \right)}{\text{tr}\{(\bar{x}_{g_{ij}})^T \bar{x}_{g_{ij}}\}},$$

determines the magnitude of inner and outer modified weight *hopping* respectively, which as explained in Sect. 3.2.2.1 has to be two times or $\kappa^{l,\text{outer}}_{g_{ij}}$—th times the distance of the current weight vector to the forbidden hyper-plane as shown in Fig. 3.5. Therefore, the **existence** of the control signal is ensured because the weights never reach the forbidden hyperplanes.

Regarding the **negativity** of $\dot{V}$ we proceed as follows. Assume that $W^{\star l}_{g_{ij}}$ contains the actual unknown weight values of $W^l_{g_{ij}}$ such that $\left| \bar{x}_{g_{ij}} \cdot W^{\star l}_{g_{ij}} \right| >> \theta^l_{ij}$ and $\tilde{W}^l_{g_{ij}} = W^l_{g_{ij}} - W^{\star l}_{g_{ij}}$. Then, according to Sect. 3.2.2.1, the weight hopping can be equivalently written with respect to $\tilde{W}^l_{g_{ij}}$ as: $-2\theta^l_{ij} \frac{\tilde{W}^l_{g_{ij}}}{\|\tilde{W}^l_{g_{ij}}\|}$ when the inner hopping condition is activated or $-\kappa^{l,\text{outer}}_{g_{ij}} \rho^l_{ij} \frac{\tilde{W}^l_{g_{ij}}}{\|\tilde{W}^l_{g_{ij}}\|}$ when the outer hopping condition is activated. Under this consideration, the modified updating law is rewritten as:

$$\dot{W}^l_{g_{ij}} = -\left(\bar{x}_{g_{ij}} \right)^T p_i e_i u_j s_l(x) d_{g_l} - 2\sigma^l_{ij} \theta^l_{ij} \frac{\tilde{W}^l_{g_{ij}}}{\|\tilde{W}^l_{g_{ij}}\|} - (1 - \sigma^l_{ij})\kappa^{l,\text{outer}}_{g_{ij}} \frac{\tilde{W}^l_{g_{ij}}}{\|\tilde{W}^l_{g_{ij}}\|}.$$

With this updating law it can be easily verified that:

$$\dot{V} \leq -L\left(e, \hat{x}, \eta\right) - \gamma_1 \Theta^l_{g_{ij}} \frac{(\tilde{W}^l_{g_{ij}})^T \tilde{W}^l_{g_{ij}}}{\|\tilde{W}^l_{g_{ij}}\|}, \tag{3.55}$$

with $L\left(e, \hat{x}, \eta\right)$ given by the right-hand side of (3.54) and $\Theta^l_{g_{ij}}$ is a positive number expressed as:

$$\Theta^l_{g_{ij}} = \sigma^l_{ij} \sum 2\theta^l_{ij} + (1 - \sigma^l_{ij})\kappa^{l,\text{outer}}_{g_{ij}} \sum \rho^l_{ij} \geq 0, \tag{3.56}$$

for all time, $\frac{(\tilde{W}^l_{g_{ij}})^T \tilde{W}^l_{g_{ij}}}{\|\tilde{W}^l_{g_{ij}}\|} > 0$ and thus $\dot{V} \leq 0$, where the summation includes all weight vectors which require hopping. Therefore, the negativity of $\dot{V}$ is actually strengthened due to the last negative term.   □

3.3 Simulation Results on the Speed Regulation of a DC Motor

We present the test case of a DC motor speed regulation, focusing mainly on simulations that examine the effectiveness of the NF controller under the presence of manipulated additional unknown dynamics terms. For comparison reasons, we run the same experiment using the well-known RHONN approach (Rovithakis and M.A.Christodoulou 2000), which is of comparable structure. Other simulations demonstrating the approximation and control abilities of the NF approach under only parametric uncertainties can be found in Boutalis et al. (2009), Theodoridis et al. (2010).

We consider the speed regulation of a 1 kW DC motor with a normalized model described by the following dynamical equations (Leonhard 1985):

$$T_a \frac{\mathrm{d}I_a}{\mathrm{d}t} = -I_a - \Phi\Omega + V_a,$$

$$T_m \frac{\mathrm{d}\Omega}{\mathrm{d}t} = \Phi I_a - K_0\Omega - m_L,$$

$$T_f \frac{\mathrm{d}\Phi}{\mathrm{d}t} = -I_f + V_f,$$

$$\Phi = \frac{aI_f}{1 + bI_f}. \tag{3.57}$$

The states are chosen to be the armature current, the angular speed, and the stator flux, $x = \begin{bmatrix} I_a & \Omega & \Phi \end{bmatrix}^T$. As control inputs the armature and the field voltages, $x = \begin{bmatrix} V_a & V_f \end{bmatrix}^T$, are used. With this choice, we have:

$$
\begin{bmatrix} \dot{x}_1 \\ \dot{x}_2 \\ \dot{x}_3 \end{bmatrix} = \begin{bmatrix} -\frac{1}{T_a}x_1 - \frac{1}{T_a}x_2 x_3 \\ \frac{1}{T_m}x_1 x_3 - \frac{K_0}{T_m}x_2 - \frac{m_L}{T_m} \\ -\frac{1}{T_f}\frac{x_3}{a-\beta x_3} \end{bmatrix} + \begin{bmatrix} \frac{1}{T_a} & 0 \\ 0 & 0 \\ 0 & \frac{1}{T_f} \end{bmatrix} \begin{bmatrix} u_1 \\ u_2 \end{bmatrix}, \tag{3.58}
$$

which is of a nonlinear, affine in the control form. The regulation problem of a DC motor is translated as follows: Find a state feedback to force the angular velocity and the armature current to go to zero, while the magnetic flux varies.

When Φ is considered constant, the above nonlinear third-order system can be linearized and reduced to a second-order form having 2 states ($x_1 = I_a$ and $x_2 = \Omega$), with the value Φ being included as a constant parameter. However, this method of linearization and reduction cannot be used when one has to alter the field excitation to fulfill additional requirements imposed to the control system, as for example, in the loss minimization problem (Margaris et al. 1991).

Inspired by that, we first assume that the system is described, within a degree of accuracy, by a second-order ($n = 2$) nonlinear NF system of the form (3.9), where $x_1 = I_a$ and $x_2 = \Phi$. Coefficients a_i in matrix A of (3.9) were chosen to be $a_i = 15$. The number of fuzzy partitions in X_f was chosen to be $m = 5$ and the range of $f_1 = [-182.5667, 0]$, $f_2 = [-19.3627, 30.0566]$. The depth of high order terms was $k = 2$ (only first-order sigmoidal terms $s(x_1)$, $s(x_2)$ were used). The number of fuzzy partitions of each g_{ii} in X_g is $m = 3$ and the range of g_{11} is $[147 , 151]$ and of g_{22} is $[0.029 , 0.033]$ assuming that the constant real nonzero coefficients of the control inputs in the real system are unknown. The NF model is initially trained using artificial inputs assuming the persistently exciting form $u = 1 + 0.8\sin (3t)$. The parameters of the sigmoidals that have been used are: $\alpha_1 = 0.4$, $\alpha_2 = 5$, $\beta_1 = \beta_2 = 1$, and $\gamma_1 = \gamma_2 = 0$, with the fuzzy partitions of the output membership functions selected as:

$$
\begin{aligned}
\bar{x}_{f_1} &= \begin{bmatrix} -149.04 & -173.78 & -103.31 & -24.45 & -164.39 \end{bmatrix}, \\
\bar{x}_{f_2} &= \begin{bmatrix} -15.24 & 23.08 & 6.51 & -3.72 & 17.06 \end{bmatrix},
\end{aligned}
$$

and

$$
\begin{aligned}
\bar{x}_{g_1} &= \begin{bmatrix} 148 & 149 & 150 \end{bmatrix}, \\
\bar{x}_{g_2} &= \begin{bmatrix} 0.030 & 0.031 & 0.032 \end{bmatrix}.
\end{aligned}
$$

In the simulations carried out, the actual system is simulated by using the complete set of equations (3.58), while the 1 kW DC motor parameter values are listed in Table 3.1.

The initial state values during the simulations were set: $\Omega = 0.6$, $I_a = 0.6$, and $\Phi = 0.98$. Also, in order to include an additional manipulated by us dynamics uncertainty term, the actual system is simulated by using the following set of equations:

Table 3.1 Parameter values for the DC motor

Parameter	Value
$1/T_a$	148.88 s^{-1}
$1/T_m$	42.91 s^{-1}
K_0/T_m	0.0129 N · m/rad
T_f	31.88 s
m_L	0.0
a	2.6
β	1.6

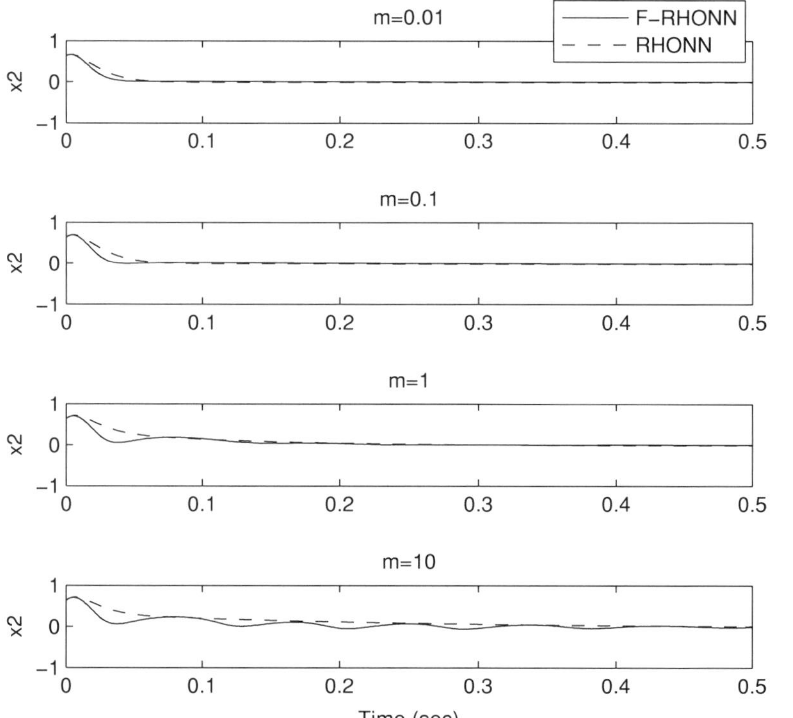

Fig. 3.6 Convergence of motor speed x_2 to zero , with dynamic uncertainties (when $\eta-$ initial $= 2$ and $\mu = 0.01, 0.1, 1, 10$)

$$\begin{bmatrix} \dot{x}_1 \\ \dot{x}_2 \\ \dot{x}_3 \end{bmatrix} = \begin{bmatrix} -\frac{1}{T_a}x_1 - \frac{1}{T_a}x_2x_3 \\ \frac{1}{T_m}x_1x_3 - \frac{K_0}{T_m}x_2 - \frac{m_L}{T_m} \\ -\frac{1}{T_f}\frac{x_3}{a-\beta x_3} \end{bmatrix} + \begin{bmatrix} \frac{1}{T_a} & 0 \\ 0 & 0 \\ 0 & \frac{1}{T_f} \end{bmatrix} \begin{bmatrix} u_1 \\ u_2 \end{bmatrix} + \begin{bmatrix} 5 \\ 10 \\ 0 \end{bmatrix} \eta. \qquad (3.59)$$

The produced control law (3.43), is applied on this system, which in turn produces states x_1, x_2, which in the sequel are used for the computation of the estimation errors that are employed by the updating laws. The parameters of the control law (3.43)

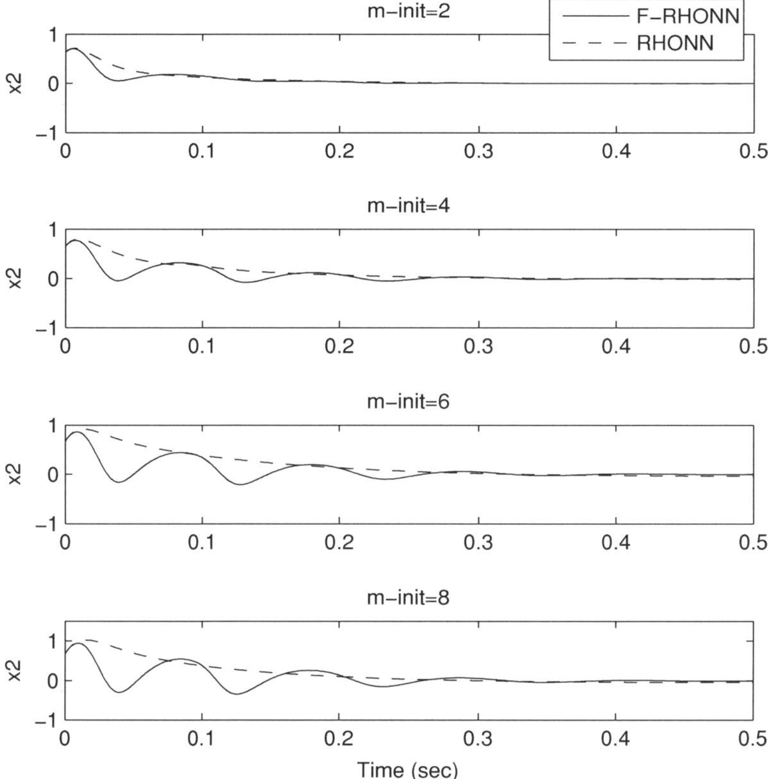

Fig. 3.7 Convergence of motor speed x_2 to zero, with dynamic uncertainties (when $\mu = 1$ and $\eta - \text{initial} = 2, 4, 6, 8$)

are: $K = \text{diag}\,(2, 3)$ and $a = 0.01$ of u_p. As concerning the dynamic uncertainties part, we used the following quasi-steady-state:

$$\dot{h} = \dot{x}^T x = \begin{bmatrix} \dot{x}_1 & \dot{x}_2 \end{bmatrix} \begin{bmatrix} x_1 \\ x_2 \end{bmatrix}, \tag{3.60}$$

in conjunction with Eq. (3.37) using different values from 0.01 to 10 for constant μ and value -10 for the diagonal elements of A_0.

According to the above equations, we simulated the Dc Motor model with $\mu = 0.01, 0.1, 1, 10$ when $\eta_{\text{initial}} = 2$ and all the initial values referred before. So, we take Fig. 3.6 for the evolution of states under the presence of unmodeled dynamics.

In the sequel, we simulated the Dc Motor model with $\eta_{\text{initial}} = 2, 4, 6, 8$ when $\mu = 1$ and we take Fig. 3.7 for the evolution of states under the presence of unmodeled dynamics.

Table 3.2 Comparison of F-RHONNs and RHONN approaches for the regulation of Dc motor when $\eta_{initial} = 2$

	$\mu = 0.01$	$\mu = 0.1$	$\mu = 1$	$\mu = 10$
$MSEx_2$, F-RHONN	0.0134	0.0149	0.0204	0.0320
$MSEx_2$, RHONN	0.0171	0.0206	0.0310	0.0549

Table 3.3 Comparison of F-RHONNs and RHONN approaches for the regulation of Dc motor when $\mu = 1$

	$\eta_{in} = 2$	$\eta_{in} = 4$	$\eta_{in} = 6$	$\eta_{in} = 8$
$MSEx_2$, F-RHONNs	0.0204	0.0303	0.0442	0.0620
$MSEx_2$, RHONN	0.0313	0.0560	0.0840	0.1196

We also measured the mean squared errors for the two different simulations and we take the following Tables 3.2 and 3.3, where it can be seen that our two stage algorithm performs well under the presence of unmodeled dynamics and much better compared to simple RHONNs, even if the $\eta_{initial}$ or the μ changes.

3.4 Summary

An indirect adaptive control scheme aiming at the regulation of nonlinear unknown plants under the presence of parametric and dynamic uncertainties was presented in this chapter. The proposed scheme was based on a new NF dynamical systems modeling, which uses the concept of fuzzy dynamical systems (FDS) operating in conjunction with recurrent high order neural networks (F-RHONNs). Once the system is identified around an operation point is adaptively regulated to zero. Weight updating laws derived from the derivative of Lyapunov candidate function, guarantee that both the identification error and the system states reach zero exponentially fast, while keeping all signals in the closed-loop bounded. Another significant aspect of this chapter was the issue of existence of the control signal which is always ensured by employing a novel method of parameter hopping instead of the conventional projection method. It is demonstrated from the simulations that, by following the proposed procedure, one can obtain asymptotic regulation quite well in the presence of parametric uncertainties and unmodeled dynamics.

References

Boutalis, Y., Christodoulou, M., & Theodoridis, D. (2008). A new neuro-fuzzy intelligent method for indirect adaptive control of nonlinear systems via a novel approach of parameter hopping. In, *Proceedings of Joint International Conferences in Soft Computing Intelligent Systems and Symposium on Advanced Intelligent Systems* (pp. 918–923). Nagoya, Japan.

Boutalis, Y. S., Theodoridis, D. C., & Christodoulou, M. A. (2009). A new neuro fds definition for indirect adaptive control of unknown nonlinear systems using a method of parameter hopping. *IEEE Transactions on Neural Networks, 20*, 609–625.

Boutalis, Y. S., Christodoulou, M.A. & Theodoridis, D. C. (2013). Indirect adaptive control of nonlinear systems based on bilinear neurofuzzy approximation. International Journal of Neural Systems, *23*, 17

Cortes, J. (2008). Discontinuous dynamical systems: A tutorial on solutions, nonsmooth analysis, and stability. IEEE Control Systems Magazine, *28*, 36–73.

Filippov, A. (1998). *Differential equations with discontinuous righthand sides*. Norwell: Kluwer.

Ioannou, P., & Fidan, B. (2006). *Adaptive control tutorial*. USA: SIAM, Advances in Design and Control Series.

Kokotovic, P. V., Khalil, H. K., & O'Reilly, J., (1986). *Singular perturbation methods in control: Analysis and design*. New York: Academic Press.

Kosmatopoulos, E. B., & Ioannou, P. A. (1999). A switching adaptive controller for feedback linearizable systems. *IEEE Transactions on Automatic Control, 44*, 742–750.

Kosmatopoulos, E. B., & Ioannou, P. A. (2002). Robust switching adaptive control of multi-input nonlinear systems. *IEEE Transactions on Automatic Control, 47*, 610–624.

Leonhard, W. (1985). Control of electrical drives. Berlin: Springer.

Mandic, D. P. (2004). A generalized normalized gradient descent algorithm. *IEEE Signal Processing Letters, 11*, 115–118.

Margaris, N., Goutas, T., Doulgeri, Z., & Paschali, A. (1991). Loss minimization in dc drives. *IEEE Transactions on Industrial Electronics, 38*, 328–336.

Morse, A., Mayne, D., & Goodwin, G. (1992). Applications of hysteresis switching in parameter adaptive control. *IEEE Transactions on Automatic Control, 34*, 1343–1354.

Polycarpou, M., & Ioannou, P. (1993). On the existence and uniquness of solutions in adaptive control systems. *IEEE Transactions on Automatic Control, 38*, 474–479.

Popov, V. M. (1973). *Hyperstability of Control Systems*. NY: Springer.

Rovithakis, G. & M. A.Christodoulou. (2000). Adaptive control with recurrent high order neural networks (theory and industrial applications). In *Advances in Industrial Control*. London: Springer Verlag London Limited.

Theodoridis, D., Boutalis, Y. & Christodoulou, M. (2009a). A new neuro-fuzzy dynamical system definition based on high order neural network function approximators. In, *European control conference ECC-09*, Budapest, Hungary.

Theodoridis, D., Boutalis, Y., & Christodoulou, M. (2010). Indirect adaptive control of unknown multi variable nonlinear systems with parametric and dynamic uncertainties using a new neuro-fuzzy system description. *International Journal of Neural Systems, 20*, 129–148.

Theodoridis, D. C., Christodoulou, M. A., & Boutalis, Y. S. (2008). Indirect adaptive neuro—fuzzy control based on high order neural network function approximators. *Proccedings of the 16th mediterranean conference on control and automation—MED08* (pp. 386–393). Corsica, France: Ajaccio.

Theodoridis, D. C., Boutalis, Y. S. & Christodoulou, M. A. (2009b). Direct adaptive control of unknown nonlinear systems using a new neuro-fuzzy method together with a novel approach of parameter hopping. Kybernetica, *45*, 349–386.

Vamvoudakis, K. G. & Lewis, F. L. (2009). Online actor critic algorithm to solve the continuous-time infinite horizon optimal control problem. In Proc. Int. Joint Conf. on Neural Networks, 3180–3187, Atlanta.

Weller, S., & Goodwin, G. (1994). Hysteresis switching adaptive control of linear multivariable systems. *IEEE Transactions on Automatic Control, 39*, 1360–1375.

Yesildirek, A., & Lewis, F. L. (1995). Feedback linearization using neural networks. *Automatica, 31*, 1659–1664.

Chapter 4
Direct Adaptive Neurofuzzy Control of SISO Systems

4.1 Direct Adaptive Regulation

The direct adaptive regulation of unknown nonlinear SISO dynamical systems assuming a Brunovsky form with modeling imperfections, is considered in this section (Theodoridis et al. 2010). Since the plant is considered unknown, we propose its approximation by a special form of a Brunovsky-type NF dynamical system (B-NFDS) assuming also the existence of modeling error terms, which are expressed as disturbance depending on both input and system states plus a not-necessarily-known constant value. The development is combined with a sensitivity analysis of the closed-loop with the presence of modeling imperfections and provides a comprehensive and rigorous analysis of the stability properties of the closed-loop system.

Consider nonlinear dynamical systems of the Brunovsky canonical form:

$$\dot{x} = A_c x + b_c [f(x) + g(x) \cdot u], \tag{4.1}$$

where the state $x \in R^n$ is assumed to be completely measured, the control input $u \in R$, f and g are scalar nonlinear functions of the state being only involved in the dynamic equation of x_n. Also, $A_c = \begin{bmatrix} 0 & 1 & 0 & \cdots & 0 \\ 0 & 0 & 1 & 0 & \cdots \\ \cdots\cdots & 0 & \cdots & 0 \\ 0 & 0 & \cdots & 0 & 1 \\ 0 & 0 & \cdots & 0 & 0 \end{bmatrix}$ and $b_c = \begin{bmatrix} 0 & \cdots & 0 & 1 \end{bmatrix}^T$.

The state regulation problem is known as our attempt to force the state to zero from an arbitrary initial value by applying appropriate feedback control to the plant input. However, the problem as it is stated above for the system (4.1), is very difficult or even impossible to be solved since functions f, g are assumed to be completely unknown. To overcome this problem, we assume that the unknown plant can be described by the following model arriving from a NF representation described in Sect. 4.1.1 below, plus a modeling error term $\omega_n(x, u)$:

Y. Boutalis et al., *System Identification and Adaptive Control*,
Advances in Industrial Control, DOI: 10.1007/978-3-319-06364-5_4,
© Springer International Publishing Switzerland 2014

$$\dot{x} = A_c x + b_c [\bar{x}_f W_f^* s_f(x) + \bar{x}_g W_g^* s_g(x)u + \omega_n(x, u)], \qquad (4.2)$$

where $\bar{x}_f, \bar{x}_g$ are vectors determined by the designer and contain the centers of the partitions of the fuzzy output membership functions. W_f^*, W_g^* are unknown weight matrices and s_f, s_g are sigmoidal vectors containing high order sigmoidal terms with parameters which are also decided appropriately by the designer.

Therefore, the state regulation problem is analyzed for the system (4.2) instead of (4.1). Since, W_f^* and W_g^* are unknown, our solution consists of designing a control law $u(W_f, W_g, x)$ and appropriate update laws for W_f and W_g to guarantee convergence of the state to zero or to a residual set, something that will be analyzed in the following sections. Moreover, the boundedness of x and of all signals in the closed-loop has to be ensured.

The following assumptions guarantee the existence and uniqueness of solution for any finite initial condition of (4.1) and $u \in U_c$.

Assumption 6 Given a class $U_c \subset R$ of admissible inputs, then for any $u \in U_c$ and any finite initial condition, the state trajectories are uniformly bounded for any finite $T > 0$. Meaning that we do not allow systems processing trajectories which escape at infinite, in finite time T, T being arbitrarily small. Hence, $|x(T)| < \infty$.

Assumption 7 Functions f, g are continuous with respect to their arguments and satisfy a local Lipschitz condition so that the solution $x(t)$ of (4.1) is unique for any finite initial condition and $u \in U_c$.

4.1.1 Neurofuzzy Modeling

Following the development made in Sect. 2.1, we are using a fuzzy approximation of the system in (4.1), which uses two fuzzy subsystem blocks for the description of $f(x)$ and $g(x)$ as follows:

$$\hat{f}(x) = \sum \bar{x}_{f_p} \times (I')_f^p(x), \qquad (4.3)$$

$$\hat{g}_j(x) = \sum \bar{x}_{g_{jp}} \times (I')_{g_j}^p(x), \qquad (4.4)$$

where the summation is carried out over the number of all available fuzzy rules and $(I')_f$, $(I')_{g_j}$ are appropriate weighted fuzzy rule indicator functions.

In order to simplify the model structure, since some rules result to the same output partition, we could replace the NNs associated to the rules having the same output with one NN and therefore the summations in (4.3), (4.4) are carried out over the number of the corresponding output partitions. Therefore, the system of (4.1) is replaced by the following equivalent Brunovsky-type F-HONN, which depends on the centers of the fuzzy output partitions $\bar{x}_{f_p}$ and $\bar{x}_{g_{jp}}$.

$$\dot{\hat{x}} = A_c\hat{x} + b_c\left[\sum_{p=1}^{q}\bar{x}_{f_p}\cdot\sum_{l=1}^{k}w_f^{pl}\cdot s_{f_l}(x) + \sum_{j=1}^{m}\left(\sum_{p=1}^{q}\bar{x}_{g_{jp}}\cdot\sum_{l=1}^{k}w_{g_j}^{pl}\cdot s_{g_l}(x)\right)\cdot u_j\right],$$

(4.5)

where $j = 1,\ldots,m$ are the number of control inputs, $p = 1,\ldots,q$ are the number of fuzzy centers of f and g, respectively, and finally $l = 1,\ldots,k$ are the number of high order terms of the sigmoidal functions. In a more compact form, the above NF model can be written as:

$$\dot{\hat{x}} = A_c\hat{x} + b_c[\bar{x}_f W_f s_f(x) + \bar{x}_g W_g s_g(x)u],$$

(4.6)

where, similar to (4.2), $\bar{x}_f$, $\bar{x}_g$ are vectors containing the centers of the partitions of the fuzzy output variables corresponding to $f(x)$ and $g(x)$, respectively, $s_f(x)$, $s_g(x)$ are vectors containing high order combinations of sigmoid functions of the state x and W_f, W_g are matrices containing respective neural weights according to (4.5) and (4.6). The dimensions and the contents of all the above matrices are chosen so that both $\bar{x}_f W_f s_f(x)$ and $\bar{x}_g W_g s_g(x)$ are scalar. For notational simplicity, we also assume that all output fuzzy variables are partitioned to the same number, q, of partitions. Under these specifications $\bar{x}_f$ is a $1 \times q$ row vector of the form:

$$\bar{x}_f = \left[\bar{x}_{f_1}\bar{x}_{f_2}\cdots\bar{x}_{f_q}\right],$$

where $\bar{x}_{f_p}$ denotes the center or the fuzzy p-th partition of f. These centers can be determined manually or automatically with the help of a fuzzy c-means clustering algorithm as a part of an off-line fuzzy system structural identification procedure. Also, $s_f(x) = \left[s_{f_1}(x) \ldots s_{f_k}(x)\right]^T$, where each $s_{f_l}(x)$ with $l = 1, 2, \ldots, k$, is a high order combination of sigmoid functions of the state variables and W_f is a $q \times k$ matrix with neural weights. W_f can be also written as a collection of column vectors $W_f^l, l = 1, 2, \ldots, k$, that is $W_f = \left[W_f^1 W_f^2 \cdots W_f^k\right]$ or in a more detailed form as:

$$W_f = \begin{bmatrix} w_f^{11} & w_f^{12} & \cdots & w_f^{1k} \\ w_f^{21} & w_f^{22} & \cdots & w_f^{2k} \\ \vdots & \vdots & \vdots & \vdots \\ w_f^{q1} & w_f^{q2} & \cdots & w_f^{qk} \end{bmatrix}.$$

Similarly, $\bar{x}_g$ is a $1 \times q$ row vector of the form:

$$\bar{x}_g = \left[\bar{x}_{g_1}\ \bar{x}_{g_2}\ \cdots\ \bar{x}_{g_q}\right],$$

where $\bar{x}_{g_k}$ denotes the center or the k-th partition of g. W_g, $s_g(x)$ have the same dimensions as W_f, $s_f(x)$, respectively.

4.1.2 Adaptive Regulation with Modeling Error Effects

In this section, we present a solution to the adaptive regulation problem and investigate the modeling error effects. Assuming the presence of modeling errors the unknown system can be written as (4.2), where the dimensions of the involved matrices and vectors have been given in Sect. 4.1.1.

Define now v as:

$$v \stackrel{\Delta}{=} b_c \left[\bar{x}_f W_f s_f(x) + \bar{x}_g W_g s_g(x)u \right] - \dot{x} + A_c x, \tag{4.7}$$

substituting Eq. (4.2) to (4.7) we have:

$$v \stackrel{\Delta}{=} b_c \left[\bar{x}_f \tilde{W}_f s_f(x) + \bar{x}_g \tilde{W}_g s_g(x)u - \omega_n(x, u) \right], \tag{4.8}$$

where $\tilde{W}_f = W_f - W_f^*$ and $\tilde{W}_g = W_g - W_g^*$. W_f and W_g are estimates of W_f^* and W_g^*, respectively, and are obtained by updating laws which are to be designed in the sequel. v cannot be measured since $\dot{x}$ is unknown. To overcome this problem, we use the following filtered version of v:

$$v = \dot{\xi} + K_c \xi,$$

where $K_c = \begin{bmatrix} 0 & -1 & 0 & \cdots & 0 \\ 0 & 0 & -1 & 0 & \cdots \\ \cdots & \cdots & 0 & \cdots & 0 \\ 0 & 0 & \cdots & 0 & -1 \\ k_n & k_{n-1} & \cdots & k_2 & k_1 \end{bmatrix}$ is a positive definite matrix. Also, $[k_n \cdots k_2 \ k_1] \in R^n$ is a vector such that all roots of the polynomial $h(s) = s^n + k_1 s^{n-1} + \cdots + k_n$ are in the open left half-plane. In the sequel, using Eq. (4.8) we have that:

$$\dot{\xi} = -K_c \xi + b_c \left[\bar{x}_f \tilde{W}_f s_f(x) + \bar{x}_g \tilde{W}_g s_g(x)u - \omega_n(x, u) \right]. \tag{4.9}$$

To implement Eq. (4.9), we take:

$$\xi \stackrel{\Delta}{=} \zeta - x. \tag{4.10}$$

Employing Eq. (4.10), Eq. (4.9) can be written as:

$$\dot{\zeta} + K_c \zeta = K_c x + A_c x + b_c \left[\bar{x}_f W_f s_f(x) + \bar{x}_g W_g s_g(x)u \right], \tag{4.11}$$

with state $\zeta \in \Re^n$. This method is referred to as error filtering.

The regulation of the system can be achieved by selecting the control input to be:

$$u = - \left[\bar{x}_g W_g s_g (x) \right]^{-1} \left[\bar{x}_f W_f s_f (x) + \upsilon \right], \tag{4.12}$$

with

$$\upsilon = kx, \tag{4.13}$$

where k is a vector of the form $k = [k_n \cdots k_2 \, k_1] \in R^n$.

Thus, substituting Eq. (4.12), Eq. (4.11) becomes:

$$\dot{\zeta} = \Lambda_c \zeta, \tag{4.14}$$

where $\Lambda_c = -K_c = A_c - b_c k$. Λ_c is a matrix with its eigenvalues on the left half plane.

To continue, consider the Lyapunov candidate function:

$$V = \frac{1}{2}\xi^T P \xi + \frac{1}{2}\zeta^T P \zeta + \frac{1}{2\gamma_1}\mathrm{tr}\left\{ \tilde{W}_f^T \tilde{W}_f \right\} + \frac{1}{2\gamma_2}\mathrm{tr}\left\{ \tilde{W}_g^T \tilde{W}_g \right\}, \tag{4.15}$$

where $P > 0$ is chosen to satisfy the Lyapunov equation:

$$P \Lambda_c + \Lambda_c^T P = -Q, \tag{4.16}$$

where Q is a positive definite diagonal matrix chosen appropriately by the designer. If we take the derivative of Eq. (4.15) with respect to time we obtain:

$$\dot{V} = -\xi^T Q \xi - \zeta^T Q \zeta + \xi^T P b_c \bar{x}_f \tilde{W}_f s_f (x) + \xi^T P b_c \bar{x}_g \tilde{W}_g s_g (x) u$$
$$- \xi^T P b_c \omega_n (x, u) + \frac{1}{\gamma_1}\mathrm{tr}\left\{ \dot{W}_f^T \tilde{W}_f \right\} + \frac{1}{\gamma_2}\mathrm{tr}\left\{ \dot{W}_g^T \tilde{W}_g \right\}. \tag{4.17}$$

Hence, if we choose:

$$\mathrm{tr}\left\{ \dot{W}_f^T \tilde{W}_f \right\} = -\gamma_1 \xi^T P b_c \bar{x}_f \tilde{W}_f s_f (x), \tag{4.18}$$

$$\mathrm{tr}\left\{ \dot{W}_g^T \tilde{W}_g \right\} = -\gamma_2 \xi^T P b_c \bar{x}_g \tilde{W}_g s_g (x) u, \tag{4.19}$$

$\dot{V}$ becomes:

$$\dot{V} \leq -\lambda_{\min}(Q) \|\xi\|^2 - \lambda_{\min}(Q) \|\zeta\|^2 + \|\xi\| \, \|P\| \, \|\omega_n (x, u)\|. \tag{4.20}$$

It can be easily verified that Eqs. (4.18) and (4.19), after using matrix trace properties, result in the following weight updating laws:

$$\dot{W}_f = -\gamma_1 \bar{x}_f^T b_c^T P \xi s_f^T (x), \tag{4.21}$$

$$\dot{W}_g = -\gamma_2 \bar{x}_g^T b_c^T P \xi s_g^T(x) u, \tag{4.22}$$

where ξ is the vector defined in (4.10), u is a scalar and γ_1, γ_2 are positive constants expressing the learning rates.

The above equations can be also element-wise written as:

(A) for the elements of W_f:

$$\dot{w}_{fpl} = -\gamma_1 \bar{x}_{fp} p(n,:) \xi s_{fl}(x) \tag{4.23}$$

or equivalently:

$$\dot{W}_f^l = -\gamma_1 p(n,:) \xi \left(x_f\right)^T s_{fl}(x), \tag{4.24}$$

for all $l = 1, 2, \ldots, k$, $p = 1, 2, \ldots, q$ and $p(n,:)$ is the n-th row vector of matrix P.

(B) for the elements of W_g:

$$\dot{w}_{gpl} = -\gamma_2 \bar{x}_{gp} p(n,:) \xi s_{gl}(x) u \tag{4.25}$$

or equivalently:

$$\dot{W}_g^l = -\gamma_2 \left(\bar{x}_g\right)^T p(n,:) \xi s_{gl}(x) u, \tag{4.26}$$

for all $l = 1, 2, \ldots, k$ and $p = 1, 2, \ldots, q$.

Two issues that need special consideration is the existence and the boundedness of the control signal. From (4.12), (4.13) it can be easily seen that the existence of the control signal is ensured if the inverse of $\bar{x}_g W_g s_g(x)$ exists. Apart from that, in order the control law to be physically implementable the quantity $\bar{x}_f W_f s_f(x)$ and $\bar{x}_g W_g s_g(x)$ should remain within pre-specified bounds. In adaptive control literature, these issues are usually tackled by introducing a "projection" method in the weight updating algorithm. However, as has already been explained in previous chapters, the weight updating laws presented above are directional. Therefore, the projection procedure is replaced by the so-called "hopping" method, which can give more detailed insight in the various subvector components of the weight matrices. The idea of "hopping" was first introduced in Chap. 2 and was explained better in Chap. 3, it is however in the sequel further specialized.

4.1.2.1 Weight Hopping in SISO Systems

Consider now the control law in (4.12), (4.13). To ensure its existence, one should ensure that $\bar{x}_g W_g s_g(x) \neq 0$. Taking into account that $\bar{x}_g$ is a $1 \times q$ vector and that $W_g = [W_g^1, W_g^1, \ldots, W_g^k,]$ is a $q \times k$ matrix with each W_g^l, $l = 1, \ldots, k$ being a column vector, one may rewrite $\bar{x}_g W_g s_g(x) \neq 0$ as $\bar{x}_g W_g^1 s_g^1(x) + \bar{x}_g W_g^2 s_g^2(x) + \cdots + \bar{x}_g W_g^k s_g^k(x) \neq 0$, where $s_g^i(x)$ is the ith element of vector $s_g(x)$. Therefore:

$$\bar{x}_g W_g^l \neq -\sum_{i \neq l} \frac{\bar{x}_g W_g^i s_g^i(x)}{s_g^l(x)}, \tag{4.27}$$

where the right-hand side is a scalar quantity. It is obvious that the equation $\bar{x}_g W_g^l = -\sum_{i \neq j} \frac{\bar{x}_g W_g^i s_g^i(x)}{s_g^l(x)}$ determines the forbidden hyperplane. Moreover, the weight updating law (4.26), expressed in subvector form determines that the weight updating direction (depending on the vector $\bar{x}_g$) is perpendicular to $\bar{x}_g W_g^l = 0$ and all its parallel hyperplanes. Therefore, the procedure of parameter hopping described in previous chapters can be applied here in order to drive the required weight vectors to the other side of the hyper-space. According to (4.27), the forbidden hyper-plane is determined taking into account all the other weight vectors W_g^i. The procedure that will be followed is to check at each iteration condition (4.27) and perform the "hopping" by appropriately modifying the weight updating law in all vectors W_g^l, which call for "hopping." As it will be shown in the next theorems, the introduction of hopping ensures the existence of the control signal and it does not threaten the stability of the system. For implementation reasons, condition (4.27) can be also written as:

$$\left| \bar{x}_g W_g^l + \sum_{i \neq l} \frac{\bar{x}_g W_g^i s_g^i(x)}{s_g^l(x)} \right| > \theta_l, \tag{4.28}$$

where θ_l is a small positive number. Hopping may occur when this condition is violated. Apparently, one may apply hopping either to one of the weight subvectors W_g^i or to all of them simultaneously by determining each hopping magnitude according to the contribution of each vector to (4.28). For simplicity we will use the first approach.

Regarding the term $\bar{x}_f W_f s_f(x)$ of (4.12), in order the control law to be physically implementable, one has to ensure that $|\bar{x}_f W_f s_f(x)| \leq \rho$, where ρ is an upper bound. Since $|\bar{x}_f W_f s_f(x)| \leq \sum_{i=1}^{k} |\bar{x}_f W_f^i s_f^i(x)|$, one is allowed to have a better insight in the algorithm if puts specialized bounds, each one for each component of $\bar{x}_f W_f s_f(x)$. Each specialized bound may be expressed by:

$$|\bar{x}_f W_f^i s_f^i(x)| \leq \rho_i. \tag{4.29}$$

Since the weight updating of W_f^i given by (4.24) is also directional, the above inequality is also sufficient for the boundedness of $\|W_f^i\|$, provided that the initial values of W_f^i lie in a hyper-sphere determined by this bound. One strategy that can be followed is to apply again a "hopping" to the weight updating equation whenever a vector is approaching the forbidden outer hyper-plane and is directed toward it. The "hopping" could send the weight back to the desired hyper-space allowing thus the algorithm to search the entire space for a better weight solution. This strategy

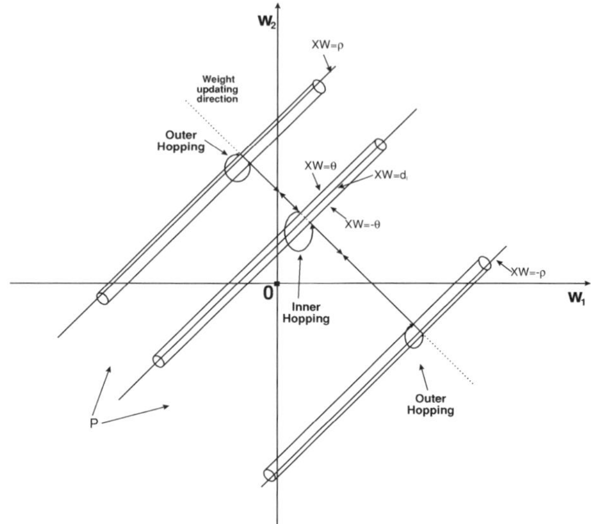

Fig. 4.1 Pictorial representation of inner and outer parameter hopping for systems having Brunovsky-type form

is depicted in Fig. 4.1. The same rationale can be also applied if one wants to keep $\bar{x}_g W_g s_g(x)$ also bounded.

Lemma 4.1 below introduces this *hopping* in the weight updating law.

Lemma 4.1 *The updating law for the elements of W_g^l given by (4.26) and modified according to the hopping method:*

$$
\dot{W}_g^l = \begin{cases}
-\gamma_2 \left(\bar{x}_g\right)^T p(n,:) \xi s_{g_l}(x) u & \text{if } \left|\sum \bar{x}_g \cdot W_g^i s_g^i\right| > \theta_l \\
& \text{or } \left|\sum \bar{x}_g \cdot W_g^i s_g^i\right| = \theta_l \\
& \qquad \text{and } \bar{x}_g \cdot \dot{W}_g^l \leq 0 \\
\\
-\gamma_2 \left(\bar{x}_g\right)^T p(n,:) \xi s_{g_l}(x) u + & \text{otherwise} \\
\dfrac{2\left(d_l - \bar{x}_g W_g^l\right)(\bar{x}_g)^T}{\operatorname{tr}\{(\bar{x}_g)^T \bar{x}_g\}}
\end{cases},
$$

ensures the existence of the control signal.

Proof The first part of the weight updating equation is used when the weights are at a certain distance (condition if $\left|\sum \bar{x}_g^i \cdot W_g^i s_g^i\right| > \theta_l$) from the forbidden plane or at the safe limit (condition $\left|\sum \bar{x}_g^i \cdot W_g^i s_g^i\right| = \theta_l$) but with the direction of updating moving the weights far from the forbidden plane (condition $\bar{x}_g \cdot \dot{W}_g^l < 0$).

In the second part of $\dot{W}_g^l$, term $\dfrac{2\left(d_l - \bar{x}_g W_g^l\right)(\bar{x}_g)^T}{\mathrm{tr}\{(\bar{x}_g)^T \bar{x}_g\}}$ determines the magnitude of weight *hopping* and $d_l = -\sum\limits_{i \neq j} \dfrac{\bar{x}_g W_g^i s_g^i(x)}{s_g^l(x)}$. As explained in Chap. 3 (see Sect. 3.2.2), this magnitude has to be two times the distance of the current weight vector to the forbidden hyper-plane. Therefore, the *existence* of the control signal is ensured because the weights never reach the forbidden plane. Moreover, as it will be shown in Lemma 4.3, this hopping does not threaten the stability of the overall control scheme.

The inclusion of weight hopping in the weights updating law guarantees that the control signal does not go to infinity. Apart from that, it is also of practical use to ensure that $\bar{x}_g W_g s_g(x)$ does not approach even temporarily infinity because in this case the method may become algorithmically unstable driving at the same time the control signal to zero, failing to control the system. To ensure that this situation does not happen we have again to ensure that $\left| \bar{x}_g \cdot W_g^l \right| < \rho_l$ with ρ_l being again a design parameter determining an external limit for $\bar{x}_g \cdot W_g^l$.

Following the same lines of thought regarding weight hopping as were introduced in Chaps. 2, 3, we could again consider the forbidden hyper-planes being defined by the equation $\left| \bar{x}_g \cdot W_g^l \right| = \rho_l$. When the weight vector reaches one of the forbidden hyper-planes $\bar{x}_g \cdot W_g^l = \rho_l$ and the direction of updating is toward the forbidden hyper-plane, a new *modified hopping* is introduced which moves the weights to the interior of the desired space, away from the forbidden hyper-plane.

This procedure is depicted in Fig. 4.1, in a simplified 2-dimensional representation. The magnitude of hopping is $-\dfrac{\kappa\left(\bar{x}_g W_g^l (\bar{x}_g)^T\right)}{\mathrm{tr}\{(\bar{x}_g)^T \bar{x}_g\}}$, being determined by following again the same vectorial proof given in Chap. 3 (Sect. 3.2.2) and depicted in Fig. 3.2. By performing *hopping* when $\bar{x}_g \cdot W_g^l$ reaches either the inner or outer forbidden planes, $\bar{x}_g \cdot W_g^l$ is confined in space $P_g = \left\{ \bar{x}_g \cdot W_g^l : \left| \bar{x}_g \cdot W_g^l \right| \leq \rho_l \text{ and } \left| \sum \bar{x}_g \cdot W_g^i s_g^i \right| > \theta_l \right\}$ lying between these hyper-planes. The weight updating law for W_g^l incorporating the two hopping conditions can now be expressed as:

$$
\dot{W}_g^l =
\begin{cases}
-\gamma_2 \left(\bar{x}_g\right)^T p(n,:)\xi s_{g_l}(x)u &
\begin{aligned}
&\text{if } \bar{x}_g \cdot W_g^l \in P_g \\
&\text{or } \bar{x}_g \cdot W_g^l = (\pm \theta_l \text{ or } \pm \rho_l) \\
&\text{and } \bar{x}_g \cdot W_g^l <> \text{ or } >< 0
\end{aligned} \\[2em]
\begin{aligned}
&-\gamma_2 \left(\bar{x}_g\right)^T p(n,:)\xi s_{g_l}(x)u+ \\
&2\sigma_i \frac{\left(d_l - \bar{x}_g W_g^l\right)(\bar{x}_g)^T}{\mathrm{tr}\{(\bar{x}_g)^T \bar{x}_g\}} - \\
&(1-\sigma_i)\frac{\kappa\left(\bar{x}_g W_g^l (\bar{x}_g)^T\right)}{\mathrm{tr}\{(\bar{x}_g)^T \bar{x}_g\}}
\end{aligned} & \text{otherwise}
\end{cases}
\qquad (4.30)
$$

where

$$
\sigma_i = \begin{cases}
0 & \begin{array}{l} \text{if } \bar{x}_g W_g^l = \pm\rho_l \\ \text{and } \bar{x}_g \dot{W}_g^l <> 0 \end{array} \\[2ex]
1 & \begin{array}{l} \text{if } \sum \bar{x}_g W_g^i s_g^i = \pm\theta_l \\ \text{and } \bar{x}_g \dot{W}_g^l >< 0 \end{array}
\end{cases}
\tag{4.31}
$$

and κ a small positive constant value chosen by the designer according to Chap. 3 (Fig. 3.5). In the current notation, the "$\pm$" symbol has a one to one correspondence with the "$<>$" one, meaning that "$+$" case corresponds to "$<$" case and the "$-$" case corresponds to "$>$" case.

To continue the analysis, regarding the modeling error term $\omega_n(x, u)$ we can distinguish two possible cases. The complete model matching at zero case and the modeling error at zero case.

4.1.2.2 Complete Model Matching at Zero Case

We can make the following assumption.

Assumption 8 The modeling error term satisfies:

$$
\mid \omega_n(x, u) \mid \leq \ell_1' \, |x| + \ell_1''|u|,
$$

where ℓ_1' and ℓ_1'' are known positive constants.

Also, we can find an *a priori* known constant $\ell_u > 0$, such that:

$$
|u| \leq \ell_u |x|,
\tag{4.32}
$$

and Assumption 8 becomes equivalent to:

$$
|\omega_n(x)| \leq \ell_1 |x|,
\tag{4.33}
$$

where

$$
\ell_1 = \ell_1' + \ell_1'' \ell_u,
$$

is a positive constant. Assumption 8 together with $\ell_1'' = 0$ tell us that at zero we have no modeling error in the controlled vector fields since then $|\omega(0, u)| \leq 0$.

One can easily verify that (4.32) is valid provided that $\bar{x}_f W_f$ is uniformly bounded by a known positive constant ε_l so that $\bar{x}_f W_f(t)$ is confined to the set $P_f = \{\bar{x}_f \cdot W_f^l : \left|\bar{x}_f \cdot W_f^l\right| \leq \varepsilon_l\}$ through the use of a hopping action. In particular, the standard updating law (4.24) is modified to:

$$\dot{W}_f^l = \begin{cases} -\gamma_1 \left(\bar{x}_f \right)^T p(n,:) \xi s_{f_l}(x) & \begin{array}{l} \text{if } \bar{x}_f \cdot W_f^l \in P_f \\ \text{or } \bar{x}_f \cdot W_f^l = \pm \varepsilon_l \\ \text{and } \bar{x}_f \cdot \dot{W}_f^l >< 0 \end{array} \\[2em] \dfrac{-\gamma_1 \left(\bar{x}_f \right)^T p(n,:) \xi s_{f_l}(x) - \kappa \left(\bar{x}_f W_f^l (\bar{x}_f)^T \right)}{\text{tr} \left\{ (\bar{x}_f)^T \bar{x}_f \right\}} & \text{otherwise} \end{cases} \qquad (4.34)$$

Therefore, we have the following lemma.

Lemma 4.2 *If the initial weights are chosen such that $\bar{x}_f \cdot W_f^l(0) \in P_f$ and $\bar{x}_f \cdot W_f^{l\star} \in P_f$ then, we have $\bar{x}_f \cdot W_f^l \in P_f$, for all $t \geq 0$.*

Proof The above lemma can be readily established by noting that whenever $|\bar{x}_f \cdot (W_f^l)^-| \geq \varepsilon_l$ then:

$$\frac{d}{dt} \left(|\bar{x}_f \cdot (W_f^l)^+|^2 \right) \leq 0, \qquad (4.35)$$

which implies that after modified hopping occurs, the weights $(W_f^l)^+$, are directed towards the interior of P_f. For simplicity, since we will be working from now on with the time $(\cdot)^+$, we omit the $+$ sign from the exponent. It is true that:

$$\frac{d}{dt} \left(|\bar{x}_f \cdot W_f^l|^2 \right) = W_f^{l\,T} \dot{W}_f^l \bar{x}_f \bar{x}_f^T. \qquad (4.36)$$

Since $\bar{x}_f \bar{x}_f^T > 0$, only $W_f^{l\,T} \dot{W}_f^l$ determines the sign of the above derivative. Employing the modified adaptive law (4.34), we obtain:

$$\left(W_f^l \right)^T \dot{W}_f^l = -\gamma_1 \left(W_f^l \right)^T (\bar{x}_f)^T p(n,:) \xi s_{f_l}(x) - \kappa \varepsilon_l \frac{\left(W_f^l \right)^T W_f^l}{\left\| W_f^l \right\|}, \qquad (4.37)$$

where $\varepsilon_l = \dfrac{\left(\bar{x}_f W_f^l (\bar{x}_f)^T \right)}{\text{tr} \left\{ (\bar{x}_f)^T \bar{x}_f \right\}}$. As concerning the second part of the above equation it is obvious that $\kappa \varepsilon_l > 0$ and $\dfrac{\left(W_f^l \right)^T W_f^l}{\left\| W_f^l \right\|} > 0$. So:

$$-\kappa \varepsilon_l \frac{\left(W_f^l \right)^T W_f^l}{\left\| W_f^l \right\|} < 0.$$

Now, regarding the first part of Eq. (4.37), we can distinguish two cases:

Case 1: $\bar{x}_f \cdot W_f^l = \varepsilon_l$ and $\bar{x}_f \cdot \dot{W}_f^l < 0$.
From the above notation we have that:

$$\bar{x}_f \dot{W}_f^l = -\gamma_1 x_f \left(\bar{x}_f\right)^T p(n,:)\xi s_{f_l}(x) < 0 \Rightarrow \gamma_1 p(n,:)\xi s_{f_l}(x) > 0. \qquad (4.38)$$

Also, $\bar{x}_f W_f^l \geq \varepsilon_l$ and so the first part of Eq. (4.37) becomes:

$$-\gamma_1 \left(W_f^l\right)^T \left(\bar{x}_f\right)^T p(n,:)\xi s_{f_l}(x) \leq -\gamma_1 \varepsilon_l p(n,:)\xi s_{f_l}(x).$$

According to Eq. (4.38):

$$-\gamma_1 \left(W_f^l\right)^T \left(\bar{x}_f\right)^T p(n,:)\xi s_{f_l}(x) < 0.$$

Case 2: $\bar{x}_f \cdot W_f^l \leq -\varepsilon_l$ and $\bar{x}_f \cdot \dot{W}_f^l > 0$.
From the above notation we have that:

$$\bar{x}_f \dot{W}_f^l = -\gamma_1 \bar{x}_f \left(\bar{x}_f\right)^T p(n,:)\xi s_{f_l}(x) > 0 \Rightarrow \gamma_1 p(n,:)\xi s_{f_l}(x) < 0. \qquad (4.39)$$

Also, $\bar{x}_f W_f^l \leq -\varepsilon_l$ and so the first part of Eq. (4.37) becomes:

$$-\gamma_1 \left(W_f^l\right)^T \left(\bar{x}_f\right)^T p(n,:)\xi s_{f_l}(x) \geq -\gamma_1 (-\varepsilon_l) p(n,:)\xi s_{f_l}(x).$$

According to Eq. (4.39):

$$-\gamma_1 \left(W_f^l\right)^T \left(\bar{x}_f\right)^T p(n,:)\xi s_{f_l}(x) < 0.$$

Therefore, we finally obtain:

$$\frac{\mathrm{d}}{\mathrm{d}t}\left(|\bar{x}_f \cdot W_f^l(t)|^2\right) \leq 0.$$

$\square$

Employing Assumption 8, Eq. (4.20) becomes:

$$\begin{aligned}
\dot{V} &\leq -\lambda_{\min}(Q)\|\xi\|^2 - \lambda_{\min}(Q)\|\zeta\|^2 + \ell_1 \|\xi\|\|P\|\|x\| \\
&\leq -\lambda_{\min}(Q)\left(\|\xi\|^2 + \|\zeta\|^2\right) + \ell_1\|P\|\|\xi\|^2 + \ell_1\|P\|\|\xi\|\|\zeta\| \\
&\leq -\left[\|\xi\|\ \|\zeta\|\right] \cdot \begin{bmatrix} \lambda_{\min}(Q) - l_1\|P\| & -l_1\|P\| \\ 0 & \lambda_{\min}(Q) \end{bmatrix} \cdot \begin{bmatrix} \|\xi\| \\ \|\zeta\| \end{bmatrix}.
\end{aligned} \qquad (4.40)$$

Hence, if we chose $\lambda_{\min}(Q) \geq \ell_1\|P\|$ then Eq. (4.40) becomes negative semi-definite. Thus, we have:

$$\dot{V} \leq 0. \tag{4.41}$$

Regarding the *negativity* of $\dot{V}$ when weight hopping is applied, we proceed with the following lemma.

Lemma 4.3 *Based on the adaptive laws* (4.30), (4.34) *the additional terms introduced in the expression for $\dot{V}$, can only make $\dot{V}$ more negative.*

Proof In the second part of W_f^l, term $-\dfrac{\kappa\left(\bar{x}_f W_f^l (\bar{x}_f)^T\right)}{\mathrm{tr}\{(\bar{x}_f)^T \bar{x}_f\}}$ determines the magnitude of weight *modified hopping*, which as explained in Chap. 3 (Fig. 3.5) has to be κ times the distance of the current weight vector, with $0 < \kappa < 1$. In order to analyze the effect of the additional terms introduced by the hopping condition in the *negativity* of $\dot{V}$, we proceed as follows.

Following similar analysis with proof of Theorem 3.3 in Sect. 3.2.2, the modified updating law concerning W_f^l is rewritten as:

$$\dot{W}_f^l = -\gamma_1 \left(\bar{x}_f\right)^T p(n, :)\xi s_{f_l}(x) - \kappa \varepsilon_l \frac{\tilde{W}_f^l}{\|\tilde{W}_f^l\|}.$$

With this updating law, it can be easily verified that Eq. (4.40) becomes:

$$\dot{V} \leq -\left[\|\xi\| \ \|\zeta\|\right] \cdot \begin{bmatrix} \lambda_{\min}(Q) - l_1 \|P\| & -l_1 \|P\| \\ 0 & \lambda_{\min}(Q) \end{bmatrix} \cdot \begin{bmatrix} \|\xi\| \\ \|\zeta\| \end{bmatrix} - I_{f_l} \Theta_{f_l}, \tag{4.42}$$

where I_{f_l} is an indicator function defined as $I_{f_l} = 1$, if the hopping condition is applied, otherwise $I_{f_l} = 0$. Also, $\Theta_{f_l} = \sum \kappa \varepsilon_l \dfrac{(\tilde{W}_f^l)^T \tilde{W}_f^l}{\|\tilde{W}_f^l\|} \geq 0$ is a positive real constant for all time, where the summation includes all weight vectors which require hopping. Therefore, the negativity of $\dot{V}$ is actually strengthened due to the last negative terms.

By using the modified updating law for W_g^l, the negativity of the Lyapunov function is also not compromised. Indeed, the second part of the modified form of $\dot{W}_g^l$ shown in Eq. (4.30), is rewritten as:

$$\dot{W}_g^l = -\gamma_2 \left(\bar{x}_g\right)^T p(n, :)\xi s_{g_l}(x)u - 2\theta_l \frac{\tilde{W}_g^l}{\|\tilde{W}_g^l\|},$$

when the inner hopping condition is activated or

$$\dot{W}_g^l = -\gamma_2 \left(\bar{x}_g\right)^T p(n, :)\xi s_{g_l}(x)u - k\rho_l \frac{\tilde{W}_g^l}{\|\tilde{W}_g^l\|},$$

when the outer hopping condition is activated. With this updating law, it can be easily verified that (4.42) becomes:

$$\dot{V} \le - \begin{bmatrix} \|\xi\| & \|\zeta\| \end{bmatrix} \cdot \begin{bmatrix} \lambda_{\min}(Q) - l_1\|P\| & -l_1\|P\| \\ 0 & \lambda_{\min}(Q) \end{bmatrix} \cdot \begin{bmatrix} \|\xi\| \\ \|\zeta\| \end{bmatrix} - I_{f_l}\Theta_{f_l} - I_{g_l}\Theta_{g_l} \quad (4.43)$$

where I_{g_l} is an indicator function defined as $I_{g_l} = 1$, if the hopping condition is applied, otherwise $I_{g_l} = 0$. Also, Θ_{g_l} is a positive real constant expressed as:
$\Theta_{g_l} = \sum 2\theta_l \dfrac{(\tilde{W}_g^l)^T \tilde{W}_g^l}{\|\tilde{W}_g^l\|} \ge 0$ or $\Theta_{g_l} = \sum \kappa \varepsilon_l \dfrac{(\tilde{W}_g^l)^T \tilde{W}_g^l}{\|\tilde{W}_g^l\|} \ge 0$ when we have inner or
outer modified hopping for all time, respectively, where the summation includes all weight vectors which perform hopping. Therefore, once again the negativity of $\dot{V}$ is actually strengthened due to the last negative term.

Therefore, it is obvious that Lemma 4.3 implies that the hopping modifications (4.30), (4.34) guarantee the boundedness of the weights, without affecting the rest of the stability properties established in the absence of hopping.

Hence, we can now prove the following theorem.

Theorem 4.1 *The control law (4.12) and (4.13) together with the updating laws (4.30) and (4.34) guarantee the following properties:*

- $\xi, x, W_f, W_g \in L_\infty, \quad |\xi| \in L_2,$

- $\lim_{t\to\infty} \xi(t) = 0, \quad \lim_{t\to\infty} x(t) = 0,$

- $\lim_{t\to\infty} \dot{W}_f(t) = 0, \quad \lim_{t\to\infty} \dot{W}_g(t) = 0,$

provided that $\lambda_{\min}(Q) > \ell_1\|P\|$ *and* Assumption 8 *is satisfied.*

Proof From Eq. (4.41), we have that $V \in L_\infty$ which implies that $\xi, \tilde{W}_f, \tilde{W}_g \in L_\infty$. Furthermore, $W_f = \tilde{W}_f + W_f^* \in L_\infty$ and $W_g = \tilde{W}_g + W_g^* \in L_\infty$. Since $\xi \in L_\infty$, this also implies that $x \in L_\infty$. Moreover, since V is a monotone decreasing function of time and bounded from below, the $\lim_{t\to\infty} V(t) = V_\infty$ exists; so by integrating $\dot{V}$ from 0 to ∞ we have:

$$(\lambda_{\min}(Q) - l_1\|P\|) \int_0^\infty \|\xi\|^2\, dt + \lambda_{\min}(Q) \int_0^\infty \|\zeta\|^2\, dt - \ell_1\|P\| \int_0^\infty \|\xi\|\,\|\zeta\|\, dt + I_g$$

$$\times \int_0^\infty \Theta_{g_l}\, dt + I_f \int_0^\infty \Theta_{f_l}\, dt \le [V(0) - V_\infty] < \infty,$$

which implies that $|\xi| \in L_2$. We also have that:

$$\dot{\xi} = \Lambda_c\xi + b_c[\bar{x}_f \tilde{W}_f s_f(x) + \bar{x}_g \tilde{W}_g s_g(x)u - \omega_n(x, u)].$$

Hence and since $u \in L_\infty$, the sigmoidals and center vectors are bounded by definition, $\tilde{W}_f, \tilde{W}_g \in L_\infty$ and Assumption 8 holds, so since $\xi \in L_2 \cap L_\infty$ and $\dot{\xi} \in L_\infty$, applying Barbalat's Lemma we conclude that $\lim_{t \to \infty} \xi(t) = 0$.

Now, using the boundedness of $u, s_f(x), s_g(x), x$ and the convergence of $\xi(t)$, $\zeta(t)$ to zero, we have that $\dot{W}_f, \dot{W}_g$ also converge to zero. Hence, we have that:

$$\lim_{t \to \infty} x(t) = \lim_{t \to \infty} \zeta(t) - \lim_{t \to \infty} \xi(t) = 0.$$

Thus:

$$\lim_{t \to \infty} x(t) = 0.$$

Remark 4.1 Inequality $\lambda_{\min}(Q) \geq \ell_1 \|P\|$ shows how the design constant matrix Q should be selected, in order to guarantee convergence of the state x to zero, even in the presence of modeling error terms which are not uniformly bounded *a priori*. The values of Q become large as we allow for large model imperfections but Q is implemented as a gain in the construction of $\dot{\zeta}$ and for practical reasons it cannot take arbitrarily large values. This leads to a compromise between the value of Q and the maximum allowable modeling error terms.

4.1.2.3 The Modeling Error at Zero Case.

In Sect. 4.1.2.2, we have assumed that the modeling error term satisfies the following condition:

$$|\omega_n(x, u)| \leq l_1' |x| + l_1'' |u|,$$

which implies that the modeling error becomes zero when $|x| = 0$ and we have proven convergence of the state x to zero, plus boundedness of all signals in the closed-loop. In this section however, we examine the more general case which is described by the following assumption.

Assumption 9 The modeling error term satisfies:

$$|\omega_n(x, u)| \leq l_0 + l_1' |x| + l_1'' |u|.$$

Having made this assumption, we now allow a not-necessarily-known modeling error $l_0 \neq 0$ at zero. Furthermore, as stated previously, we can find *a priori* known constant $l_u > 0$, such that:

$$|u| \leq l_u |x|,$$

thus making:

$$|\omega_n(x, u)| \equiv |\omega_n(x)|,$$

and Assumption 9 equivalent to:

$$|\omega_n(x)| \leq l_0 + l_1|x|, \tag{4.44}$$

where

$$l_1 = l_1' + l_1''l_u, \tag{4.45}$$

is a positive constant. Employing (4.44), Eq. (4.20) becomes:

$$\begin{aligned}
\dot{V} &\leq -\lambda_{\min}(Q)\|\xi\|^2 - \lambda_{\min}(Q)\|\zeta\|^2 + \|\xi\|\,\|P\|\,(l_0 + l_1\,\|x\|) \\
&\leq -\lambda_{\min}(Q)\|\xi\|^2 - \lambda_{\min}(Q)\|\zeta\|^2 + l_1\,\|P\|\,\|\xi\|^2 + l_1\,\|P\|\,\|\xi\|\,\|\zeta\| \\
&\quad + l_0\,\|P\|\,\|\xi\|.
\end{aligned} \tag{4.46}$$

To continue, we need to state and prove the following lemma.

Lemma 4.4 *The control law:*

$$u = -\left[\bar{x}_g W_g s_g(x)\right]^{-1}\left[\bar{x}_f W_f s_f(x) + \upsilon\right], \tag{4.47}$$

$$\upsilon = kx, \tag{4.48}$$

where the synaptic weight estimates W_f and W_g, are adjusted according to equations (4.30), (4.34) *guarantee the following properties:*

- $\zeta(t) \leq 0, \quad \forall t \geq 0,$

- $\lim_{t \to \infty} \zeta(t) = 0$ *exponentially fast provided that* $\zeta(t) < 0.$

Proof that if we use the control laws (4.47), (4.48), Eq. (4.11) becomes:

$$\dot{\zeta} = -K_c\zeta, \quad \forall t \geq 0$$

which is a homogeneous differential equation with solution

$$\zeta(t) = \zeta(0)e^{-K_c t}.$$

Hence, if $\zeta(0)$ which represents the initial value of $\zeta(t)$, is chosen negative, we obtain:

$$\zeta(t) \leq 0, \quad \forall t \geq 0.$$

Moreover, $\zeta(t)$ converges to zero exponentially fast. Hence, we can distinguish the following cases:

Case 1: If $x \geq 0$, we have that $\zeta(t) \geq \xi(t)$ but $\zeta(t) \leq 0, \forall t \geq 0$ which implies that $\|\zeta(t)\| \leq \|\xi(t)\|$. So, we have:

$$\|x\| \leq \|\zeta\| + \|\xi\| \leq 2\|\xi\|. \tag{4.49}$$

Therefore, Eq. (4.46) becomes:

$$
\begin{aligned}
\dot{V} &\leq -\lambda_{\min}(Q)\,\|\xi\|^2 - \lambda_{\min}(Q)\,\|\zeta\|^2 + 2l_1\,\|P\|\,\|\xi\|^2 + l_0\,\|P\|\,\|\xi\| \\
&\leq -\left[\lambda_{\min}(Q)\,\|\xi\| - 2l_1\,\|P\|\,\|\xi\| - l_0\,\|P\|\right]\|\xi\| - \lambda_{\min}(Q)\,\|\zeta\|^2 \\
&\leq 0,
\end{aligned}
\tag{4.50}
$$

provided that:

$$
\|\xi\| > \frac{l_0\,\|P\|}{\lambda_{\min}(Q) - 2l_1\,\|P\|},
\tag{4.51}
$$

with $\lambda_{\min}(Q) > 2l_1\,\|P\|$.

Case 2: If $x < 0$, we have that $\zeta(t) < \xi(t)$ but $\zeta(t) \leq 0$, $\forall t \geq 0$ which implies that $\|\zeta(t)\| > \|\xi(t)\|$. So, we have:

$$
\|x\| \leq \|\zeta\| + \|\xi\| \leq 2\,\|\zeta\|.
\tag{4.52}
$$

Therefore, Eq. (4.46) becomes:

$$
\begin{aligned}
\dot{V} &\leq -\lambda_{\min}(Q)\,\|\xi\|^2 - \lambda_{\min}(Q)\,\|\zeta\|^2 + 2l_1\,\|P\|\,\|\xi\|\,\|\zeta\| + l_0\,\|P\|\,\|\xi\| \\
&\leq -\left(\lambda_{\min}(Q)\,\|\xi\| - l_0\,\|P\|\right)\|\xi\| - \left(\lambda_{\min}(Q) - 2l_1\,\|P\|\right)\|\zeta\|^2 \leq 0 \\
&\leq 0,
\end{aligned}
\tag{4.53}
$$

provided that:

$$
\|\xi\| > \frac{l_0\,\|P\|}{\lambda_{\min}(Q)}
\tag{4.54}
$$

and $\lambda_{\min}(Q) > 0$.

Conclusively, $\forall x \in R^n$ the Lyapunov candidate function becomes negative when $\|\xi\| > \frac{l_0\,\|P\|}{\lambda_{\min}(Q) - 2l_1\,\|P\|}$ and $\lambda_{\min}(Q) > 2l_1\,\|P\|$.

In the sequel, inequality (4.51) together with (4.49), (4.52) demonstrate that the trajectories of $\xi(t)$ and $x(t)$ are uniformly bounded with respect to the arbitrarily small, (since $\lambda_{\min}(Q)$ can be chosen sufficiently large), sets Ξ and X shown below:

$$
\Xi = \left\{ \xi(t) : \|\xi(t)\| \leq \frac{2l_0\,\|P\|}{\lambda_{\min}(Q) - 2l_1\,\|P\|}, \ \lambda_{\min}(Q) > 2l_1\,\|P\| > 0 \right\}
$$

and

$$
X = \left\{ x(t) : \|x(t)\| \leq \frac{4l_0\,\|P\|}{\lambda_{\min}(Q) - 2l_1\,\|P\|}, \ \lambda_{\min}(Q) > 2l_1\,\|P\| > 0 \right\}.
$$

Thus, we have proven the following theorem:

Theorem 4.2 *Consider the system* (4.2) *with the modeling error term satisfying* (4.44). *Then, the control law* (4.12), (4.13) *together with the updating laws* (4.30) *and* (4.34) *guarantees the uniform ultimate boundedness with respect to the sets:*

- $\Xi = \left\{ \xi(t) : \|\xi(t)\| \leq \frac{2l_0\|P\|}{\lambda_{\min}(Q) - 2l_1\|P\|}, \ \lambda_{\min}(Q) > 2l_1\|P\| > 0 \right\}$,

- $X = \left\{ x(t) : \|x(t)\| \leq \frac{4l_0\|P\|}{\lambda_{\min}(Q) - 2l_1\|P\|}, \ \lambda_{\min}(Q) > 2l_1\|P\| > 0 \right\}$.

Furthermore,

$$\dot{\xi} = \Lambda_c \xi + b_c [\bar{x}_f \tilde{W}_f s_f(x) + \bar{x}_g \tilde{W}_g s_g(x) u - \omega_n(x, u)].$$

Hence, since the boundedness of $\tilde{W}_f$ and $\tilde{W}_g$ is ensured by the use of the hopping algorithm and $\omega_n(x)$ owing to (4.44) and Theorem 4.2, we conclude that $\dot{\xi} \in L_\infty$.

Remark 4.2 The previous analysis reveals that in the case where we have a modeling error different from zero at $\|x\| = 0$, our adaptive regulator can guarantee at least uniform ultimate boundedness of all signals in the closed-loop. In particular, Theorem 4.2 shows that if l_0 is sufficiently small, or if the design constants q_i (diagonal elements of matrix Q) are chosen such that $\lambda_{\min}(Q) > 2l_1\|P\|$, then $\|x(t)\|$ can be arbitrarily close to zero and in the limit as $q_i \to \infty$, actually becomes zero but as we stated in Remark 4.1, implementation issues constrain the maximum allowable value of q_i.

4.2 Adaptive Tracking

In this section, we investigate trajectory tracking problems based on the fuzzy-high order neural networks (F-HONNs) controller algorithm developed for systems having Brunovsky-type form, Theodoridis et al. (2009).

Tracking is known as our attempt to force the state of the actual system to follow the state of a given stable dynamical system.

A large class of single-input-single-output nonlinear systems can be modeled by Eq. (4.1) which is in Brunovsky-type form as described in Sect. 4.1.

The Assumptions 6, 7 together with the following Assumption 10 are also imposed on (4.1), to guarantee the existence and uniqueness of solution for any finite initial condition and $u \in U_c$.

Assumption 10 In order the system to be universally controllable, assume that the input function $g(x)$ satisfies $|g(x)| > 0$, for all x in a certain controllability compact region $U_c \in R$.

The control objective is to force the state vector x to follow a desired trajectory x_d of a stable model having the following form:

$$\dot{x}_d = A_c x_d + b_c \cdot f_n(x_d, u_d, t). \tag{4.55}$$

where A_c, b_c were defined in Sect. 4.1.

Define the tracking error e as:

$$e = x_d - x = \begin{bmatrix} e_1\, e_2\, \cdots\, e_{n-1} \end{bmatrix}^T. \tag{4.56}$$

However, the problem as it is stated above for the system (4.1), is very difficult or even impossible to be solved since the f, g are assumed to be completely unknown. To overcome this problem we assume that the unknown plant can be described by the following model arriving from the NF model described in Sect. 4.1.1, plus a modeling error term $\mu(x, u)$ such as:

$$\dot{x} = A_c x + b_c[\bar{x}_f W_f^* s_f(x) + \bar{x}_g W_g^* s_g(x)u + \mu(x, u)], \tag{4.57}$$

where $\bar{x}_f$, $\bar{x}_g$ are vectors determined by the designer, which contain the centers of the partitions of the fuzzy output variables, being the fuzzy approximations of $f(x)$ and $g(x)$, respectively. Also, W_f^*, W_g^* are unknown weight matrices and s_f, s_g are sigmoidal vectors containing high order sigmoidal terms with parameters which are also determined by the designer.

Therefore, the state trajectory problem is analyzed for the system (4.57) instead of (4.1). Since, W_f^* and W_g^* are unknown, our solution consists of designing a control law $u(W_f, W_g, e, \dot{x}_{dn})$ and appropriate update laws for W_f and W_g to guarantee convergence of the state to the desired trajectory and in some cases, which will be analyzed in the following sections, boundedness of x and of all signals in the closed-loop.

4.2.1 Complete Matching Case

In this section, we investigate the adaptive model reference control problem when the modeling error term is zero, or in other words, when we have complete model matching. Thus, the unknown system can be written as:

$$\dot{x} = A_c x + b_c[\bar{x}_f W_f^* s_f(x) + \bar{x}_g W_g^* s_g(x)u], \tag{4.58}$$

where the meaning and the dimensions of the involved matrices and vectors have been given in Sect. 4.1.1.

Taking the derivative of Eq. (4.56) and substituting Eq. (4.58), we have:

$$\dot{e} = A_c e + b_c \cdot f_n(x_d, u_d, t) - b_c[\bar{x}_f W_f^* s_f(x) + \bar{x}_g W_g^* s_g(x)u]. \tag{4.59}$$

The trajectory tracking of the system can be achieved by selecting the control input to be:

$$u = \frac{-\bar{x}_f W_f s_f(x) + K_s s + \lambda_1 e_{(n-1)} + \cdots + \lambda_{n-1} e_{(1)} + \dot{x}_{dn} + u_r}{\bar{x}_g W_g s_g(x)}, \tag{4.60}$$

where $\dot{x}_{dn} = f_n(x_d, u_d, t)$, $K_s s$ is a linear PD type compensator (Ioannou and Fidan 2006) with K_s a positive constant and $s = \lambda e$ with λ being a vector of the form $\lambda = [1 \ \lambda_1 \cdots \lambda_{n-1}] \in R^n$ such that all roots of the polynomial $h(s) = s^n + \lambda_1 s^{n-1} + \cdots + \lambda_{n-1}$ are in the open left half-plane. Also, $e_{(n-1)}$ is the $(n-1)$th element of vector e and $u_r = -K_r sgn\left(e^T P b_c\right)$ is a sliding control term, (Sanchez and Bernal 2000; Kung et al. 2005; Abid et al. 2007) to overcome system uncertainties. In Eq. (4.60) we can substitute the equality $K_s s + \lambda_1 e_{(n-1)} + \cdots + \lambda_{n-1} e_{(1)} = k^T e$, where $k = \left[(K_s + \lambda_1) \ (K_s \lambda_1 + \lambda_2) \ \cdots \ (K_s \lambda_{n-2} + \lambda_{n-1}) \ (K_s \lambda_{n-1})\right]^T \in R^n$. Assuming for the moment that there are not any uncertainties, we may also assume that $u_r = 0$ and Eq. (4.60) becomes:

$$u_c = \frac{-\bar{x}_f W_f s_f(x) + k^T e + \dot{x}_{dn}}{\bar{x}_g W_g s_g(x)}. \tag{4.61}$$

Solving Eq. (4.61) in respect to $\dot{x}_{dn}$ we get the following equation:

$$\dot{x}_{dn} = \bar{x}_f W_f s_f(x) + \bar{x}_g W_g s_g(x) u_c - k^T e. \tag{4.62}$$

Next, substituting Eq. (4.62) to Eq. (4.59) we have:

$$\begin{aligned}
\dot{e} &= A_c e + b_c[\bar{x}_f W_f s_f(x) + \bar{x}_g W_g s_g(x) u_c - k^T e] \\
&\quad - b_c[\bar{x}_f W_f^* s_f(x) + \bar{x}_g W_g^* s_g(x) u_c] \\
&= \Lambda_c e + b_c[\bar{x}_f \tilde{W}_f s_f(x) + \bar{x}_g \tilde{W}_g s_g(x) u_c],
\end{aligned} \tag{4.63}$$

where $\Lambda_c = \begin{bmatrix} 0 & 1 & 0 & \cdots & 0 \\ 0 & 0 & 1 & 0 & \cdots \\ \cdots & \cdots & 0 & \cdots & 0 \\ 0 & 0 & \cdots & 0 & 1 \\ -k_n & -k_{n-1} & \cdots & -k_2 & -k_1 \end{bmatrix}$ with $k_n = K_s + \lambda_1$, $k_{n-1} = K_s \lambda_1 + \lambda_2, \ldots, k_2 = K_s \lambda_{n-2} + \lambda_{n-1}, k_1 = K_s \lambda_{n-1}$ is a Hurwitz matrix.

To continue, consider the Lyapunov candidate function:

$$V = \frac{1}{2} e^T P e + \frac{1}{2} \text{tr} \left\{ \tilde{W}_f^T D_f^{-1} \tilde{W}_f \right\} + \frac{1}{2} \text{tr} \left\{ \tilde{W}_g^T D_g^{-1} \tilde{W}_g \right\}, \tag{4.64}$$

where $P > 0$ is chosen to satisfy the Lyapunov equation:

$$P \Lambda_c + \Lambda_c^T P = -Q, \tag{4.65}$$

where Q is a positive definite diagonal matrix chosen appropriately by the designer. If we take the derivative of Eq. (4.64) with respect to time and take into account Eq. (4.63) we obtain:

$$\dot{V} = -e^T Q e + e^T P b_c \bar{x}_f \tilde{W}_f s_f + e^T P b_c \bar{x}_g \tilde{W}_g s_g u_c$$
$$+ \mathrm{tr} \left\{ \dot{W}_f^T D_f^{-1} \tilde{W}_f \right\} + \mathrm{tr} \left\{ \dot{W}_g^T D_g^{-1} \tilde{W}_g \right\}. \tag{4.66}$$

Hence, if we choose:

$$\mathrm{tr} \left\{ \dot{W}_f^T D_f^{-1} \tilde{W}_f \right\} = -e^T P b_c \bar{x}_f \tilde{W}_f s_f(x), \tag{4.67}$$

$$\mathrm{tr} \left\{ \dot{W}_g^T D_g^{-1} \tilde{W}_g \right\} = -e^T P b_c \bar{x}_g \tilde{W}_g s_g(x) u_c, \tag{4.68}$$

$\dot{V}$ becomes:

$$\dot{V} \leq -\lambda_{\min}(Q) \|e\|^2. \tag{4.69}$$

It can be easily verified that Eqs. (4.67) and (4.68) after using matrix trace properties, result to the following weight updating laws:

$$\dot{W}_f = -\bar{x}_f^T b_c^T P e s_f^T(x) D_f, \tag{4.70}$$

$$\dot{W}_g = -\bar{x}_g^T b_c^T P e s_g^T(x) u_c D_g. \tag{4.71}$$

As it is proved in the following Theorem 4.3, the weight updating laws for the column submatrices of W_f, W_g given by:

$$\dot{W}_f^l = -\left(\bar{x}_f\right)^T p(:, n)^T e s_l(x) d_{f_l}, \tag{4.72}$$

and

$$\dot{W}_g^l = -\left(\bar{x}_g\right)^T p(:, n)^T e u_c s_l(x) d_{g_l}, \tag{4.73}$$

where $p(:, n)$ is the n-th column of matrix P, can serve the purpose of driving e to zero.

The problem of the existence of the control law arises also in this problem. It is well known that, in adaptive control schemes it is possible the identifier to become uncontrollable, although the physical system is controllable (Kosmatopoulos and Ioannou 1999, 2002). This is actually because that the control law of (4.60) may become uncomputable if $\left[\bar{x}_g W_g s_g(x)\right]^{-1}$ does not exist. Therefore, the existence of $\left[\bar{x}_g W_g s_g(x)\right]^{-1}$ has to be ensured. In this case, we perform the same analysis as in Chap. 3 to include a weight hopping term in the updating laws.

This way, the learning law (4.73) for the adaptation of the weights W_g is modified and becomes:

$$\dot{W}_g^l = -\left(\bar{x}_g\right)^T p(:,n)^T e u_c s_l(x) d_{g_l} - \frac{2\sigma_{g_l}\kappa_{g_l}^{\text{inner}}\left(\bar{x}_g W_g^l(\bar{x}_g)^T\right)}{\left|\bar{x}_g\right|^2}, \qquad (4.74)$$

where the last term determines the magnitude of weight hopping and

$$\sigma_{g_l} = \begin{cases} 0 & \begin{aligned} &\text{if } |\bar{x}_g \cdot W_g^l| > \theta_l \\ &\text{or } |\bar{x}_g \cdot W_g^l| = \theta_l \\ &\text{and } \bar{x}_g \cdot \dot{W}_g^l \leq 0 \end{aligned} \\[1em] 1 & \text{otherwise} \end{cases} \qquad (4.75)$$

expresses the hopping condition. Also, $\kappa_{g_l}^{\text{inner}}$ is a positive number constraint in the interval $(1, 1.5)$ in order to avoid the infinite hopping inside the forbidden hyperplanes according to Chap. 3, Remark 3.2 and Fig. 3.5. Due to the hopping condition, the control law (4.60) actually indirectly performs like a switching controller (Kosmatopoulos and Ioannou 1999, 2002), where the switching here is determined by the weight hopping.

The next theorem gives the control scheme for trajectory tracking without modeling errors.

Theorem 4.3 *The closed-loop system given by* (4.55), (4.58), (4.59), (4.60) *together with the updating laws* (4.72), (4.74), *guarantees the following properties:*

1. $e, \tilde{W}_f, \tilde{W}_g \in L_\infty, \quad e \in L_2,$
2. $\lim_{t\to\infty} e(t) = 0,$
3. $\lim_{t\to\infty} \dot{W}_f(t) = 0, \quad \lim_{t\to\infty} \dot{W}_g(t) = 0.$

Proof (1) Since $V > 0$ and $\dot{V} \leq 0$, we conclude that $V \in L_\infty$, which implies that $e, \tilde{W}_f, \tilde{W}_g \in L_\infty$. Furthermore, $W_f = \tilde{W}_f + W_f^*, W_g = \tilde{W}_g + W_g^*$ are also bounded. Since V is a non-increasing function of time and bounded from below, the $\lim_{t\to\infty} V = V_\infty$ exists; therefore, by integrating $\dot{V}$ from 0 to ∞ we have:

$$\lambda_{\min}(Q) \int_0^\infty \|e\|^2 \, dt \leq [V(0) - V_\infty] < \infty,$$

which implies that $e \in L_2$.

(2) Since $e \in L_2 \cap L_\infty$, using Barbalat's Lemma we conclude that $\lim_{t\to\infty} e(t) = 0$.

(3) Furthermore, using the boundedness of u, s_f, s_g and the convergence of $e(t)$ to zero, we have that $\dot{W}_f, \dot{W}_g$ also converges to zero (Ioannou and Fidan 2006).

Remark 4.3 In order the properties of above Theorem 4.3 to be valid, it suffices to show that by using the modified updating law for W_g^l when the hopping condition is activated, the negativity of the time derivative of Lyapunov function is not compromised. This can be shown by considering the additional terms introduced in $\dot{V}$

(due to the modification in the weight updating of W_g^l) and verifying that they are negative. A detailed proof can be found in Theodoridis et al. (2012).

4.2.2 Inclusion of a Non-zero Approximation Error

Let us assume now that we have a non-zero NF approximation error $\mu(x, u)$ which is bounded by a small positive number $\bar{\mu}_n$. Thus, we use Eq. (4.57) in addition with the control law:

$$u = u_c + u_s = \frac{-\bar{x}_f W_f s_f(x) + k^T e + \dot{x}_{dn} + u_r}{\bar{x}_g W_g s_g(x)}, \tag{4.76}$$

which takes into account the sliding control term u_r with $u_s = \left[\bar{x}_g W_g s_g(x)\right]^{-1} u_r$. Now, we can distinguish two possible cases.

Case 1: Without the sliding control term u_r.

The time derivative of the Lyapunov function (4.64) taking into account the modeling error term of the NF approximator (4.57) becomes:

$$\begin{aligned}
\dot{V} &\le -\lambda_{\min}(Q)\,\|e\|^2 - e^T P b_c \mu \\
&\le -\lambda_{\min}(Q)\,\|e\|^2 + \|e P b_c \mu\| \\
&\le -\|e\|\left(\lambda_{\min}(Q)\,\|e\| - \|p(:, n)\|\,\bar{\mu}_n\right).
\end{aligned} \tag{4.77}$$

The above expression will be negative if $\|e\| > \frac{\|p(:,n)\|\bar{\mu}_n}{\lambda_{\min}(Q)}$. In other words, this implies that the trajectory of $e(t)$ is uniformly bounded with respect to the arbitrarily small, (since $\lambda_{\min}(Q)$ can be chosen sufficiently large), set E shown below:

$$E = \left\{e(t) : \|e(t)\| \le \frac{\|p(:, n)\|\,\bar{\mu}_n}{\lambda_{\min}(Q)},\ \lambda_{\min}(Q) > 0\right\}.$$

Case 2: With the sliding control term u_r.

The time derivative of the Lyapunov function (4.64) taking into account the modeling error term of the NF approximator (4.57) once again becomes:

$$\begin{aligned}
\dot{V} &\le -\lambda_{\min}(Q)\,\|e\|^2 - e^T P b_c \mu - K_r\left(e^T P b_c\right) sgn\left(e^T P b_c\right) \\
&\le -\lambda_{\min}(Q)\,\|e\|^2 + \|e P b_c \mu\| - K_r\,\|e P b_c\| \\
&\le -\lambda_{\min}(Q)\,\|e\|^2 - \|e\|\,\|p(:, n)\|\,(K_r - \bar{\mu}_n).
\end{aligned} \tag{4.78}$$

The above expression will be negative by appropriately selecting K_r, so that $K_r > \bar{\mu}_n$.

Following the above analysis two issues need special consideration, the existence and the boundedness of the control signal in the presence of modeling errors. From

(4.76), it can be easily seen that the existence of the control signal is ensured if the inverse of $\bar{x}_g W_g s_g(x)$ exists. Apart from that, in order the control law to be physically implementable the quantities $\bar{x}_f W_f s_f(x)$ and $\bar{x}_g W_g s_g(x)$ should remain within pre-specified bounds. As has already been analyzed in previous sections, we tackle this issues by employing the "hopping" method, as it was first introduced in Boutalis et al. (2009) and is presented in detail in Sect. 3.2.2.

Conclusively, one can easily verify that (4.76) is valid provided that $\bar{x}_f W_f, \bar{x}_g W_g$ are uniformly bounded by known positive constants ε_l, ρ_l and additionally $\bar{x}_g W_g$ always exists. Thus, $\bar{x}_f W_f(t)$ is confined to the set:

$$P_{f,\mathrm{SISO}} = \{\bar{x}_f \cdot W_f^l : \left|\bar{x}_f \cdot W_f^l\right| \le \varepsilon_l\},$$

and $\bar{x}_g W_g(t)$ is confined to the set:

$$P_{g,\mathrm{SISO}} = \left\{\bar{x}_g \cdot W_g^l : \left|\bar{x}_g \cdot W_g^l\right| \le \rho_l \text{ and } \left|\bar{x}_g \cdot W_g^l\right| > \theta_l\right\},$$

through the use of a hopping algorithm. In particular, the standard update law (4.70) is modified to:

$$\dot{W}_f^l = -\left(\bar{x}_f\right)^T p(:,n)^T es_l(x)d_{fl} - \frac{\sigma_{fl}\kappa_{fl}^{\mathrm{outer}}\left(\bar{x}_f W_f^l(\bar{x}_f)^T\right)}{\left|\bar{x}_f\right|^2}, \tag{4.79}$$

where

$$\sigma_{gl} = \begin{cases} 0 & \begin{array}{l} \text{if } \bar{x}_f \cdot W_f^l \in P_f \\ \text{or } \bar{x}_f \cdot W_f^l = \pm\varepsilon_l \\ \text{and } \bar{x}_f \cdot \dot{W}_f^l >< 0 \end{array} \\[2em] 1 & \text{otherwise} \end{cases} \tag{4.80}$$

and $\kappa_{fl}^{\mathrm{outer}}$ a positive constant $(0 < \kappa_{fl}^{\mathrm{outer}} < 1)$ decided from the designer according to Chap. 3, Fig. 3.5. The weight updating law (4.74) incorporating the two hopping conditions (outer and inner hopping) can now be rewritten as:

$$\dot{W}_g^l = -\left(\bar{x}_g\right)^T p(:,n)^T eus_l(x)d_{gl} - \frac{2\sigma_{gl}\kappa_{gl}^{\mathrm{inner}}\left(\bar{x}_g W_g^l(\bar{x}_g)^T\right)}{\left|\bar{x}_g\right|^2}$$

$$- \frac{(1-\sigma_{gl})\kappa_{gl}^{\mathrm{outer}}\left(\bar{x}_g W_g^l(\bar{x}_g)^T\right)}{\left|\bar{x}_g\right|^2}, \tag{4.81}$$

where

$$
\sigma_{gl} =
\begin{cases}
0 & \begin{aligned} &\text{if } \bar{x}_g W_g^l = \pm\rho_l \\ &\text{and } \bar{x}_g \dot{W}_g^l <> 0 \end{aligned} \\[2em]
1 & \begin{aligned} &\text{if } \bar{x}_g W_g^l = \pm\theta_l \\ &\text{and } \bar{x}_g \dot{W}_g^l >< 0 \end{aligned}
\end{cases}
\tag{4.82}
$$

and $\kappa_{gl}^{\text{inner}}$, $\kappa_{gl}^{\text{outer}}$ are small positive constant values ($1 < \kappa_{gl}^{\text{inner}} < 1.5$ and $0 < \kappa_{gl}^{\text{outer}} < 1$) for the inner and outer hopping, respectively, chosen by the designer according to Fig. 3.5. The above hopping conditions are depicted in Figs. 2.3, 3.4 where simplified 2-dimensional representations are given.

Remark 4.4 In order the properties of above Theorem 4.3 to be valid, it suffices to show that by using the modified updating laws for W_f^l, W_g^l when the outer hopping condition is activated, the negativity of the time derivative of Lyapunov function is not compromised. Similar to Remark 4.3, this can be shown by considering the additional terms introduced in $\dot{V}$ (due to the modification in the weight updating of W_g^l) and verifying that they are negative. A detailed proof can be found in Theodoridis et al. (2012).

4.3 Simulation Results

We present simulation results of the proposed NF controllers on the control of a well-known benchmark systems. First, we consider the "Inverted Pendulum" system, where we control it to track a periodic wave trajectory under the presence of modeling errors. The second example is the "Van der pol" oscillator where we regulate it's states and make comparisons, against the well-established approach on the use of HONNs (Rovithakis and M.A.Christodoulou 2000).

4.3.1 Inverted Pendulum

A highly unstable benchmark system in the area of nonlinear control is the well-known inverted pendulum on a cart. In this system, an inverted pendulum is attached to a cart equipped with a motor that drives it along a horizontal track. Taking into account only angular position and velocity, the dynamical model of the inverted pendulum in Fig.4.2 may assume the following Brunovsky canonical form (Slontine and Li 1991):

$$
\dot{x}_1 = x_2,
$$

$$
\dot{x}_2 = \frac{g \sin x_1 - \dfrac{mlx_2^2 \cos x_1 \sin x_1}{m_C+m}}{l\left(\dfrac{4}{3} - \dfrac{m \cos^2 x_1}{m_C+m}\right)} + \frac{\dfrac{\cos x_1}{m_C+m}}{l\left(\dfrac{4}{3} - \dfrac{m \cos^2 x_1}{m_C+m}\right)} u + \mathrm{d}(t),
\tag{4.83}
$$

Fig. 4.2 Inverted Pendulum

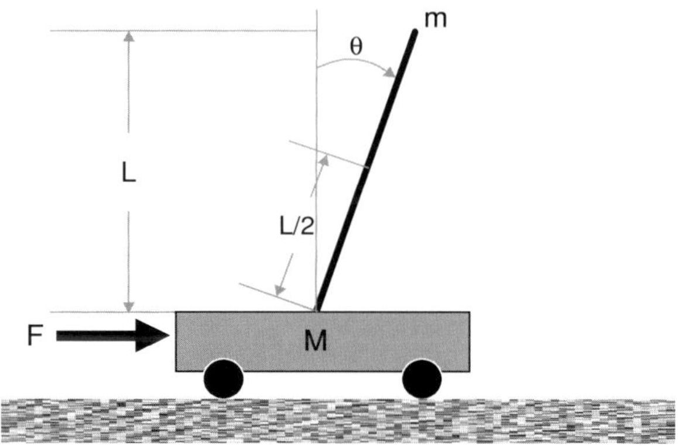

with $x_1 = \theta$ and $x_2 = \dot{\theta}$ are the angle from the vertical position and the angular velocity, respectively and $d(t)$ is a disturbance term. Also, $g = 9.8$ m/s^2 is the acceleration due to gravity, $m_c = M$ is the mass of the cart, m is the mass of the pole, $l = L/2$ is the half-length of the pole and $u = F$ is the force applied to the cart. We choose $m_c = 1$ kg, $m = 0.1$ kg, and $l = 0.5$ m in the following simulation.

It is our intention to track a reference signal. For that reason we use the adaptive laws described by Eqs. (4.79), (4.81) and the control law (4.60). The parameters used in the updating laws for the external and internal bounds chosen as $\varepsilon_l = 56, \theta_l = 0.12$ and $\rho_l = 44$ while the constant values used for the appropriate hopping $\kappa_{fl}^{\mathrm{inner}} = 1.02$, $\kappa_{gl}^{\mathrm{inner}} = 1.14$ and $\kappa_{gl}^{\mathrm{outer}} = 0.58$. Numerical training data were obtained by using Eq. (4.83) with initial conditions $\left[x_1(0)\ x_2(0)\right] = \left[\frac{\pi}{4}\ 0\right]$,disturbance term $d(t) = 0.1\,(1 + \sin(0.25t))$ and sampling time 10^{-2} sec. Also, the tracking desired signal was chosen as:

$$\left[x_{d_1}\ x_{d_2}\right] = \left[\frac{\pi}{9}\sin(t) + 0.25\sin(3t)\ \ \frac{\pi}{9}\cos(t) + 0.75\cos(3t)\right],$$

with initial condition $\left[x_{d_1}(0)\ x_{d_2}(0)\right] = \left[0\ 1.1\right]$.Our NF model was chosen to use five output partitions of f:

$$\bar{x}_f = \left[-4.91\ -1.00\ 0.09\ 4.63\ 5.04\right],$$

and 5 output partitions of g:

$$\bar{x}_g = \left[-0.14\ -0.06\ 0.01\ 0.08\ 0.13\right].$$

The number of high order sigmoidal terms (HOST) used in HONNs was chosen to be five $(s(x_1), s(x_2), s(x_1)\cdot s(x_2), s^2(x_1), s^2(x_2))$ up to second order for f and two $(s(x_1), s(x_2))$ up to first order for g. We select the initial weights as: $W_f(0) = 0$, $W_g(0) = 0.34$, we have chosen the learning rates as $D_f = \mathrm{diag}(1.48, 1.48)$ and $D_g = \mathrm{diag}(2.99, 2.99)$, and the parameters of the sigmoidal terms such as: $a_1 =$

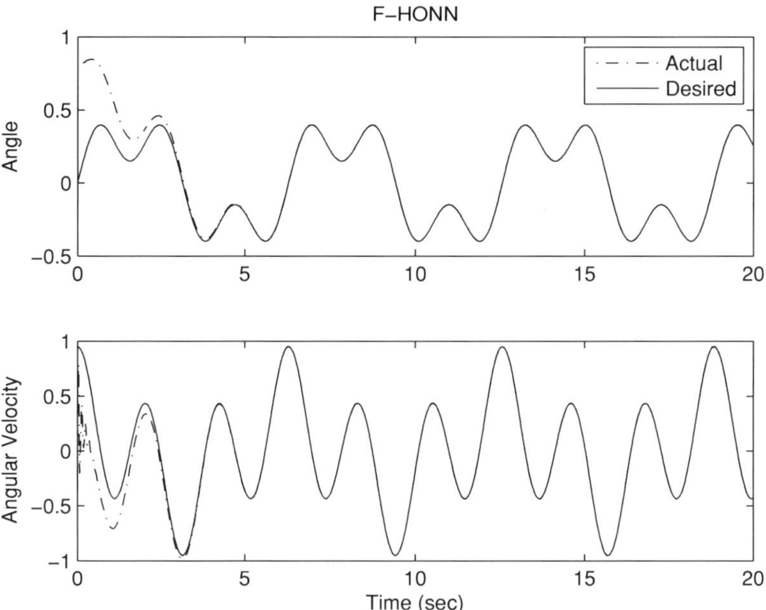

Fig. 4.3 Tracking of angle x_1 and angular velocity x_2, for the inverted pendulum

0.42, $a_2 = 0.11$, $\beta_1 = 0.94$, $\beta_2 = 1.98$ and $\gamma_1 = 2.64$, $\gamma_2 = 4.79$. Also, after careful selection $k_1 = 6.3$, $k_2 = 5.1$ and $Q = \mathrm{diag}(1, 1)$.

Figure 4.3 shows the convergence of the actual (dotted line) to the desired (solid line) trajectory while Fig. 4.4 shows the evolution of the control signal. As we can clearly remark, the controller has an excellent tracking behavior.

4.3.2 Van der Pol Oscillator

Another well-known benchmark problem is the regulation of Van der Pol oscillator. Its dynamical equations are given by the following equations (Lewis et al. 1999):

$$\dot{x}_1 = x_2,$$

$$\dot{x}_2 = x_2 \cdot \left(a - x_1^2\right) \cdot b - x_1 + u + d(x_2, u),$$

where we have included the disturbance term:

$$d(x_2, u) = 2x_2 + \sin(10x_2) + \sin(2u) + 0.5. \tag{4.84}$$

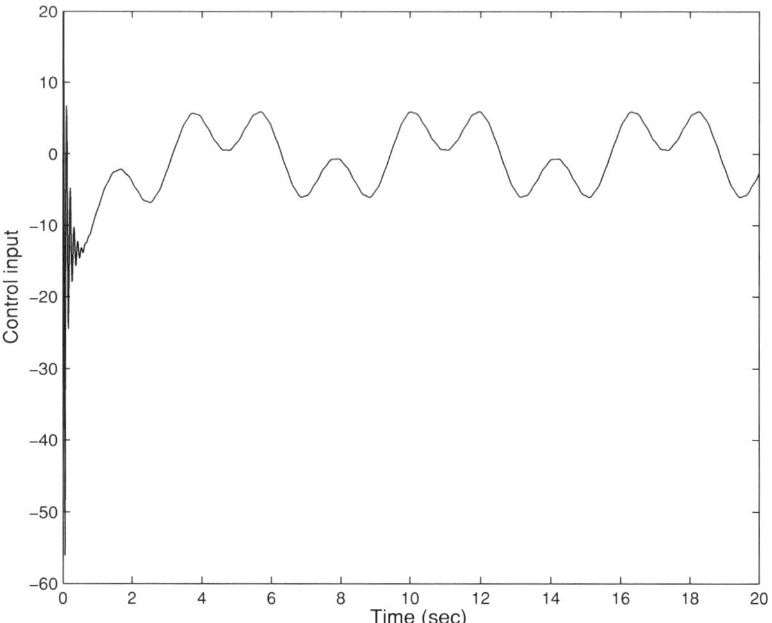

Fig. 4.4 Evolution of the tracking control input u

The NF controller has the following parameter values: $\bar{x}_f$ is a row vector of $q = 5$ centers of fuzzy partitions covering the range $[-23, -11]$, $s_f(x)$ is a column vector with $k = 5$ high order sigmoidal terms and W_f is a $q \times k$ matrix with neural weights. Similarly, $\bar{x}_g$ is a row vector of $q = 5$ centers of fuzzy partitions covering the range $[0.8, 1.2]$, s_g is a column vector with $k = 5$ high order sigmoidal terms and W_g is a $q \times k$ matrix containing the corresponding neural weights. Also, $k_1 = 2$ and $k_2 = 30$ are the elements of vector k in (4.13).

The proposed NF model was chosen to have state initial conditions:

$$\begin{bmatrix} x_1(0) & x_2(0) \end{bmatrix} = \begin{bmatrix} 0.4 & 0.5 \end{bmatrix},$$

and sampling time 10^{-3} s. Furthermore, regarding the other parameters we have chosen the initial weights to be $W_f(0) = 0$ and $W_g(0) = 0.5$ and the learning rates $\gamma_1 = 0.1$ and $\gamma_2 = 0.01$. Also, the parameters of the sigmoidal terms were selected as: $a_1 = 0.1$, $a_2 = 0.5$, $\beta_1 = \beta_2 = 1$, $\gamma_1 = \gamma_2 = 0$.

Figure 4.5 shows the regulation of states x_1 and x_2 while Fig. 4.6 gives the evolution of control input u and modeling error d. It can be seen that the NF approach seems to be superior compared to simple HONNs.

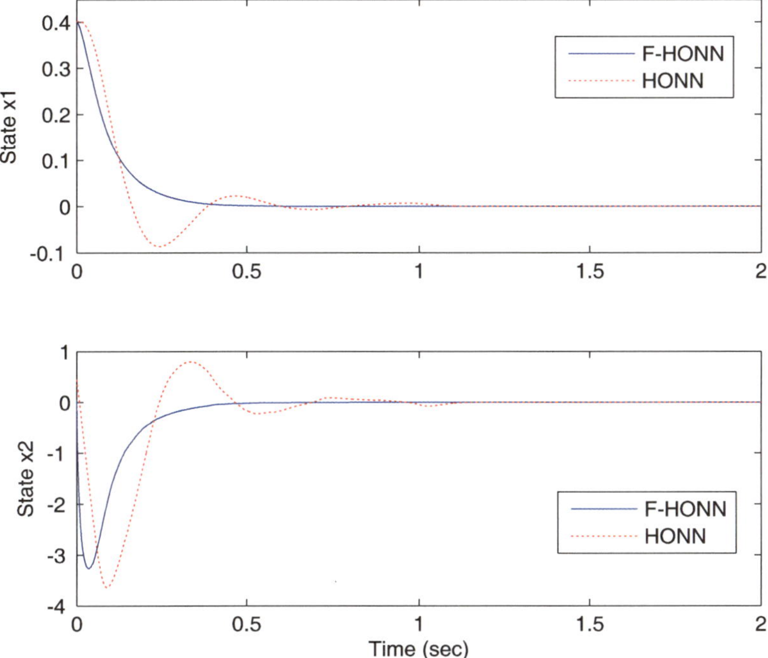

Fig. 4.5 Evolution of state variables x_1 and x_2 of the regulated Van der pol oscillator

The mean squared error (MSE) of HONNs and F-HONNs approaches was measured and is shown in Table 4.1, demonstrating a significant (order of magnitude) increase in the control performance.

Conclusively, the comparison between HONNs and F-HONNs leads to a superiority of F-HONNs regarding the control abilities. Especially, if we choose smaller learning rate and reduced number of high order terms, then the difference between the two methods becomes very noticeable, emphasizing the approximation superiority of the NF model over the simple neural one.

Finally, as concerning the hopping condition on the adaptive law of weights, the value $\vartheta_i = 0.01$ was set to determine the "safe" distance from the forbidden hyperplane and therefore to ensure the existence of control signal. Figures 4.7, 4.8 demonstrate time instants where hopping occurs. It has to be noted that, observing also Fig. 4.5, this hopping does not affect the continuity in the evolution of the states.

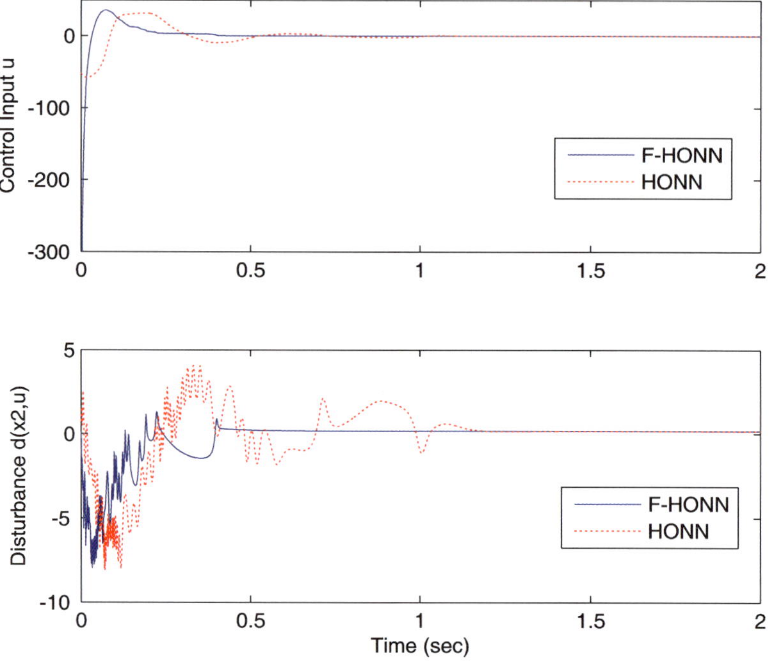

Fig. 4.6 Evolution of control input u, disturbance $d(x_2, u)$ of the regulated Van der pol oscillator

Table 4.1 Comparison of HONN and F-HONNs approaches for Van der pol oscillator simulations

Examples		HONN	F-HONNs
Van der pol	MSEx_1	0.0075	0.0042
	MSEx_2	0.7265	0.3915

Fig. 4.7 Evolution of $\bar{x}_{g_{1,2}} W_{g_{1,2}}$ with time which shows the activation of hopping condition

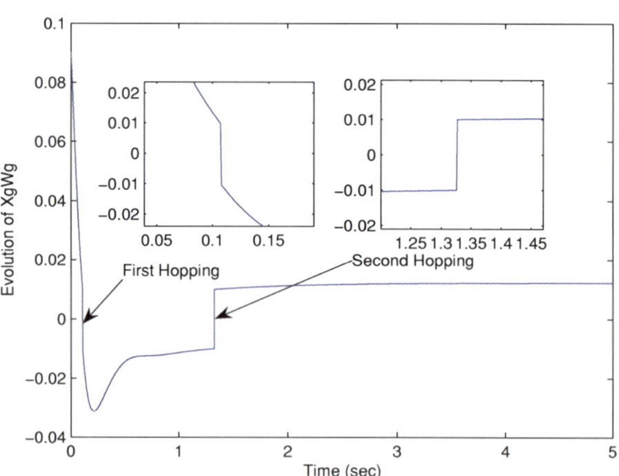

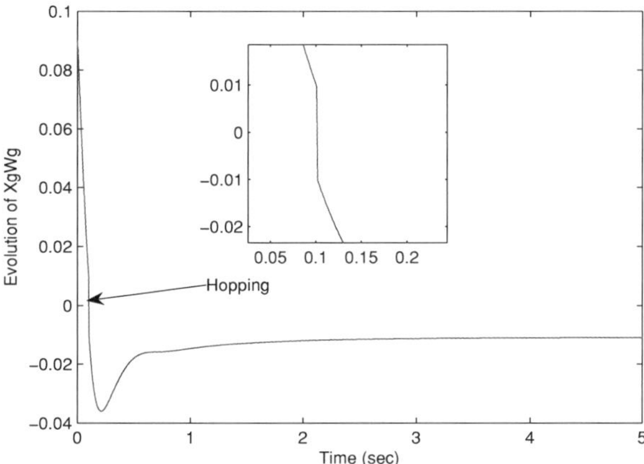

Fig. 4.8 Evolution of $\bar{x}_{g_{2,1}} W_{g_{2,1}}$ with time which shows the activation of hopping condition

4.4 Summary

Based on the Neurofuzzy approximation of unknown nonlinear functions, this chapter covered the direct adaptive regulation of Brunovsky-type SISO systems assuming the presence of modeling errors. Moreover, an adaptive trajectory tracking controller for the same type of systems was presented.

Weight updating laws were given and it was proved that when the structural identification is appropriate and the modeling error terms are within a certain region depending on the input and state values, then the error reaches zero very fast. In case the modeling error term depends also on a not necessarily known constant value, then it was proved that the error remains bounded.

The main contribution of this chapter was that it presented an effective NF approach for the approximation and in the sequel control of SISO uncertain nonlinear systems, and provided with mathematical proofs of the stability and convergence of the control system using the Lyapunov stability theorem.

Simulations illustrate the potency of the method where it was shown that by following the proposed procedure despite the presence of modeling errors, one can obtain asymptotic regulation or trajectory tracking quite well. Moreover, compared to simple HONN-based direct regulation, the proposed method proved to be superior.

References

Abid, H., Chtourou, M., & Toumi, A. (2007). Fuzzy indirect adaptive control scheme for nonlinear systems based on lyapunov approach and sliding mode. *International Journal of Computational Cognition, 5*, 1–8.

Boutalis, Y. S., Theodoridis, D. C., & Christodoulou, M. A. (2009). A new neuro fds definition for indirect adaptive control of unknown nonlinear systems using a method of parameter hopping. *IEEE Transactions on Neural Networks, 20*, 609–625.

Ioannou, P., & Fidan, B. (2006). *Adaptive control tutorial*. Advances in Design and Control Series. Pennsylvania: SIAM.

Kosmatopoulos, E. B., & Ioannou, P. A. (1999). A switching adaptive controller for feedback linearizable systems. *IEEE Transactions on Automatic Control, 44*, 742–750.

Kosmatopoulos, E. B., & Ioannou, P. A. (2002). Robust switching adaptive control of multi-input nonlinear systems. *IEEE Transactions on Automatic Control, 47*, 610–624.

Kung, C. C., Chen, T. H., & Kung, L. H. (2005). Modified adaptive fuzzy sliding mode controller for uncertain nonlinear systems. *IEICE Transactions on Fundamentals, E88A*, 1328–1334.

Lewis, F., Jaganathan, S. & Yesildirek, A. (1999). *Neural network control of robot manipulators and nonlinear systems*. Pennsylvania: Taylor and Francis.

Rovithakis, G. & Christodoulou, M. A. (2000). Adaptive control with recurrent high order neural networks (Theory and Industrial Applications). In *Advances in industrial control*. London: Springer.

Sanchez, E. N., & Bernal, M. A. (2000). Adaptive recurrent neural control for nonlinear system tracking. *IEEE Transactions on SMC-Part B, 30*, 886–889.

Slontine, J. J. E., & Li, W. (1991). *Applied Nonlinear Control*. Englewood Cliffs, New Jersey: Prentice-Hall.

Theodoridis, D., Boutalis, Y. & Christodoulou, M. (2009). Direct adaptive regulation and robustness analysis for systems in brunovsky form using a new neuro-fuzzy method. In *European Control Conference ECC-09*. Budapest, Hungary.

Theodoridis, D., Boutalis, Y., & Christodoulou, M. (2010). A new neuro-fuzzy approach with robustness analysis for direct adaptive regulation of systems in brunovsky form. *International Journal of Neural Systems, 20*, 319–339.

Theodoridis, D., Boutalis, Y., & Christodoulou, M. (2012). Direct adaptive neuro-fuzzy trajectory tracking of uncertain nonlinear systems. *International Journal of Adaptive Control and Signal Processing, 26*, 660–688.

Chapter 5
Direct Adaptive Neurofuzzy Control of MIMO Systems

5.1 Regulation of MIMO Systems Assuming Only Parametric Uncertainties

We consider affine in the control, nonlinear dynamical systems of the form

$$\dot{x} = f(x) + G(x) \cdot u, \tag{5.1}$$

where the state $x \in R^n$ is assumed to be completely measured, the control u is in R^q, f is an unknown smooth vector field called the drift term, and G is a matrix with rows containing the unknown smooth controlled vector fields g_{ij}.

Following the analysis presented in Sect. 3.1.1 (Boutalis et al. 2009; Theodoridis et al. 2009), we use an affine in the control fuzzy dynamical system, which approximates the system in (5.1) and uses two fuzzy subsystem blocks for modeling $f(x)$ and $G(x)$ as follows:

$$\dot{\hat{x}} = A\hat{x} + X_f W_f s_f(x) + X_g W_g S_g(x)u, \tag{5.2}$$

where X_f, X_g are matrices containing the centers of the partitions of every fuzzy output variable of $f(x)$ and $G(x)$, respectively, $s_f(x)$, $S_g(x)$ are vector and matrix, respectively, containing high-order combinations of sigmoid functions of the state x, and W_f, W_g are matrices containing respective neural weights according to (5.2).

The state regulation problem is known as our attempt to force the state to zero from an arbitrary initial value by applying appropriate feedback control to the plant input. However, since the plant is considered unknown, we assume that the unknown plant can be described by the following model arriving from the NF representation of (5.2), where the weight values W_f^* and W_g^* are unknown:

$$\dot{x} = Ax + X_f W_f^\star s_f(x) + X_g W_g^\star S_g(x)u. \tag{5.3}$$

Y. Boutalis et al., *System Identification and Adaptive Control*,
Advances in Industrial Control, DOI: 10.1007/978-3-319-06364-5_5,
© Springer International Publishing Switzerland 2014

Therefore, the state regulation problem is analyzed for the system (5.3) instead of (5.1). Since W_f^* and W_g^* are unknown, our solution consists of designing a control law $u(W_f, W_g, x)$ and appropriate update laws for W_f and W_g to guarantee convergence of the state to zero and in some cases, which will be analyzed in the following sections, boundedness of x and of all signals in the closed-loop.

The following mild assumptions are also imposed on (5.1), to guarantee the existence and uniqueness of solution for any finite initial condition and $u \in U_c$:

Assumption 11 In order for the system to be universally controllable, assume that the input function $G(x)$ satisfies diag $\left(G(x) \cdot G^T(x)\right) > 0$, for all x in a certain controllability compact region $\Omega_x \in R^n$. The assumption simply denotes that every state variable can be affected by at least one input variable.

Assumption 12 Given a class $U_c \subset R^q$ of admissible inputs, then for any $u \in U_c$ and any finite initial condition, the state trajectories are uniformly bounded for any finite $T > 0$. Meaning that we do not allow systems processing trajectories that escape at infinite, in finite time T, T being arbitrarily small. Hence, $|x(T)| < \infty$.

Assumption 13 Functions f_i, g_{ij}, $i = 1, 2, \ldots, n$ are continuous with respect to their arguments and satisfy a local Lipschitz condition so that the solution $x(t)$ of (5.1) is unique for any finite initial condition and $u \in U_c$.

Next, we present a solution to the adaptive regulation problem of multivariable multi-input systems and investigate the error effects assuming only parameter uncertainties, that is, there is complete model matching (Theodoridis et al. 2009). Under this assumption the unknown system can be written as (5.3), where $x \in \Re^n$ is the system state vector, $u \in \Re^q$ are the control inputs, X_f, X_g are $n \times n \cdot q$, and $n \times n \cdot m \cdot q$ block diagonal matrices, respectively, W_f^* is a $n \cdot q \times k$ matrix of synaptic weights and W_g^* is a $n \cdot m \cdot q \times n \cdot m$ block diagonal matrix. Finally, $s_f(x)$ is a k-dimensional vector and $S_g(x)$ is a $n \cdot m \times m$ block diagonal matrix with each diagonal element $s_i(x)$ being a high-order combination of sigmoid functions of the state variables.
The above parameter dimensions and description of Eqs. (5.2), (5.3) have been extensively explained in Sect. 3.1.1.
Define now ν as

$$\nu \triangleq X_f W_f s_f(x) + X_g W_g S_g(x)u - \dot{x} + Ax, \tag{5.4}$$

substituting Eq. (5.3) into Eq. (5.4) we have

$$\nu \triangleq X_f \tilde{W}_f s_f(x) + X_g \tilde{W}_g S_g(x)u, \tag{5.5}$$

where $\tilde{W}_f = W_f - W_f^*$ and $\tilde{W}_g = W_g - W_g^*$. W_f and W_g are estimates of W_f^* and W_g^*, respectively, and are obtained by updating laws that are to be designed in the sequel. ν cannot be measured since $\dot{x}$ is unknown. To overcome this problem, we use the following filtered version of ν:

$$\nu = \dot{\xi} + K\xi, \tag{5.6}$$

where $K = \begin{bmatrix} k_1 & 0 & \cdots & 0 \\ 0 & k_2 & 0 & \cdots \\ \cdots & 0 & \cdots & 0 \\ 0 & \cdots & 0 & k_n \end{bmatrix}$ is a diagonal positive definite matrix.

In the sequel, according to Eq. (5.4) we have that

$$\dot{\xi} + K\xi = -\dot{x} + Ax + X_f W_f s_f(x) + X_g W_g S_g(x)u, \tag{5.7}$$

and after substituting Eq. (5.3) we have

$$\dot{\xi} = -K\xi + X_f \tilde{W}_f s_f(x) + X_g \tilde{W}_g S_g(x)u. \tag{5.8}$$

To implement Eq. (5.8), we take

$$\xi \triangleq \zeta - x. \tag{5.9}$$

Employing Eq. (5.9), Eq. (5.7) can be written as

$$\dot{\zeta} + K\zeta = Kx + Ax + X_f W_f s_f(x) + X_g W_g S_g(x)u, \tag{5.10}$$

with state $\zeta \in \Re^n$. This method is referred to as error filtering.

The regulation of the system can be achieved by selecting the control input to be

$$u = -\left[X_g W_g S_g(x)\right]^+ \left[X_f W_f s_f(x) + v\right], \tag{5.11}$$

where $[\cdot]^+$ means pseudo-inverse in Moore–Penrose sense (Moore 1920; Penrose 1955) and

$$v = (K + A)x. \tag{5.12}$$

Thus, substituting Eq. (5.11), Eq. (5.10) becomes

$$\dot{\zeta} = -K\zeta. \tag{5.13}$$

To continue, consider the Lyapunov candidate function

$$V = \xi^T \xi + \zeta^T \zeta + \frac{1}{2}\mathrm{tr}\left\{\tilde{W}_f^T D_f^{-1} \tilde{W}_f\right\} + \frac{1}{2}\mathrm{tr}\left\{\tilde{W}_g^T D_g^{-1} \tilde{W}_g\right\}. \tag{5.14}$$

If we take the derivative of Eq. (5.14) with respect to time and substituting Eqs. (5.8), (5.13) we obtain

$$\dot{V} = -\xi^T K \xi - \zeta^T K \zeta + \xi^T X_f \tilde{W}_f s_f(x) + \xi^T X_g \tilde{W}_g S_g(x) u$$
$$+ \, \mathrm{tr} \left\{ \dot{W}_f^T D_f^{-1} \tilde{W}_f \right\} + \mathrm{tr} \left\{ \dot{W}_g^T D_g^{-1} \tilde{W}_g \right\}.$$

Hence, if we choose

$$\mathrm{tr} \left\{ \dot{W}_f^T D_f^{-1} \tilde{W}_f \right\} = -\xi^T X_f \tilde{W}_f s_f(x), \tag{5.15}$$

$$\mathrm{tr} \left\{ \dot{W}_g^T D_g^{-1} \tilde{W}_g \right\} = -\xi^T X_g \tilde{W}_g S_g(x) u, \tag{5.16}$$

$\dot{V}$ becomes

$$\dot{V} \leq -\lambda_{\min}(K) \|\xi\|^2 - \lambda_{\min}(K) \|\zeta\|^2. \tag{5.17}$$

Using matrix trace properties, it can be easily verified that Eqs. (5.15) and (5.16) can be written as (5.20), (5.21). These can be element-wise written as

(a) For the elements of W_f:

$$\dot{w}_{f_i}^{pl} = -\bar{x}_{f_i}^p \xi_i s_l(x) d_{f_l} \tag{5.18}$$

or subvector-wise as $\dot{W}_{f_i}^l = -\left(\bar{x}_{f_i} \right)^T \xi_i s_l(x) d_{f_l}$, for all $i = 1, 2, \ldots, n$, $p = 1, 2, \ldots, q$ and $l = 1, 2, \ldots, k$.

(b) For the elements of W_g:

$$\dot{w}_{g_{ij}}^p = -\bar{x}_{g_{ij}}^p \xi_i u_j s_i(x) d_{g_i} \tag{5.19}$$

or subvector-wise as $\dot{W}_{g_{ij}} = -\left(\bar{x}_{g_{ij}} \right)^T \xi_i u_j s_i(x) d_{g_i}$, for all $i = 1, 2, \ldots, n$, $j = 1, 2, \ldots, m$ and $p = 1, 2, \ldots, q$.

Equations (5.18) and (5.19) can be finally written in a compact form as

$$\dot{W}_f = -X_f^T \xi s_f^T(x) D_f, \tag{5.20}$$

$$\dot{W}_g = -X_g^T \xi u^T S_g^T(x) D_g, \tag{5.21}$$

where $\xi = (\xi_1, \xi_2, \ldots, \xi_n)^T$ and $u = (u_1, u_2, \ldots, u_m)^T$.

Based on the derived updating laws, we cannot conclude anything about the weight convergence if the existence and boundedness of signal u are not ensured. So the weight updating laws (5.18), (5.19) have to be modified by employing the method of parameter "Modified Hopping" that has already been analyzed in the previous chapters. In our case, the existence of $\left[X_g W_g S_g(x) \right]^+$ has to be ensured. Since the submatrices of $S_g(x)$ are diagonal to the diagonal elements $s_i(x) \neq 0$ and X_g, W_g are block diagonal, the existence of the pseudo-inverse is ensured when $\bar{x}_{g_{i,j+m(i-1)}} \cdot W_{g_{j+m(i-1)}} \neq 0$, $\forall i = 1, \ldots, n$ and $\forall j = 1, \ldots, m$. Therefore, $W_{g_{j+m(i-1)}}$ has to be confined such that $\left| \bar{x}_{g_{i,j+m(i-1)}} \cdot W_{g_{j+m(i-1)}} \right| \geq \theta_{j+m(i-1)} > 0,$

with $\theta_{j+m(i-1)}$ being a small positive design parameter (usually in the range of $[0.001, 0.01]$). For notational simplicity, we can denote with $a = i$, $j+m(i-1)$ and $b = j + m(i-1)$. The analysis of hopping condition follows the same theoretical lines as in Sect. 3.2.2 and the procedure is depicted in Fig. 3.1, where a simplified two-dimensional representation is given. Theorem 5.1 below introduces this *modified hopping* in the weight updating law.

Theorem 5.1 *The updating law for the elements of W_{gb} given by* (5.19) *and modified according to the hopping method:*

$$
\dot{W}_{gb} = \begin{cases} -\left(\bar{x}_{ga}\right)^T \xi_i u_j s_i(x) d_{gi} & \begin{aligned} &\text{if } \left|\bar{x}_{ga} \cdot W_{gb}\right| > \theta_b \\ &\text{or } \bar{x}_{ga} \cdot W_{gb} = \pm\theta_b \\ &\text{and } \bar{x}_{ga} \cdot \dot{W}_{gb} <> 0 \end{aligned} \\[2em] -\left(\bar{x}_{ga}\right)^T \xi_i u_j s_i(x) d_{gi} - \\ \dfrac{2\kappa_{gb}^{\mathrm{inner}}\left(\bar{x}_{ga} W_{gb}(\bar{x}_{ga})^T\right)}{\mathrm{tr}\{(\bar{x}_{ga})^T \bar{x}_{ga}\}} & \text{otherwise} \end{cases}
$$

ensures the existence of the control signal.

Proof The first part of the weight updating equation is used when the weights are at a certain distance (condition if $\left|\bar{x}_{ga} \cdot W_{gb}\right| > \theta_b$) from the forbidden plane or at the safe limit (condition $\left|\bar{x}_{ga} \cdot W_{gb}\right| = \pm\theta_b$) but with the direction of updating moving the weights far from the forbidden plane (condition $\bar{x}_{ga} \cdot \dot{W}_{gb} <> 0$).

In the second part of $\dot{W}_{gb}$, term $-\dfrac{2\kappa_{gb}^{\mathrm{inner}}\left(\bar{x}_{ga} W_{gb}(\bar{x}_{ga})^T\right)}{\mathrm{tr}\{(\bar{x}_{ga})^T \bar{x}_{ga}\}}$ determines the magnitude of weight *hopping*, which as explained in the vectorial proof of "hopping" in Chap. 3 and Fig. 3.5, has to be at least two times the distance of the current weight vector to the forbidden hyperplane. Therefore the **existence** of the control signal is ensured, because the weights never reach the forbidden plane, as well as the case of infinite hopping never occurs.

The inclusion of weight hopping in the weights updating law guarantees that the control signal does not go to infinity. Apart from that, it is also of practical use to ensure that $X_g W_g S_g(x)$ does not approach even temporarily at very large values because in this case the method may become algorithmically unstable driving at the same time the control signal to zero, failing to control the system. To ensure that this situation does not happen we have again to ensure that $\left|\bar{x}_{ga} \cdot W_{gb}\right| < \rho_b$ with ρ_b being again a design parameter determining an external limit for $\bar{x}_{ga} \cdot W_{gb}$.

Following the same lines of thought with the case of weight hopping introduced above we could again consider the forbidden hyperplanes being defined by the equation $\left|\bar{x}_{ga} \cdot W_{gb}\right| = \rho_b$. When the weight vector reaches one of the forbidden hyperplanes $\bar{x}_{ga} \cdot W_{gb} = \rho_b$ and the direction of updating is toward the forbidden hyperplane, a new *modified hopping* is introduced that moves the weights inside the restricting area.

This procedure is depicted in Fig. 2.3, in a simplified two-dimensional representation. The magnitude of hopping is $-\dfrac{\kappa_{gb}^{\mathrm{outer}}\left(\bar{x}_{ga} W_{gb}(\bar{x}_{ga})^T\right)}{\mathrm{tr}\{(\bar{x}_{ga})^T \bar{x}_{ga}\}}$ being determined by following again the same vectorial proof as explained in Chap. 3, with $\kappa_{gb}^{\text{outer}}$ a small positive number decided appropriately by the designer according to Fig. 3.5. By performing *modified hopping* when $\bar{x}_{ga} \cdot W_{gb}$ reaches either the inner or outer forbidden hyperplanes, $\bar{x}_{ga} \cdot W_{gb}$ are confined to lie in space:

$$P_{g,\text{MIMO}} = \left\{ \bar{x}_{ga} \cdot W_{gb} : \left| \bar{x}_{ga} \cdot W_{gb} \right| \le \rho_b \text{ and } \left| \bar{x}_{ga} \cdot W_{gb} \right| > \theta_b \right\}.$$

The weight updating law for W_{gb} incorporating the two hopping conditions can now be expressed as

$$\dot{W}_{gb} = \begin{cases} -\left(\bar{x}_{ga}\right)^T \xi_i u_j s_i(x) d_{gi} & \begin{array}{l} \text{if } \bar{x}_{ga} \cdot W_{gb} \in P_{g,\text{MIMO}} \\ \text{or } \bar{x}_{ga} \cdot W_{gb} = (\pm\theta_b \text{ or } \pm\rho_b) \\ \text{and } \bar{x}_{ga} \cdot \dot{W}_{gb} <> \text{ or } >< 0 \end{array} \\[2em] -\left(\bar{x}_{ga}\right)^T \xi_i u_j s_i(x) d_{gi} - \\ \dfrac{2\sigma_i \kappa_{gb}^{\text{inner}} \left(\bar{x}_{ga} W_{gb} (\bar{x}_{ga})^T\right)}{\text{tr}\{(\bar{x}_{ga})^T \bar{x}_{ga}\}} - & \text{otherwise} \\ \dfrac{(1-\sigma_i)\kappa_{gb}^{\text{outer}} \left(\bar{x}_{ga} W_{gb} (\bar{x}_{ga})^T\right)}{\text{tr}\{(\bar{x}_{ga})^T \bar{x}_{ga}\}} \end{cases} , \quad (5.22)$$

where

$$\sigma_i = \begin{cases} 0 & \text{if } \bar{x}_{ga} \cdot W_{gb} = \pm\rho_b \\ & \text{and } \bar{x}_{ga} \cdot \dot{W}_{gb} <> 0 \\ 1 & \text{if } \bar{x}_{ga} \cdot W_{gb} = \pm\theta_b \\ & \text{and } \bar{x}_{ga} \cdot \dot{W}_{gb} >< 0 \end{cases} . \quad (5.23)$$

Regarding the *negativity* of $\dot{V}$ we proceed with the following lemma:

Lemma 5.1 *Based on the adaptive law* (5.22) *the additional term introduced in the expression for* $\dot{V}$, *can only make* $\dot{V}$ *more negative.*

Proof Following the same lines of analysis with the proof of Theorem 3.3 in Sect. 3.2.2 and the expression of hopping magnitude in respect to $\tilde{W}_{gb}$, the modified updating law is rewritten as

$$\dot{W}_{gb} = -\gamma_2 \left(\bar{x}_{ga}\right)^T \xi_i u_j s_i(x) - 2\sigma_i \kappa_{gb}^{\text{inner}} \theta_b \frac{\tilde{W}_{gb}}{\|\tilde{W}_{gb}\|} - (1 - \sigma_i) \kappa_{gb}^{\text{outer}} \rho_b \frac{\tilde{W}_{gb}}{\|\tilde{W}_{gb}\|}.$$

With this updating law, it can be easily verified that (5.17) becomes

$$\dot{V} \le -\lambda_{\min}(K) \|\xi\|^2 - \lambda_{\min}(K) \|\zeta\|^2 - \Theta_{gb} \frac{\tilde{W}_{gb}^T \tilde{W}_{gb}}{\|\tilde{W}_{gb}\|}, \quad (5.24)$$

with Θ_{g_b} being a positive constant expressed as

$$\Theta_{g_b} = 2\sigma_i \sum \kappa_{g_b}^{\text{inner}} \theta_b + (1 - \sigma_i) \sum \kappa_{g_b}^{\text{outer}} \rho_b \geq 0,$$

for all time, where the summation includes all weight vectors that require hopping. Therefore, the negativity of $\dot{V}$ is actually strengthened due to the last negative term.

5.2 Robustness Analysis Assuming the Presence of Modeling Errors

In this section, we examine the problem of direct regulation of unknown multivariable affine in the control nonlinear systems and its robustness analysis (Theodoridis et al. 2011b). The development is combined with a sensitivity analysis of the closed-loop in the presence of modeling imperfections and provides a comprehensive and rigorous analysis showing that the adaptive regulator can guarantee the convergence of states to zero or at least uniform ultimate boundedness of all signals in the closed-loop when a not-necessarily-known modeling error is applied. The existence and boundedness of the control signal are always ensured by employing the method of parameter "modified hopping," which appears in the weight updating laws. We present the analysis in a progressive way, considering first the case of modeling error that depends on system states. Then we examine the more general case, where the modeling error depends also on a unknown constant term.

5.2.1 Modeling Errors Depending on System States

The analysis proceed as follows. Due to the approximation capabilities of the dynamic NF networks, we can assume with no loss of generality that the unknown plant (5.1) can be completely described by a dynamical NF network plus a modeling error term $\omega(x, u)$. In other words, there exist weight values $W_f^\star$ and $W_g^\star$ such that the system (5.1) can be written as

$$\dot{x} = Ax + X_f W_f^\star s_f(x) + X_g W_g^\star S_g(x)u + \omega(x, u) \tag{5.25}$$

By including the modeling errors, Eq. (5.5) becomes

$$\nu \triangleq X_f \tilde{W}_f s_f(x) + X_g \tilde{W}_g S_g(x)u - \omega(x, u). \tag{5.26}$$

After substituting Eq. (5.6) into Eq. (5.26), we get

$$\dot{\xi} = -K\xi + X_f \tilde{W}_f s_f(x) + X_g \tilde{W}_g S_g(x)u - \omega(x, u). \qquad (5.27)$$

The regulation of the system can be achieved by selecting the same control input as in Sect. 5.2

$$u = -\left[X_g W_g S_g(x)\right]^+ \left[X_f W_f s_f(x) + v\right], \qquad (5.28)$$

where $[\cdot]^+$ means pseudo-inverse in the Moore–Penrose sense and

$$v = (K + A)x. \qquad (5.29)$$

Then, taking into account that ξ is implemented by using Eq. (5.9), selecting the weight updating laws as in Eqs. (5.20), (5.21), and considering modeling errors, Eq. (5.17) which is the derivative of Lyapunov candidate function becomes

$$\dot{V} \leq -\lambda_{\min}(K)\,\|\xi\|^2 - \lambda_{\min}(K)\,\|\zeta\|^2 + \|\xi\|\,\|\omega(x, u)\|. \qquad (5.30)$$

We are now ready to make the following assumption:

Assumption 14 The modeling error term satisfies

$$\|\omega(x, u)\| \leq \ell_1'\,\|x\| + \ell_1''\,\|u\|,$$

where ℓ_1' and ℓ_1'' are known positive constants.

Also, we can find an *a priori* known constant $\ell_u > 0$, such that:

$$\|u\| \leq \ell_u\,\|x\|, \qquad (5.31)$$

and Assumption 14 becomes equivalent to

$$\|\omega(x)\| \leq \ell_1\,\|x\|, \qquad (5.32)$$

where

$$\ell_1 = \ell_1' + \ell_1''\ell_u, \qquad (5.33)$$

is a positive constant.

Since $X_g W_g$ is bounded, if Eq. (5.22) is used one can easily verify that (5.28) is valid provided that $X_f W_f$ is uniformly bounded by a known positive constant ε_i. So $X_f W_f(t)$ should be confined to the set $P_{f,\text{MIMO}} = \{\bar{x}_{f_i} \cdot W_{f_i}^l : \left|\bar{x}_{f_i} \cdot W_{f_i}^l\right| \leq \varepsilon_i\}$ through the use of a modified hopping algorithm. In particular, the standard updating law (5.18) is modified to

$$
\dot{W}^l_{f_i} =
\begin{cases}
- \left(\bar{x}_{f_i}\right)^T \xi_i s_l(x) d_{f_l} &
\begin{array}{l}
\text{if } \bar{x}_{f_i} \cdot W^l_{f_i} \in P_{f,\text{MIMO}} \\
\text{or } \bar{x}_{f_i} \cdot W^l_{f_i} = \pm \varepsilon_i \\
\text{and } \bar{x}_{f_i} \cdot \dot{W}^l_{f_i} >< 0
\end{array} \\[2em]
- \left(\bar{x}_{f_i}\right)^T \xi_i s_l(x) d_{f_l} - \\
\dfrac{\kappa^{\text{outer}}_{f_i}\left(\bar{x}_{f_i} W^l_{f_i} (\bar{x}_{f_i})^T\right)}{\operatorname{tr}\left\{(\bar{x}_{f_i})^T \bar{x}_{f_i}\right\}} &
\text{otherwise}
\end{cases}
, \qquad (5.34)
$$

therefore we have the following lemma:

Lemma 5.2 *If the initial weights are chosen such that $\bar{x}_{f_i} \cdot W^l_{f_i}(0) \in P_f$ and $\bar{x}_{f_i} \cdot W^{*l}_{f_i} \in P_f$, then we have $\bar{x}_{f_i} \cdot W^l_{f_i} \in P_f$ for all $t \geq 0$.*

Proof The proof is similar to that of Lemma 4.2 in Chap. 4.

In the sequel, employing Assumption 14, Eq. (5.30) becomes

$$
\begin{aligned}
\dot{V} &\leq -\lambda_{\min}(K) \|\xi\|^2 - \lambda_{\min}(K) \|\zeta\|^2 + \ell_1 \|\xi\| \|x\| \\
&\leq -\lambda_{\min}(K) \left(\|\xi\|^2 + \|\zeta\|^2\right) + \ell_1 \|\xi\|^2 + \ell_1 \|\xi\| \|\zeta\| \\
&\leq -\begin{bmatrix} \|\xi\| & \|\zeta\| \end{bmatrix}
\begin{bmatrix} \lambda_{\min}(K) - \ell_1 & -\ell_1 \\ 0 & \lambda_{\min}(K) \end{bmatrix}
\begin{bmatrix} \|\xi\| \\ \|\zeta\| \end{bmatrix}.
\end{aligned} \qquad (5.35)
$$

Hence, if we choose $\lambda_{\min}(K) \geq \ell_1$ then Eq. (5.35) becomes negative. Thus, we have

$$
\dot{V} \leq 0. \qquad (5.36)
$$

Regarding the **negativity** of $\dot{V}$ we proceed with the following lemma:

Lemma 5.3 *Based on the adaptive laws (5.22), (5.34) the additional terms introduced in the expression for $\dot{V}$, can only make $\dot{V}$ more negative.*

Proof The proof which refers to the adaptive law (5.22) regarding the negativity of Lyapunov is exactly the same as that of Lemma 5.1.

Using the modified updating law (5.34), term $- \dfrac{\kappa^{\text{outer}}_{f_i}\left(\bar{x}_{f_i} W^l_{f_i} (\bar{x}_{f_i})^T\right)}{\operatorname{tr}\{(\bar{x}_{f_i})^T \bar{x}_{f_i}\}}$, the second part of $W^l_{f_i}$ determines the magnitude of weight *hopping*, which according to Chap. 3 and Fig. 3.5, has to be $\kappa^{\text{outer}}_{f_i}$ times the distance of the current weight vector. Once again, regarding the **negativity** of $\dot{V}$ we proceed as follows.

Following the same lines of analysis as the proof of Theorem 3.3 in Sect. 3.2.2 the modified updating law is rewritten as

$$
\dot{W}^l_{f_i} = - \left(\bar{x}_{f_i}\right)^T \xi_i s_l(x) d_{f_l} - \kappa^{\text{outer}}_{f_i} \varepsilon_i \frac{\tilde{W}^l_{f_i}}{\|\tilde{W}^l_{f_i}\|}.
$$

With this updating law and Eq. (5.22), it can be easily verified that Eq. (5.35) becomes

$$
\dot{V} \leq -\begin{bmatrix} \|\xi\| & \|\zeta\| \end{bmatrix}\begin{bmatrix} \lambda_{\min}(K) - \ell_1 & -\ell_1 \\ 0 & \lambda_{\min}(K) \end{bmatrix}\begin{bmatrix} \|\xi\| \\ \|\zeta\| \end{bmatrix}
$$
$$
-\Theta_{f_i}\frac{(\tilde{W}_{f_i}^l)^T \tilde{W}_{f_i}^l}{\|\tilde{W}_{f_i}^l\|} - \Theta_{gb}\frac{(\tilde{W}_{gb})^T \tilde{W}_{gb}}{\|\tilde{W}_{gb}\|}, \tag{5.37}
$$

with Θ_{f_i} being a positive constant expressed as $\Theta_{f_i} = \sum \kappa_{f_i}^{\mathrm{outer}}\varepsilon_i \geq 0$ for all time, where the summation includes all weight vectors that require hopping. Therefore, the negativity of $\dot{V}$ is actually strengthened due to the last negative terms.

Lemma 5.3 implies that the hopping modifications (5.22), (5.34) guarantee boundedness of the weights, without affecting the rest of the stability properties established in the absence of modified hopping.

Hence, we can prove the following theorem:

Theorem 5.2 *The control law* (5.28) *and* (5.29) *together with the updating laws* (5.22) *and* (5.34) *guarantee the following properties:*

- $\xi, x, W_f, W_g, \zeta, \dot{\xi} \in L_\infty, \quad \|\xi\| \in L_2,$
- $\lim_{t\to\infty} \xi(t) = 0, \quad \lim_{t\to\infty} x(t) = 0,$
- $\lim_{t\to\infty} \dot{W}_f(t) = 0, \quad \lim_{t\to\infty} \dot{W}_g(t) = 0,$

provided that $\lambda_{\min}(K) \geq \ell_1.$

Proof From Eq. (5.36) we have that $V \in L_\infty$, which implies $\xi, \tilde{W}_f, \tilde{W}_g \in L_\infty$. Furthermore, $W_f = \tilde{W}_f + W_f^* \in L_\infty$ and $W_g = \tilde{W}_g + W_g^* \in L_\infty$. Since, $\xi = \zeta - x$ and $\zeta, \xi \in L_\infty$ this in turn implies that $x \in L_\infty$. Moreover, since V is a monotone decreasing function of time and bounded from below, $\lim_{t\to\infty} V(t) = V_\infty$ exists; so by integrating $\dot{V}$ from 0 to ∞ we have

$$
(\lambda_{\min}(K) - \ell_1)\int_0^\infty \|\xi\|^2\, dt + \lambda_{\min}(K)\int_0^\infty \|\zeta\|^2\, dt - \ell_1\int_0^\infty \|\xi\|\,\|\zeta\|\, dt \leq [V(0) - V_\infty] < \infty,
$$

which implies that $\|\xi\| \in L_2$. We also have that

$$
\dot{\xi} = -K\xi + X_f\tilde{W}_f s_f(x) + X_g\tilde{W}_g S_g(x)u - \omega(x).
$$

Hence and since $u, x, \dot{\xi} \in L_\infty$, the sigmoidals are bounded by definition, $\tilde{W}_f, \tilde{W}_g \in L_\infty$ and Assumption 14 holds, so since $\xi \in L_2 \cap L_\infty$ and $\dot{\xi} \in L_\infty$, applying Barbalat's Lemma (Popov 1973), we conclude that $\lim_{t\to\infty} \xi(t) = 0$.

Now, using the boundedness of $u, s_f(x), S_g(x), x$ and the convergence of $\xi(t)$ to zero, we have that $\dot{W}_f, \dot{W}_g$ also converges to zero. Hence and since $\zeta(t)$ also converges to zero, we have that

$$\lim_{t\to\infty} x(t) = \lim_{t\to\infty} \zeta(t) - \lim_{t\to\infty} \xi(t) = 0.$$

Thus,

$$\lim_{t\to\infty} x(t) = 0.$$

Remark 5.1 Inequality $\lambda_{\min}(K) \geq \ell_1$ shows how the design constant K should be selected, in order to guarantee convergence of the state x to zero, even in the presence of modeling error terms that are not uniformly bounded *a priori*, as Assumption 14 implies. To cope with large model imperfections, the value of K should become large. But K is implemented as a gain in the construction of $\dot{\zeta}$ and for practical reasons it cannot take arbitrarily large values. This leads to a compromise between the value of K and the maximum allowable modeling error terms.

5.2.2 Modeling Errors Depending on System States and a Not-Necessarily-Known Constant Value

In Sect. (5.2.1), we have assumed that the modeling error term satisfies the following condition:

$$\|\omega(x, u)\| \leq \ell_1' \|x\| + \ell_1'' \|u\|$$

which, using Eq. (5.32), implies that the modeling error becomes zero when $\|x\| = 0$ and we have proven convergence of the state x to zero, plus boundedness of all signals in the closed-loop. In this subsection however, we examine the more general case which is described by the following assumption:

Assumption 15 The modeling error term satisfies

$$\|\omega(x, u)\| \leq \ell_0 + \ell_1' \|x\| + \ell_1'' \|u\|.$$

Having made this assumption, we now allow a not-necessarily-known modeling error $l_0 \neq 0$ at zero. Furthermore, as stated previously, we can find an *a priori* known constant $l_u > 0$, such that

$$\|u\| \leq \ell_u \|x\|,$$

thus making

$$\|\omega(x, u)\| \equiv \|\omega(x)\|,$$

and Assumption 15 equivalent to

$$\|\omega(x)\| \leq \ell_0 + \ell_1 \|x\|, \tag{5.38}$$

where

$$\ell_1 = \ell_1' + \ell_1'' \ell_u, \tag{5.39}$$

is a positive constant. Employing (5.38), Eq. (5.30) becomes

$$\begin{aligned}
\dot{V} &\leq -\lambda_{\min}(K) \|\xi\|^2 - \lambda_{\min}(K) \|\zeta\|^2 + \|\xi\| (l_0 + l_1 \|x\|) \\
&\leq -\lambda_{\min}(K) \|\xi\|^2 - \lambda_{\min}(K) \|\zeta\|^2 + l_1 \|\xi\|^2 + l_1 \|\xi\| \|\zeta\| + l_0 \|\xi\|
\end{aligned} \tag{5.40}$$

To continue, we need to state and prove the following lemma:

Lemma 5.4 *The control law* (5.28), (5.29), *where the synaptic weight estimates* W_f *and* W_g, *are adjusted according to Eqs.* (5.22), (5.34) *guarantee the following properties:*

- $\zeta(t) \leq 0, \quad \forall t \geq 0$

- $\lim_{t \to \infty} \zeta(t) = 0$ *exponentially fast provided that* $\zeta(t) < 0$.

Proof The proof is exactly the same as that of Lemma 4.4 in Chap. 4.

Hence, we can distinguish the following cases:

Case 1: If $x \geq 0$ we have that $\zeta(t) \geq \xi(t)$ but $\zeta(t) \leq 0$, $\forall t \geq 0$ which implies that $\|\zeta(t)\| \leq \|\xi(t)\|$. So, we have

$$\|x\| \leq \|\zeta\| + \|\xi\| \leq 2 \|\xi\|. \tag{5.41}$$

Therefore, Eq. (5.40) becomes

$$\begin{aligned}
\dot{V} &\leq -\lambda_{\min}(K) \|\xi\|^2 - \lambda_{\min}(K) \|\zeta\|^2 + 2\ell_1 \|\xi\|^2 + \ell_0 \|\xi\| \tag{5.42} \\
&\leq -(\lambda_{\min}(K) \|\xi\| - 2\ell_1 \|\xi\| - \ell_0) \|\xi\| - \lambda_{\min}(K) \|\zeta\|^2 \leq 0, \tag{5.43}
\end{aligned}$$

provided that:

$$\|\xi\| > \frac{\ell_0}{\lambda_{\min}(K) - 2\ell_1}, \tag{5.44}$$

with $\lambda_{\min}(K) > 2\ell_1$.

Case 2: If $x < 0$ we have that $\zeta(t) < \xi(t)$ but $\zeta(t) \leq 0$, $\forall t \geq 0$ which implies that $\|\zeta(t)\| > \|\xi(t)\|$. So, we have

$$\|x\| \leq \|\zeta\| + \|\xi\| \leq 2 \|\zeta\|. \tag{5.45}$$

Therefore, Eq. (5.40) becomes

$$\begin{aligned}
\dot{V} &\leq -\lambda_{\min}(K) \|\xi\|^2 - \lambda_{\min}(K) \|\zeta\|^2 + 2\ell_1 \|\xi\| \|\zeta\| + \ell_0 \|\xi\| \tag{5.46} \\
&\leq -(\lambda_{\min}(K) \|\xi\| - \ell_0) \|\xi\| - (\lambda_{\min}(K) - 2\ell_1) \|\zeta\|^2 \leq 0, \tag{5.47}
\end{aligned}$$

provided that

$$\|\xi\| > \frac{\ell_0}{\lambda_{\min}(K)}, \tag{5.48}$$

and $\lambda_{\min}(K) > 0$.

Conclusively, $\forall x \in R^n$ the derivative of the Lyapunov candidate function becomes negative when $\|\xi\| > \frac{\ell_0}{\lambda_{\min}(K)-2\ell_1}$ and $\lambda_{\min}(K) > 2\ell_1$.

In the sequel, inequality (5.44) together with (5.41), (5.45) demonstrate that the trajectories of $\xi(t)$ and $x(t)$ are uniformly bounded with respect to the arbitrarily small, (since K can be chosen sufficiently large), sets Ξ and X shown below:

$$\Xi = \left\{ \xi(t) : \|\xi(t)\| \leq \frac{\ell_0}{\lambda_{\min}(K) - 2\ell_1}, \ \lambda_{\min}(K) > 2\ell_1 > 0 \right\},$$

and

$$X = \left\{ x(t) : \|x(t)\| \leq \frac{2\ell_0}{\lambda_{\min}(K) - 2\ell_1}, \ \lambda_{\min}(K) > 2\ell_1 > 0 \right\}.$$

Thus, we have proved the following theorem:

Theorem 5.3 *Consider the system* (5.25) *with the modeling error term satisfying* (5.38). *Then, the control law* (5.28), (5.29) *together with the update laws* (5.22) *and* (5.34) *guarantees the uniform ultimate boundedness with respect to the sets*

- $\Xi = \left\{ \xi(t) : \|\xi(t)\| \leq \frac{\ell_0}{\lambda_{\min}(K)-2\ell_1}, \ \lambda_{\min}(K) > 2\ell_1 > 0 \right\}$,
- $X = \left\{ x(t) : \|x(t)\| \leq \frac{2\ell_0}{\lambda_{\min}(K)-2\ell_1}, \ \lambda_{\min}(K) > 2\ell_1 > 0 \right\}$.

Furthermore,

$$\dot{\xi} = -K\xi + X_f \tilde{W}_f s_f(x) + X_g \tilde{W}_g S_g(x)u - \omega(x).$$

Hence, since the boundedness of $\tilde{W}_f$ and $\tilde{W}_g$ is ensured by the use of the modified hopping algorithm and $\omega(x)$ owing to (5.38) and Theorem 5.3, we conclude that $\dot{\xi} \in L_\infty$.

Remark 5.2 The previous analysis reveals that in the case where we have a modeling error different from zero at $\|x\| = 0$, the adaptive regulator can guarantee at least uniform ultimate boundedness of all signals in the closed-loop. In particular, Theorem 5.3 shows that if ℓ_0 is sufficiently small, or if the design constant K is chosen such that $\lambda_{\min}(K) > 2\ell_1$, then $\|x(t)\|$ can be arbitrarily close to zero and in the limit as $K \to \infty$, actually becomes zero but as we stated in Remark 5.1, implementation issues constrain the maximum allowable value of K.

5.3 The Model Order Problem

The NF direct adaptive regulation of unknown nonlinear multivariable multi-input dynamical systems is considered in this section and the model order problem is analyzed (Theodoridis et al. 2011a). Since the plant is considered unknown, we propose its approximation by the NF model, which however may assume *smaller number of states than the original unknown model*. The omission of states, is referred as a model order problem and is modeled by introducing a disturbance term in the approximating equations.

5.3.1 Problem Formulation

We consider an affine in the control, nonlinear dynamical system in the following form:

$$\dot{x} = f(x) + G(x) \cdot u + \phi(x, x_{ud}), \tag{5.49}$$

$$\text{and} \quad \dot{x}_{ud} = B(x, x_{ud}), \tag{5.50}$$

where the true plant is of order $N \geq n$, with the states x in $\Re^n$, the control u in $\Re^n$, f an unknown smooth vector field called the drift term, and G a diagonal matrix with elements of the unknown smooth controlled vector field $g_i, i = 1, 2, \ldots, n$ and $G = \text{diag}(g_1, g_2, \ldots, g_n)$. Also, $x_{ud} \in \Re^p$, is the state of the unmodeled dynamics and $\phi(\cdot)$, $B(\cdot)$ are unknown vector fields of x and x_{ud}. Obviously $p = N - n$.

State regulation aims at driving system states to zero from arbitrary initial values. To this end, appropriate feedback control has to be designed and applied to the plant input. However, since the plant is considered unknown and there are dynamics that have not been taken into account, we may take advantage of the NF approximator and its approximation capabilities. With no loss of generality, we assume that the unknown plant (5.49), (5.50) can be completely described by the NF model plus an unmodeled dynamics term $\phi(x, x_{ud})$. This is equivalent to assume that, there exist weight values W_f^* and W_g^* such that the system (5.49), (5.50) can be written as

$$\dot{x} = Ax + X_f W_f^{\star} s_f(x) + X_g W_g^{\star} S_g(x)u + \phi(x, x_{ud}), \tag{5.51}$$

$$\text{and} \quad \dot{x}_{ud} = B(x, x_{ud}). \tag{5.52}$$

Under this assumption, the state regulation can be analyzed and designed for the system (5.51), (5.52) instead of (5.49), (5.50). But W_f^* and W_g^* are unknown. Therefore, our solution embodies the design of a control law $u(W_f, W_g, x)$ and updating laws for W_f and W_g, which guarantee the regulation of the states and the boundedness of x and of all signals in the closed-loop.

Before proceeding further, to guarantee the existence and uniqueness of solution of (5.49) for any finite initial condition and $u \in U_c$, Assumptions 12, 13 together with the following assumption are also imposed on (5.49), (5.50):

Assumption 16 The states $x_{ud} \in R^p$ of the unmodeled dynamics, with finite initial condition are uniformly bounded for any finite time $T > 0$.

5.3.2 Adaptive Regulation

In this section, a solution to the adaptive regulation problem is presented and the model order problem is investigated. Assuming the presence of unmodeled states, the unknown system can be written as (5.51).

In the sequel, we define ν in the same way as in Eq. (5.4) and after substituting Eq. (5.51) to Eq. (5.4) we have

$$\nu \triangleq X_f \tilde{W}_f s_f(x) + X_g \tilde{W}_g S_g(x)u - \phi(x, x_{ud}), \tag{5.53}$$

where $\tilde{W}_f = W_f - W_f^\star$ and $\tilde{W}_g = W_g - W_g^\star$. W_f and W_g are estimates of $W_f^\star$ and $W_g^\star$, respectively, and can be determined by weight updating laws, which will be derived in the sequel. Alternatively, we may use the following filtered version of ν:

$$\nu = \dot{\xi} + K\xi,$$

where $K = \begin{bmatrix} k_1 & 0 & \cdots & 0 \\ 0 & k_2 & 0 & \cdots \\ \cdots & 0 & \cdots & 0 \\ 0 & \cdots & 0 & k_n \end{bmatrix}$ is a diagonal positive definite matrix and $\xi \in \Re^n$.

In the sequel, according to Eq. (5.4) we have that

$$\dot{\xi} + K\xi = -\dot{x} + Ax + X_f W_f s_f(x) + X_g W_g S_g(x)u, \tag{5.54}$$

and after substituting Eq. (5.51) we have

$$\dot{\xi} = -K\xi + X_f \tilde{W}_f s_f(x) + X_g \tilde{W}_g S_g(x)u - \phi(x, x_{ud}). \tag{5.55}$$

To implement Eq. (5.55), we take

$$\xi \triangleq \zeta - x. \tag{5.56}$$

Employing Eq. (5.56), Eq. (5.54) can be written as

$$\dot{\zeta} + K\zeta = Kx + Ax + X_f W_f s_f(x) + X_g W_g S_g(x)u, \tag{5.57}$$

with state $\zeta \in \Re^n$. This method is referred to as error filtering.

The regulation of the system can be achieved by selecting the control input to be

$$u = - \left[X_g W_g S_g(x) \right]^+ \left[X_f W_f s_f(x) + \upsilon \right], \tag{5.58}$$

where $[\cdot]^+$ means pseudo-inverse in Moore–Penrose sense and

$$\upsilon = (K + A)x. \tag{5.59}$$

After substituting Eqs. (5.58), (5.59) to Eq. (5.57), we derive

$$\dot{\zeta} = -K\zeta. \tag{5.60}$$

To continue, we consider Lemma 5.4 and the Lyapunov candidate function as

$$V = \frac{1}{2}\xi^T \xi + \frac{1}{2}\zeta^T \zeta + L(x_{\text{ud}}) + \frac{1}{2}\text{tr}\left\{ \tilde{W}_f^T D_f^{-1} \tilde{W}_f \right\} + \frac{1}{2}\text{tr}\left\{ \tilde{W}_g^T D_g^{-1} \tilde{W}_g \right\}, \tag{5.61}$$

where $L(x_{\text{ud}})$ is a strictly positive function depending on unmodeled dynamic states and D_f and D_g are positive definite diagonal gain matrices. If we take the derivative of Eq. (5.61) with respect to time and substitute Eqs. (5.55), (5.60), we obtain

$$\dot{V} \le -\xi^T K\xi - \zeta^T K\zeta + \xi^T X_f \tilde{W}_f s_f(x) + \xi^T X_g \tilde{W}_g S_g(x)u - \xi^T \phi(x, x_{\text{ud}})$$
$$+ \dot{L}(x_{\text{ud}}) + \text{tr}\left\{ \dot{W}_f^T D_f^{-1} \tilde{W}_f \right\} + \text{tr}\left\{ \dot{W}_g^T D_g^{-1} \tilde{W}_g \right\}. \tag{5.62}$$

Hence, if we choose

$$\text{tr}\left\{ \dot{W}_f^T D_f^{-1} \tilde{W}_f \right\} = -\xi^T X_f \tilde{W}_f s_f(x), \tag{5.63}$$

$$\text{tr}\left\{ \dot{W}_g^T D_g^{-1} \tilde{W}_g \right\} = -\xi^T X_g \tilde{W}_g S_g(x)u, \tag{5.64}$$

$\dot{V}$ becomes

$$\dot{V} \le -\lambda_{\min}(K) \|\xi\|^2 - \lambda_{\min}(K) \|\zeta\|^2 - \xi^T \phi(x, x_{\text{ud}}) + \dot{L}(x_{\text{ud}}). \tag{5.65}$$

It can be easily verified that Eqs. (5.63) and (5.64) after using matrix trace properties, can be subvector-wise written as

(a) For the elements of W_f:

$$\dot{W}_{f_i}^l = -\left(\bar{x}_{f_i}\right)^T \xi_i s_l(x) d_{f_l} \qquad (5.66)$$

for all $i = 1, 2, \ldots, n$ and $l = 1, 2, \ldots, k$.

(b) For the elements of W_g:

$$\dot{W}_{g_i} = -\left(\bar{x}_{g_i}\right)^T \xi_i u_i s_i(x) d_{g_i} \qquad (5.67)$$

for all $i = 1, 2, \ldots, n$.

Furthermore, we cannot conclude anything about the weight convergence if the existence and boundedness of signal u are not ensured. To this end, the weight updating laws (5.66), (5.67) have to be modified by employing the method of parameter "modified hopping."

One can easily verify that the boundedness of signal u is valid provided that $\bar{x}_{f_i} \cdot W_{f_i}^l, \bar{x}_{g_i} \cdot W_{g_i}$ are uniformly bounded by known positive constants ε_i, ρ_i and the existence of signal u when $\bar{x}_{g_i} \cdot W_{g_i} \neq 0$. Thus, $\bar{x}_{f_i} \cdot W_{f_i}^l(t)$ is confined to the set

$$P_{f,\text{MIMO}} = \{\bar{x}_{f_i} \cdot W_{f_i}^l : \left| \bar{x}_{f_i} \cdot W_{f_i}^l \right| \leq \varepsilon_i\},$$

and $\bar{x}_{g_i} \cdot W_{g_i}(t)$ is confined to the set:

$$P_{g,\text{MIMO}} = \left\{\bar{x}_{g_i} \cdot W_{g_i} : \left| \bar{x}_{g_i} \cdot W_{g_i} \right| \leq \rho_i \text{ and } \left| \bar{x}_{g_i} \cdot W_{g_i} \right| > \theta_i\right\},$$

through the use of a modified hopping algorithm. In particular, the standard updating law (5.66) is modified to

$$\dot{W}_{f_i}^l = \begin{cases} -\left(\bar{x}_{f_i}\right)^T \xi_i s_l(x) d_{f_l} & \begin{array}{l} \text{if } \bar{x}_{f_i} \cdot W_{f_i}^l \in P_{f,\text{MIMO}} \\ \text{or } \bar{x}_{f_i} \cdot W_{f_i}^l = \pm\varepsilon_i \\ \text{and } \bar{x}_{f_i} \cdot \dot{W}_{f_i}^l \gtrless 0 \end{array} \\[3ex] -\left(\bar{x}_{f_i}\right)^T \xi_i s_l(x) d_{f_l} - \\ \dfrac{\kappa_{f_i}^{\text{outer}}\left(\bar{x}_{f_i} W_{f_i}^l \left(\bar{x}_{f_i}\right)^T\right)}{\text{tr}\left\{\left(\bar{x}_{f_i}\right)^T \bar{x}_{f_i}\right\}} & \text{otherwise} \end{cases} \qquad (5.68)$$

where $\kappa_{f_i}^{\text{outer}}$ a small positive number decided from the designer according to Chap. 3 and Fig. 3.5.

By performing *modified hopping* when $\bar{x}_{g_i} \cdot W_{g_i}$ reaches either the inner or outer forbidden planes, the standard updating law (5.67) can now be expressed as

$$\dot{W}_{g_i} = \begin{cases} -\left(\bar{x}_{g_i}\right)^T \xi_i u_i s_i(x) d_{g_i} & \begin{aligned} &\text{if } \bar{x}_{g_i} \cdot W_{g_i} \in P_{g,\mathrm{MIMO}} \\ &\text{or } \bar{x}_{g_i} \cdot W_{g_i} = \pm\theta_i \\ &\text{or } \bar{x}_{g_i} \cdot W_{g_i} = \pm\rho_i \\ &\text{and } \bar{x}_{g_i} \cdot \dot{W}_{g_i} <> \text{ or } >< 0 \end{aligned} \\[2em] -\left(\bar{x}_{g_i}\right)^T \xi_i u_i s_i(x) d_{g_i} - \\ \dfrac{2\sigma_i \kappa_{g_i}^{\mathrm{inner}}\left(\bar{x}_{g_i} W_{g_i} (\bar{x}_{g_i})^T\right)}{\mathrm{tr}\{(\bar{x}_{g_i})^T \bar{x}_{g_i}\}} - & \text{otherwise} \\ \dfrac{(1-\sigma_i)\kappa_{g_i}^{\mathrm{outer}}\left(\bar{x}_{g_i} W_{g_i} (\bar{x}_{g_i})^T\right)}{\mathrm{tr}\{(\bar{x}_{g_i})^T \bar{x}_{g_i}\}} \end{cases} , \qquad (5.69)$$

where

$$\sigma_i = \begin{cases} 0 & \begin{aligned} &\text{if } \bar{x}_{g_i} \cdot W_{g_i} = \pm\rho_i \\ &\text{and } \bar{x}_{g_i} \cdot \dot{W}_{g_i} <> 0 \end{aligned} \\[1em] 1 & \begin{aligned} &\text{if } \bar{x}_{g_i} \cdot W_{g_i} = \pm\theta_i \\ &\text{and } \bar{x}_{g_i} \cdot \dot{W}_{g_i} >< 0 \end{aligned} \end{cases} . \qquad (5.70)$$

At this point we can distinguish two possible cases. The uniform asymptotic stability in the large case and the violation of the uniform asymptotic stability in the large condition.

5.3.2.1 The Uniform Asymptotic Stability in the Large Case

For completeness, we introduce from Michel and Miller (1977), the following definitions that are crucial to our discussion.

Definition 5.1 The equilibrium point $x_{\mathrm{ud}} = 0$ is said to be uniformly asymptotically stable in the large case if

1. for every $M > 0$ and any $t_0 \in \Re^+$, there exists $\bar{\alpha}(M) > 0$ such that $|x_{\mathrm{ud}}(t; x_{\mathrm{ud}}(0), t_0)| < M$ for all $t \geq t_0$ whenever $|x_{\mathrm{ud}}(0)| < \bar{\alpha}(M)$
2. for every $\bar{\alpha} > 0$ and any $t_0 \in \Re^+$, there exists $M(\bar{\alpha}) > 0$ such that $|x_{\mathrm{ud}}(t; x_{\mathrm{ud}}(0), t_0)| < M(\bar{\alpha})$ for all $t \geq t_0$ whenever $|x_{\mathrm{ud}}(0)| < \bar{\alpha}$
3. for any $\bar{\alpha}$, any $M > 0$ and $t_0 \in \Re^+$, there exists $T(M, \bar{\alpha}) > 0$, independent of t_0 such that, if $|x_{\mathrm{ud}}(0)| < \bar{\alpha}$ then $|x_{\mathrm{ud}}(t; x_{\mathrm{ud}}(0), t_0)| < M$ for all $t \geq t_0 + T(M, \bar{\alpha})$

For the state of the unmodeled dynamics we make the following assumption:

Assumption 17 The origin $x_{\mathrm{ud}} = 0$ of the unmodeled dynamics is uniformly stable in the large. More specifically, there is a C^1 function $L(x_{\mathrm{ud}})$ from $\Re^p \to \Re^+$ and continuous, strictly increasing, scalar functions $\gamma_i(|x_{\mathrm{ud}}|)$ from $\Re^+ \to \Re^+$, $i = 1, 2, 3$, which satisfy

$$\gamma_i(0) = 0, \, i = 1, 2, 3 \quad \text{and} \quad \lim_{s \to \infty} \gamma_i(s) = \infty, \quad i = 1, 2$$

such that for $x_{\mathrm{ud}} \in \Re^p$,

$$\gamma_1 (|x_{\mathrm{ud}}|) \leq L(x_{\mathrm{ud}}) \leq \gamma_2 (|x_{\mathrm{ud}}|)$$

and

$$\frac{\vartheta L}{\vartheta x_{\mathrm{ud}}} B(x, x_{\mathrm{ud}}) \leq -\gamma_3 (|x_{\mathrm{ud}}|) . \tag{5.71}$$

Employing Assumption 17, Eq. (5.65) becomes

$$
\begin{aligned}
\dot{V} &\leq -\lambda_{\min} (K) \|\xi\|^2 - \lambda_{\min} (K) \|\zeta\|^2 - \xi^T \phi (x, x_{\mathrm{ud}}) + \frac{\partial L}{\partial x_{\mathrm{ud}}} B (x, x_{\mathrm{ud}}) \\
&\leq -\lambda_{\min} (K) \|\xi\|^2 - \lambda_{\min} (K) \|\zeta\|^2 - \xi^T \phi (x, x_{\mathrm{ud}}) - \gamma_3 (|x_{\mathrm{ud}}|) \\
&\leq -\lambda_{\min} (K) \|\xi\|^2 - \lambda_{\min} (K) \|\zeta\|^2 - \xi^T \phi (x, x_{\mathrm{ud}}) \\
&\leq -\lambda_{\min} (K) \|\xi\|^2 - \lambda_{\min} (K) \|\zeta\|^2 + \|\xi\| \|\phi (x, x_{\mathrm{ud}})\| .
\end{aligned}
\tag{5.72}
$$

To continue we consider also the following assumption:

Assumption 18 Assume that the unknown vector field $\phi (x, x_{\mathrm{ud}})$ satisfies the condition

$$\|\phi (x, x_{\mathrm{ud}})\| \leq k_\phi \|x\| \|\phi_{\mathrm{ud}} (x_{\mathrm{ud}})\| \tag{5.73}$$

with $\phi_{\mathrm{ud}} (x_{\mathrm{ud}})$ an unknown vector field that depends only on x_{ud}. We further assume that $\phi_{\mathrm{ud}} (x_{\mathrm{ud}})$ is bounded uniformly by a constant θ. Hence,

$$\|\phi_{\mathrm{ud}} (x_{\mathrm{ud}})\| \leq \theta. \tag{5.74}$$

In the sequel, employing Eqs. (5.73), (5.74), Eq. (5.72) becomes

$$\dot{V} \leq -\lambda_{\min} (K) \|\xi\|^2 - \lambda_{\min} (K) \|\zeta\|^2 + k_\phi \theta \|\xi\| \|x\| . \tag{5.75}$$

Moreover, if we apply $\|x\| \leq \|\zeta\| + \|\xi\|$ then Eq. (5.75) becomes

$$\dot{V} \leq - \begin{bmatrix} \|\xi\| & \|\zeta\| \end{bmatrix} \begin{bmatrix} \lambda_{\min}(K) - k_\phi \theta & -k_\phi \theta \\ 0 & \lambda_{\min}(K) \end{bmatrix} \begin{bmatrix} \|\xi\| \\ \|\zeta\| \end{bmatrix}. \tag{5.76}$$

Hence, if we choose $\lambda_{\min}(K) \geq k_\phi \theta$ then Eq. (5.76) becomes negative. Thus, we have

$$\dot{V} \leq 0. \tag{5.77}$$

We are now ready to prove the following theorem:

Theorem 5.4 *Consider the closed-loop system*

$$\dot{x} = Ax + X_f W_f^* s_f(x) + X_g W_g^* S_g(x)u + \phi(x, x_{\mathrm{ud}})$$

$$\dot{x}_{\mathrm{ud}} = B(x, x_{\mathrm{ud}})$$

$$\dot{\zeta} = -K\zeta$$

$$u = -\left[X_g W_g S_g(x)\right]^+ \left[X_f W_f s_f(x) + v\right]$$

$$v = (K + A)x$$

$$\xi = \zeta - x$$

together with the updating laws (5.68) and (5.69) guarantee the following properties:

- $\xi, x, x_{\mathrm{ud}}, W_f, W_g \in L_\infty, \quad |\xi| \in L_2,$
- $\lim_{t\to\infty} \xi(t) = 0, \quad \lim_{t\to\infty} x(t) = 0,$
- $\lim_{t\to\infty} \dot{W}_f(t) = 0, \quad \lim_{t\to\infty} \dot{W}_g(t) = 0,$

provided that $\lambda_{\min}(K_c) > k_\phi\theta$ *and Assumption* 18 *is satisfied.*

Proof From Eq. (5.77), we have that $V \in L_\infty$ which implies $\xi, \zeta, \tilde{W}_f, \tilde{W}_g \in L_\infty$. Furthermore, $W_f = \tilde{W}_f + W_f^* \in L_\infty$ and $W_g = \tilde{W}_g + W_g^* \in L_\infty$. Since $\xi, \zeta \in L_\infty$ and $\xi = \zeta - x$, this also implies that $x \in L_\infty$. Moreover, since V is a monotone decreasing function of time and bounded from below, the $\lim_{t\to\infty} V(t) = V_\infty$ exists; so by integrating $\dot{V}$ from 0 to ∞ we have

$$\left(\lambda_{\min}(K) - k_\phi\theta\right) \int_0^\infty \|\xi\|^2 \, \mathrm{d}t + \lambda_{\min}(K) \int_0^\infty \|\zeta\|^2 \, \mathrm{d}t - k_\phi\theta \int_0^\infty \|\xi\|\,\|\zeta\| \, \mathrm{d}t \leq [V(0) - V_\infty] < \infty$$

which implies that $|\xi| \in L_2$. We also have that

$$\dot{\xi} = -K\xi + X_f \tilde{W}_f s_f(x) + X_g \tilde{W}_g S_g(x)u - \phi(x, x_{\mathrm{ud}}).$$

Hence and since $u \in L_\infty$, the sigmoidals and the fuzzy center matrices are bounded by definition, $\tilde{W}_f, \tilde{W}_g \in L_\infty$ and Assumption 18 holds, so since $\xi \in L_2 \cap L_\infty$ and $\dot{\xi} \in L_\infty$, applying Barbalat's Lemma we conclude that $\lim_{t\to\infty} \xi(t) = 0$.

Now, using the boundedness of $u, s_f(x), S_g(x), x$ and the convergence of $\xi(t)$, $\zeta(t)$ to zero, we have that $\dot{W}_f, \dot{W}_g$ also converge to zero. Hence, we have that

$$\lim_{t\to\infty} x(t) = \lim_{t\to\infty} \zeta(t) - \lim_{t\to\infty} \xi(t) = 0.$$

Thus,

$$\lim_{t\to\infty} x(t) = 0.$$

To continue, we make the following assumption:

Assumption 19 Now assume that the unknown vector field $\phi(x, x_{\mathrm{ud}})$ satisfies the condition

$$\|\phi(x, x_{\mathrm{ud}})\| \leq \theta_{\mathrm{ud}}. \tag{5.78}$$

In other words, $\phi(x, x_{\mathrm{ud}})$ is assumed to be bounded uniformly by a constant.

Thus, Eq. (5.72) becomes

$$\dot{V} \leq -\lambda_{\min}(K)\|\xi\|^2 - \lambda_{\min}(K)\|\zeta\|^2 + \theta_{ud}\|\xi\|,$$

which can be rewritten as

$$\dot{V} \leq -(\lambda_{\min}(K)\|\xi\| - \theta_{ud})\|\xi\| - \lambda_{\min}(K)\|\zeta\|^2, \tag{5.79}$$

and finally gives

$$\dot{V} \leq 0, \tag{5.80}$$

provided that

$$\|\xi\| > \frac{\theta_{ud}}{\lambda_{\min}(K)}, \tag{5.81}$$

with $\lambda_{\min}(K) > 0$. In the sequel, inequality (5.81) together with $\|x\| \leq \|\zeta\| + \|\xi\|$ leads to the conclusion that the trajectories of $\xi(t)$ and $x(t)$ lie in a small region around zero, (according to sufficiently large selection of K) and may take values from the sets Ξ and X defined as

$$\Xi = \left\{ \xi(t) : \|\xi\| \leq \frac{\theta_{ud}}{\lambda_{\min}(K)}, \ \lambda_{\min}(K) > 0 \right\},$$

and

$$X = \left\{ x(t) : \|x(t)\| \leq 2\frac{\theta_{ud}}{\lambda_{\min}(K)}, \ \lambda_{\min}(K) > 0 \right\}.$$

Thus, we have proven the following theorem:

Theorem 5.5 *Consider the approximated system* (5.51), (5.52) *and the unmodeled dynamics that satisfies Assumption 13. Assume also that Eq. (5.78) holds for the unknown vector field* $\phi(x, x_{ud})$. *Then, the control law* (5.58), (5.59) *in conjunction with the updating laws* (5.68) *and* (5.69) *guarantees the uniform ultimate boundedness with respect to the sets*

- $\Xi = \left\{ \xi(t) : \|\xi\| \leq \frac{\theta_{ud}}{\lambda_{\min}(K)}, \ \lambda_{\min}(K) > 0 \right\}$,
- $X = \left\{ x(t) : \|x(t)\| \leq 2\frac{\theta_{ud}}{\lambda_{\min}(K)}, \ \lambda_{\min}(K) > 0 \right\}$.

Furthermore,

$$\dot{\xi} = -K\xi + X_f \tilde{W}_f s_f(x) + X_g \tilde{W}_g S_g(x)u - \phi(x, x_{ud}).$$

As could be observed, since X_f, X_g, s_f, S_g are bounded by definition, the boundedness of $\tilde{W}_f$ and $\tilde{W}_g$ is ensured by the use of the hopping algorithm and

$\phi(x, x_{\mathrm{ud}})$ is bounded uniformly, we conclude that $\dot{\xi} \in L_{\infty}$. So, we are now ready to state the following remark.

Remark 5.3 The results of the above analysis denote that the appropriate selection of the design constant matrix K is very important. The wise selection of K elements makes the adaptive regulator capable of converging the system states x to zero, or at least able to ensure uniform ultimate boundedness of x and all other signals in the closed-loop. Also, by appropriate selection of small values for the constants k_{ϕ}, θ, $\bar{\theta}_{\mathrm{ud}}$ we can avoid any implementation issues that could arise if the elements of matrix K are required to have large values. Moreover, the more the NF model matches the input–output behavior of the true unknown nonlinear dynamical system, the better the control results. Finally, it is obvious that when the unknown vector field $\phi(x, x_{\mathrm{ud}})$ is uniformly bounded by a constant, any positive value for each k_i element of K, guarantees uniform boundedness of states x.

5.3.2.2 Violation of the Uniform Asymptotic Stability in the Large Condition

In this section, we investigate the effect of the unmodeled dynamics on the stability of the closed-loop system, when the uniform asymptotic stability in the large condition is violated , namely Eq. (5.71). Thus, we assume that instead of Eq. (5.71) we have

$$\frac{\vartheta L}{\vartheta x_{\mathrm{ud}}} B(x, x_{\mathrm{ud}}) \leq -\gamma_3 \left(\|x_{\mathrm{ud}}\|\right) + \rho \|x\|^2, \tag{5.82}$$

where ρ is a positive constant. Employing Eq. (5.82) within Eq. (5.65) we have

$$\dot{V} \leq -\lambda_{\min}(K) \|\xi\|^2 - \lambda_{\min}(K) \|\zeta\|^2 - \xi^T \phi(x, x_{\mathrm{ud}}) + \frac{\partial L}{\partial x_{\mathrm{ud}}} B(x, x_{\mathrm{ud}})$$

$$\leq -\lambda_{\min}(K) \|\xi\|^2 - \lambda_{\min}(K) \|\zeta\|^2 - \xi^T \phi(x, x_{\mathrm{ud}}) - \gamma_3 \left(\|x_{\mathrm{ud}}\|\right) + \rho \|x\|^2$$

$$\leq -\lambda_{\min}(K) \|\xi\|^2 - \lambda_{\min}(K) \|\zeta\|^2 + \|\xi\| \|\phi(x, x_{\mathrm{ud}})\| + \rho \|x\|^2. \tag{5.83}$$

As previously, we shall consider the following cases:

Case 1: Assume that the unknown vector field $\phi(x, x_{\mathrm{ud}})$ satisfies Eqs. (5.73), (5.74). Then Eq. (5.83) becomes

$$\dot{V} \leq -\lambda_{\min}(K) \|\xi\|^2 - \lambda_{\min}(K) \|\zeta\|^2 + k_{\phi}\theta \|\xi\| \|x\| + \rho \|x\|^2. \tag{5.84}$$

Hence, we can distinguish the following sub-cases:

A: If $x \geq 0$ we have that $\zeta(t) \geq \xi(t)$ but $\zeta(t) \leq 0$, $\forall t \geq 0$ which implies that $\|\zeta(t)\| \leq \|\xi(t)\|$. So, we have

$$\|x\| \leq \|\zeta\| + \|\xi\| \leq 2 \|\xi\|. \tag{5.85}$$

Therefore, Eq. (5.84) becomes

$$\dot{V} \leq -\lambda_{\min}(K) \|\xi\|^2 - \lambda_{\min}(K) \|\zeta\|^2 + 2k_\phi\theta \|\xi\|^2 + 4\rho \|\xi\|^2$$
$$\leq -\left(\lambda_{\min}(K) - 2k_\phi\theta - 4\rho\right) \|\xi\|^2 - \lambda_{\min}(K) \|\zeta\|^2 \leq 0, \tag{5.86}$$

provided that

$$\lambda_{\min}(K) > 2\left(k_\phi\theta + 2\rho\right). \tag{5.87}$$

B: If $x < 0$ we have that $\zeta(t) < \xi(t)$ but $\zeta(t) \leq 0$, $\forall t \geq 0$ which implies that $\|\zeta(t)\| > \|\xi(t)\|$. So, we have:

$$\|x\| \leq \|\zeta\| + \|\xi\| \leq 2 \|\zeta\|. \tag{5.88}$$

Therefore, Eq. (5.84) becomes

$$\dot{V} \leq -\lambda_{\min}(K) \|\xi\|^2 - \lambda_{\min}(K) \|\zeta\|^2 + 2k_\phi\theta \|\xi\| \|\zeta\| + 4\rho \|\zeta\|^2$$
$$\leq -\left[\|\xi\| \ \|\zeta\| \right] \begin{bmatrix} \lambda_{\min}(K) & -2k_\phi\theta \\ 0 & \lambda_{\min}(K) - 4\rho \end{bmatrix} \begin{bmatrix} \|\xi\| \\ \|\zeta\| \end{bmatrix} \leq 0, \tag{5.89}$$

provided that

$$\lambda_{\min}(K) > 4\rho. \tag{5.90}$$

Conclusively, $\forall x \in \Re^n$ the Lyapunov candidate function becomes negative when $\lambda_{\min}(K) > 2\left(k_\phi\theta + 2\rho\right)$.

Case 2: Now assume that the unknown vector field $\phi(x, x_{\text{ud}})$ satisfies Eq. (5.78). Then Eq. (5.83) becomes

$$\dot{V} \leq -\lambda_{\min}(K) \|\xi\|^2 - \lambda_{\min}(K) \|\zeta\|^2 + \theta_{\text{ud}} \|\xi\| + \rho \|x\|^2 \tag{5.91}$$

Hence, we can distinguish the following sub-cases:

A: If $x \geq 0$ we have that $\zeta(t) \geq \xi(t)$ but $\zeta(t) \leq 0$, $\forall t \geq 0$ which implies that $\|\zeta(t)\| \leq \|\xi(t)\|$. So, we have

$$\|x\| \leq \|\zeta\| + \|\xi\| \leq 2 \|\xi\|. \tag{5.92}$$

Therefore, Eq. (5.91) becomes

$$\dot{V} \leq -\lambda_{\min}(K) \|\xi\|^2 - \lambda_{\min}(K) \|\zeta\|^2 + \theta_{\text{ud}} \|\xi\| + 4\rho \|\xi\|^2$$
$$\leq -\left[(\lambda_{\min}(K) - 4\rho) \|\xi\| - \theta_{\text{ud}}\right] \|\xi\| - \lambda_{\min}(K) \|\zeta\|^2 \leq 0, \tag{5.93}$$

provided that

$$\|\xi\| > \frac{\theta_{\mathrm{ud}}}{\lambda_{\min}(K) - 4\rho}, \tag{5.94}$$

with $\lambda_{\min}(K) > 4\rho$.

B: If $x < 0$ we have that $\zeta(t) < \xi(t)$ but $\zeta(t) \leq 0$, $\forall t \geq 0$ which implies that $\|\zeta(t)\| > \|\xi(t)\|$. So, we have

$$\|x\| \leq \|\zeta\| + \|\xi\| \leq 2\|\zeta\|. \tag{5.95}$$

Therefore, Eq. (5.91) becomes

$$\begin{aligned}
\dot{V} &\leq -\lambda_{\min}(K)\|\xi\|^2 - \lambda_{\min}(K)\|\zeta\|^2 + \theta_{\mathrm{ud}}\|\xi\| + 4\rho\|\zeta\|^2 \\
&\leq -(\lambda_{\min}(K)\|\xi\| - \theta_{\mathrm{ud}})\|\xi\| - (\lambda_{\min}(K) - 4\rho)\|\zeta\|^2 \leq 0, \tag{5.96}
\end{aligned}$$

provided that

$$\|\xi\| > \frac{\theta_{\mathrm{ud}}}{\lambda_{\min}(K)}, \tag{5.97}$$

with $\lambda_{\min}(K) > 0$.

Conclusively, $\forall x \in \Re^n$ the Lyapunov candidate function becomes negative when $\|\xi\| > \frac{\theta_{\mathrm{ud}}}{\lambda_{\min}(K) - 4\rho}$ and $\lambda_{\min}(K) > 4\rho$.

In the sequel, inequalities (5.94), (5.97) together with Eqs. (5.92), (5.95) lead to the conclusion that the trajectories of $\xi(t)$ and $x(t)$ are uniformly bounded in a small region around zero, (according to sufficiently large eigenvalues of K). Therefore, sets Ξ and X can be written as

$$\Xi = \left\{ \xi(t) : \|\xi(t)\| \leq \frac{\theta_{\mathrm{ud}}}{\lambda_{\min}(K) - 4\rho}, \ \lambda_{\min}(K) > 4\rho \right\},$$

and

$$X = \left\{ x(t) : \|x(t)\| \leq \frac{2\theta_{\mathrm{ud}}}{\lambda_{\min}(K) - 4\rho}, \ \lambda_{\min}(K) > 4\rho \right\}.$$

Thus, we have proved the following theorem:

Theorem 5.6 *Consider the approximated system* (5.51), (5.52) *and the unmodeled dynamics that satisfies Eq.* (5.73). *Also, assume that Eq.* (5.74) *holds for the unknown vector field* $\phi(x, x_{\mathrm{ud}})$. *Then the control law* (5.58), (5.59) *in conjunction with the updating laws* (5.68) *and* (5.69) *guarantees the uniform ultimate boundedness with respect to the sets*

- $\Xi = \left\{ \xi(t) : \|\xi(t)\| \leq \frac{\theta_{\mathrm{ud}}}{\lambda_{\min}(K) - 4\rho}, \ \lambda_{\min}(K) > 4\rho \right\}$,
- $X = \left\{ x(t) : \|x(t)\| \leq \frac{2\theta_{\mathrm{ud}}}{\lambda_{\min}(K) - 4\rho}, \ \lambda_{\min}(K) > 4\rho \right\}$.

Furthermore,

$$\dot{\xi} = -K\xi + X_f \tilde{W}_f S_f(x) + X_g \tilde{W}_g S_g(x)u - \phi(x, x_{\mathrm{ud}}).$$

As could be observed, since X_f, X_g, s_f, S_g are bounded by definition, the boundedness of $\tilde{W}_f$ and $\tilde{W}_g$ is ensured by the use of the modified hopping algorithm and $\phi(x, x_{\mathrm{ud}})$ is bounded uniformly, we conclude that $\dot{\xi} \in L_\infty$. Thus, we are now ready to state one more remark as follows:

Remark 5.4 Theorem 5.6 demonstrates that the violation of the uniform asymptotic stability in the large condition, does not prevent the adaptive regulator from assuring the uniform ultimate boundedness of the states x, and of all signals in the closed-loop. Thus, the more careful selection of the design constant elements k_i of matrix K leads to better performance of the controller, provided that the implementation constraints are satisfied. Finally, the accuracy of the NF approximating model is a performance index of the adaptive controller.

5.4 State Trajectory Tracking

We are now ready to extend the application of the proposed controller of a multivariable multi-input system such that the system states track a reference signal. We start by writing down the general form of the dynamic state equations of an affine in the control nonlinear dynamical system

$$\dot{x} = f(x) + G(x)u, \tag{5.98}$$

where the state $x \in \mathfrak{R}^n$ is assumed to be completely measured, the control u is in R^m, $f(x) \in \mathfrak{R}^n$ is an unknown smooth vector field called the drift term, and $G(x) \in \mathfrak{R}^{q \times n}$ is a matrix with rows containing the unknown smooth controlled vector fields with elements g_{ij}.

Since the plant is considered unknown, we assume that the unknown plant can be described by the following NF model (Boutalis et al. 2009; Theodoridis et al. 2012) plus a modeling error term $\mu(x, u)$, where the weight values $W_f^\star$ and $W_g^\star$ are unknown.

$$\dot{x} = Ax + X_f W_f^\star s_f(x) + X_g W_g^\star S_g(x)u + \mu(x, u). \tag{5.99}$$

Since $W_f^\star$ and $W_g^\star$ are unknown, our solution consists of designing a control law $u(W_f, W_g, e, \dot{x}_d)$ and appropriate updating laws for W_f and W_g to guarantee convergence of states to the desired trajectory and boundedness of x and of all other signals in the closed-loop.

First, we examine the case where we have complete model matching or nonzero NF approximation error of the unknown nonlinear dynamical system.

5.4.1 The Complete Model Matching Case

In this section, we present a solution to the tracking problem without the presence of modeling error effects, assuming that the unknown system is given by Eq. (5.99) with $\mu(x, u) = 0$. The control objective is to force the state vector x to follow a desired trajectory x_d of a stable model having the following form:

$$\dot{x}_d = f(x_d, u_d, t). \tag{5.100}$$

where x_d and $\dot{x}_d$ are known quantities.

Define the tracking error e as:

$$e = x_d - x. \tag{5.101}$$

Then, taking the derivative of Eq. (5.101) and substituting Eqs. (5.99), (5.100) we obtain

$$\dot{e} = \dot{x}_d - \left(Ax + X_f W_f^\star s_f(x) + X_g W_g^\star S_g(x)u \right). \tag{5.102}$$

Define now r as

$$r \triangleq X_f W_f s_f(x) + X_g W_g S_g(x)u + \dot{e} - \dot{x}_d + Ax \tag{5.103}$$

substituting Eq. (5.102) into Eq. (5.103) we have

$$r \triangleq X_f \tilde{W}_f s_f(x) + X_g \tilde{W}_g S_g(x)u, \tag{5.104}$$

where $\tilde{W}_f = W_f - W_f^*$ and $\tilde{W}_g = W_g - W_g^*$. W_f and W_g are estimates of W_f^* and W_g^*, respectively, and are obtained by updating laws which are to be designed in the sequel. r cannot be measured since $\dot{e}$ (and $\dot{x}$, respectively) is unknown. To overcome this problem, we use the following filtered version of r:

$$r = \dot{\xi} + K\xi$$

where $K = \begin{bmatrix} k_1 & 0 & \cdots & 0 \\ 0 & k_2 & 0 & \cdots \\ \cdots & 0 & \cdots & 0 \\ 0 & \cdots & 0 & k_n \end{bmatrix}$ is a diagonal positive definite matrix and $\xi \in \Re^n$. In the sequel, according to Eq. (5.103) we have that

$$\dot{\xi} + K\xi = \dot{e} - \dot{x}_d + Ax + X_f W_f s_f(x) + X_g W_g S_g(x)u, \tag{5.105}$$

and after substituting Eq. (5.102) we have

$$\dot{\xi} = -K\xi + X_f \tilde{W}_f s_f(x) + X_g \tilde{W}_g S_g(x)u. \tag{5.106}$$

To implement Eq. (5.106), we take

$$\xi \triangleq e - \zeta. \qquad (5.107)$$

Employing Eq. (5.107) to Eq. (5.105) we have

$$-\dot{\zeta} - K\zeta = -Ke - \dot{x}_d + Ax + X_f W_f S_f(x) + X_g W_g S_g(x)u, \qquad (5.108)$$

with state $\zeta \in \Re^n$. This method is referred to as error filtering.

The trajectory tracking of the system can be achieved by selecting the control input to be

$$u = \left[X_g W_g S_g(x)\right]^+ \left[-X_f W_f s_f(x) + Ke - Ax + \dot{x}_d\right], \qquad (5.109)$$

where $[\cdot]^+$ means pseudo-inverse in Moore–Penrose sense.

Taking into account Eq. (5.109), Eq. (5.108) becomes

$$\dot{\zeta} = -K\zeta. \qquad (5.110)$$

To continue, consider the Lyapunov candidate function

$$V = \frac{1}{2}\xi^T \xi + \frac{1}{2}\zeta^T \zeta + \frac{1}{2}\mathrm{tr}\left\{\tilde{W}_f^T D_f^{-1} \tilde{W}_f\right\} + \frac{1}{2}\mathrm{tr}\left\{\tilde{W}_g^T D_g^{-1} \tilde{W}_g\right\}. \qquad (5.111)$$

If we take the derivative of Eq. (5.111) with respect to time and substitute Eqs. (5.106), (5.110), we obtain

$$\dot{V} = -\xi^T K\xi - \zeta^T K\zeta + \xi^T X_f \tilde{W}_f s_f(x) + \xi^T X_g \tilde{W}_g S_g(x)u$$
$$+ \mathrm{tr}\left\{\dot{\tilde{W}}_f^T D_f^{-1} \tilde{W}_f\right\} + \mathrm{tr}\left\{\dot{\tilde{W}}_g^T D_g^{-1} \tilde{W}_g\right\}. \qquad (5.112)$$

Hence, if we choose

$$\mathrm{tr}\left\{\dot{\tilde{W}}_f^T D_f^{-1} \tilde{W}_f\right\} = -\xi^T X_f \tilde{W}_f s_f(x), \qquad (5.113)$$

$$\mathrm{tr}\left\{\dot{\tilde{W}}_g^T D_g^{-1} \tilde{W}_g\right\} = -\xi^T X_g \tilde{W}_g S_g(x)u, \qquad (5.114)$$

$\dot{V}$ becomes

$$\dot{V} \leq -\lambda_{\min}(K) \|\xi\|^2 - \lambda_{\min}(K) \|\zeta\|^2. \qquad (5.115)$$

It can be easily verified that Eqs. (5.113) and (5.114) after making the appropriate operations according to matrix trace properties, result in the following weight updating laws:

$$\dot{W}_f = -X_f^T \xi s_f^T(x) D_f, \qquad (5.116)$$

$$\dot{W}_g = -X_g^T \xi u^T S_g^T(x) D_g. \tag{5.117}$$

As it will be proved in Theorem 5.7 below, the weight updating laws for the column submatrices of W_f, W_g are given by

$$\dot{W}_{f_i}^l = -\left(\bar{x}_{f_i}\right)^T \xi_i s_l(x) d_{f_l}, \tag{5.118}$$

and

$$\dot{W}_{g_{ij}} = -\left(\bar{x}_{g_{ij}}\right)^T \xi_i u_j s_i(x) d_{g_{ij}}, \tag{5.119}$$

can serve the purpose of driving e to zero.

Following the same lines as in the SISO case, Chap. 4, we have to ensure the existence of the control signal. Thus, the updating law concerning the weights of W_g is modified as follows:

$$\dot{W}_{g_{ij}} = -\left(\bar{x}_{g_{ij}}\right)^T \xi_i u_j s_i(x) d_{g_{ij}} - \frac{2\sigma_{g_{ij}} \kappa_{g_{ij}}^{\text{inner}} \left(\bar{x}_{g_{ij}} W_{g_{ij}} (\bar{x}_{g_{ij}})^T\right)}{\left|\bar{x}_{g_{ij}}\right|^2}, \tag{5.120}$$

where $\sigma_{g_{ij}}$ is defined as (4.7) and $\kappa_{g_{ij}}^{\text{inner}}$ is a positive constant ($1 < \kappa_{g_{ij}}^{\text{inner}} < 1.5$) decided from the designer according to Chap. 3 and Fig. 3.5. So, we are now ready to state the following theorem:

Theorem 5.7 *The closed-loop system given by (5.99), (5.100), (5.102), (5.109) together with the updating laws (5.118), (5.120), guarantees the following properties:*

1. $\xi, \zeta, e, \tilde{W}_f, \tilde{W}_g \in L_\infty, \quad \xi, e \in L_2$,
2. $\lim_{t \to \infty} e(t) = 0$,
3. $\lim_{t \to \infty} \dot{W}_f(t) = 0, \quad \lim_{t \to \infty} \dot{W}_g(t) = 0$.

Proof The proof of the above Theorem 5.7 is similar to that of Theorem 4.3 and is not repeated here.

5.4.2 Inclusion of a Non-zero Approximation Error

Let us assume that we have a nonzero NF approximation error $\mu(x, u)$ which is bounded by a small positive number $\bar{\mu}$ (Theodoridis et al. 2012). Thus, we use Eq. (5.99) in addition to the control law:

$$u = u_c + u_s = \left[X_g W_g S_g(x)\right]^+ \left[-X_f W_f s_f(x) + Ke - Ax + \dot{x}_d + u_r\right], \tag{5.121}$$

which takes into account the sliding mode control term $u_r = -K_r \frac{\xi^T sgn(\xi)}{\zeta^T \zeta + \delta} \zeta$, with δ a very small positive number $(0 < \delta \ll 1)$ and $u_s = \left[\bar{x}_g W_g s_g(x)\right]^+ u_r$. Now, we can distinguish two possible cases.

Case 1: Without the sliding control term u_r.

The time derivative of the Lyapunov function (5.115) becomes

$$
\begin{aligned}
\dot{V} &\le -\lambda_{\min}(K)\|\xi\|^2 - \lambda_{\min}(K)\|\zeta\|^2 - \xi^T \mu \\
&\le -\lambda_{\min}(K)\|\xi\|^2 - \lambda_{\min}(K)\|\zeta\|^2 + \|\xi\|\bar{\mu} \\
&\le -\|\xi\|\left(\lambda_{\min}(K)\|\xi\| - \bar{\mu}\right) - \lambda_{\min}(K)\|\zeta\|^2.
\end{aligned}
\tag{5.122}
$$

The above expression will be negative if $\|\xi\| > \frac{\bar{\mu}}{\lambda_{\min}(K)}$. In other words, this implies that the trajectory of $\xi(t)$ is uniformly bounded with respect to the arbitrarily small, (since $\lambda_{\min}(K)$ can be chosen sufficiently large), set $\varXi$ shown below:

$$
\varXi = \left\{ \xi(t) : \|\xi(t)\| \le \frac{\bar{\mu}}{\lambda_{\min}(K)}, \ \lambda_{\min}(K) > 0 \right\}.
$$

Case 2: With the sliding control term u_r.

The time derivative of the Lyapunov function (5.115) once again becomes

$$
\begin{aligned}
\dot{V} &\le -\lambda_{\min}(K)\|\xi\|^2 - \lambda_{\min}(K)\|\zeta\|^2 - \xi^T \mu - \zeta^T u_r \\
&\le -\lambda_{\min}(K)\|\xi\|^2 - \lambda_{\min}(K)\|\zeta\|^2 + \|\xi\|\bar{\mu} - K_r \frac{\|\xi\|\|\zeta\|^2}{\|\zeta\|^2 + \delta} \\
&\le -\lambda_{\min}(K)\|\xi\|^2 - \lambda_{\min}(K)\|\zeta\|^2 + \|\xi\|\left(\bar{\mu} - K_r \frac{\|\zeta\|^2}{\|\zeta\|^2 + \delta}\right).
\end{aligned}
\tag{5.123}
$$

The above expression will be negative by appropriately selecting K_r from the designer so that $K_r > \bar{\mu}$, assuming $\frac{\|\zeta\|^2}{\|\zeta\|^2 + \delta} \simeq 1$.

Continuously, one can easily verify that (5.109) is valid provided that $\bar{x}_{f_i} W^l_{f_i}$, $\bar{x}_{g_{ij}} W_{g_{ij}}$ are uniformly bounded by known positive constants ε^l_i, ρ_{ij} and additionally $\bar{x}_{g_{ij}} W_{g_{ij}}$ always exists. Thus, $\bar{x}_{f_i} W^l_{f_i}$ is confined to the set

$$
P_{f,\text{MIMO}} = \left\{ \bar{x}_{f_i} \cdot W^l_{f_i} : \left| \bar{x}_{f_i} \cdot W^l_{f_i} \right| \le \varepsilon^l_i \right\},
$$

while $\bar{x}_{g_{ij}} W_{g_{ij}}$ is confined to the set

$$
P_{g,\text{MIMO}} = \left\{ \bar{x}_{g_{ij}} \cdot W_{g_{ij}} : \left| \bar{x}_{g_{ij}} \cdot W_{g_{ij}} \right| \le \rho_{ij} \text{ and } \left| \bar{x}_{g_{ij}} \cdot W_{g_{ij}} \right| > \theta_{ij} \right\},
$$

through the use of a hopping algorithm. In particular, the standard updating law (4.79) is modified to:

$$\dot{W}_f^l = -\left(\bar{x}_{f_i}\right)^T \xi_i s_l(x) d_{f_l} - \frac{\sigma_{f_i}^l \kappa_{f_i}^{l,\text{outer}} \left(\bar{x}_{f_i} W_{f_i}^l (\bar{x}_{f_i})^T\right)}{\left|\bar{x}_{f_i}\right|^2}, \tag{5.124}$$

where

$$\sigma_{f_i}^l = \begin{cases} 0 & \begin{array}{l} \text{if } \bar{x}_{f_i} \cdot W_{f_i}^l \in P_1 \\ \text{or } \bar{x}_{f_i} \cdot W_{f_i}^l = \pm\varepsilon_i^l \\ \text{and } \bar{x}_{f_i} \cdot \dot{W}_{f_i}^l >< 0 \end{array} \\ \\ 1 & \text{otherwise} \end{cases} \tag{5.125}$$

and $\kappa_{f_i}^{l,\text{outer}}$ a positive constant decided from the designer according to Chap. 3 and Fig. 3.5. The weight updating law (5.120) incorporating the two hopping conditions (outer and inner hopping) can now be rewritten as

$$\dot{W}_{g_{ij}} = -\left(\bar{x}_{g_{ij}}\right)^T \xi_i u_j s_i(x) d_{g_{ij}} - \frac{2\sigma_{g_{ij}} \kappa_{g_{ij}}^{\text{inner}} \left(\bar{x}_{g_{ij}} W_{g_{ij}} (\bar{x}_{g_{ij}})^T\right)}{\left|\bar{x}_{g_{ij}}\right|^2}$$
$$-\frac{(1 - \sigma_{g_{ij}})\kappa_{g_{ij}}^{\text{outer}} \left(\bar{x}_{g_{ij}} W_{g_{ij}} (\bar{x}_{g_{ij}})^T\right)}{\left|\bar{x}_{g_{ij}}\right|^2}, \tag{5.126}$$

where

$$\sigma_{g_{ij}} = \begin{cases} 0 & \begin{array}{l} \text{if } \bar{x}_{g_{ij}} \cdot W_{g_{ij}} = \pm\rho_{ij} \\ \text{and } \bar{x}_{g_{ij}} \cdot \dot{W}_{g_{ij}} <> 0 \end{array} \\ \\ 1 & \begin{array}{l} \text{if } \bar{x}_{g_{ij}} \cdot W_{g_{ij}} = \pm\theta_{ij} \\ \text{and } \bar{x}_{g_{ij}} \cdot \dot{W}_{g_{ij}} >< 0 \end{array} \end{cases} \tag{5.127}$$

and $\kappa_{g_{ij}}^{\text{inner}}$, $\kappa_{g_{ij}}^{\text{outer}}$ are small positive constant values for the inner and outer hopping, respectively, chosen by the designer according to Chap. 3 and Fig. 3.5. The above hopping conditions are depicted in Fig. 3.4 where a simplified two-dimensional representation is given.

5.5 Simulation Results

In this section, we present a series of simulations demonstrating some aspects of the NF direct controllers presented in this chapter. As a test nonlinear system, we use the third-order dynamical equations of the benchmark chaotic "Lorenz system" and present its adaptive regulation in two test cases. In the first case we assume that we are confident about the order of the system, while in the second case we present results when the NF approximator assumes a reduced model order. Comparisons between the NF and a simple RHONN direct controller (Rovithakis and Christodoulou 2000) are given to demonstrate the performance superiority of the proposed NF formulation. Finally, we present state trajectory tracking simulations on the same system illustrating its performance and the stability of the closed-loop system during the tracking period. In this test case, we track several reference trajectories and we make comparisons again with simple RHONNs. The interested reader could find more results in relevant works of the authors [see for example in Theodoridis et al. (2011a, b, 2012)].

5.5.1 Exact Model Order

The Lorenz system was derived to model the two-dimensional convection flow of a fluid layer heated from below and cooled from above. The model represents the Earth's atmosphere heated by the ground's absorption of sunlight and losing heat into space. It can be described by the following dynamical equations:

$$\dot{x}_1 = \sigma(x_2 - x_1)$$
$$\dot{x}_2 = \rho x_1 - x_2 - x_1 x_3$$
$$\dot{x}_3 = -\beta x_3 + x_1 x_2 \tag{5.128}$$

where x_1, x_2 and x_3 represent measures of fluid velocity, horizontal, and vertical temperature variations, correspondingly. The parameters σ, ρ and β are positive which represent the Prandtl number, Rayleigh number, and geometric factor, correspondingly. Selecting $\sigma = 10$, $\rho = 28$ and $\beta = 8/3$ the system presents three unstable equilibrium points and the system trajectory wanders forever near a strange invariant set called strange attractor presenting thus a chaotic behavior, Yeap and Ahmed (1994). The same equations may also be applied in models of different systems like the models of lasers and dynamos and the model that describes the motion of a chaotic waterwheel (Strogatz 1994).

By incorporating control inputs, the dynamics are expressed as (Yeap and Ahmed 1994)

$$\dot{x}_1 = \sigma(x_2 - x_1) + u_1$$
$$\dot{x}_2 = \rho x_1 - x_2 - x_1 x_3 + u_2$$
$$\dot{x}_3 = -\beta x_3 + x_1 x_2 + u_3 \tag{5.129}$$

The control objective is to derive appropriate state feedback control law to regulate the system to one of its equilibria, which is $(0, 0, 0)$. In particular, we consider that (5.129) has the following initial condition:

$$x_0 = [-0.5, 0.8, 2]^T.$$

Fig. 5.1 Convergence of Lorenz system states x_1 (*red dotted line*), x_2 (*blue solid line*) and x_3 (*green dashed line*) to zero using the direct adaptive NF control

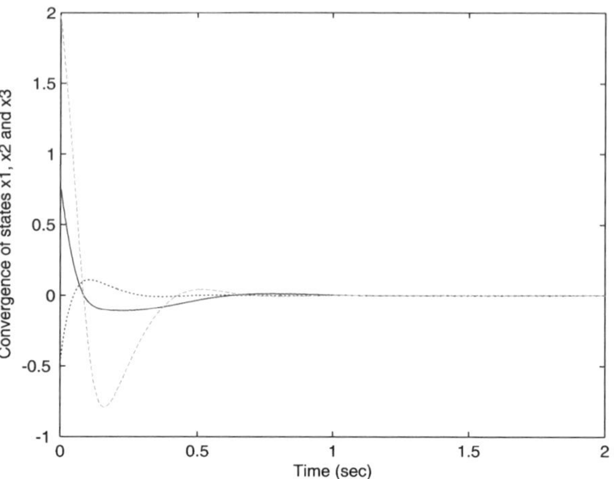

The main parameters for the control law (5.11), (5.12) and the learning laws (5.18), (5.22) are selected as

$$A = \text{diag}(-20, -10, -40)$$
$$K = \text{diag}(21, 38, 40)$$

and the parameters of the sigmoidals that have been used are $\alpha_1 = \alpha_2 = \alpha_3 = 1$, $\beta_1 = \beta_2 = \beta_3 = 1$ and $\gamma_1 = \gamma_2 = \gamma_3 = 0$.

In the sequel, we first assume that the system is described, within a degree of accuracy, by a NF system of the form (5.2) with the number of states being $n = 3$, the number of fuzzy partitions being $p = 5$ and the depth of high-order sigmoid terms $k = 9$. In this case $s_i(x)$ assume high-order connection up to the second order. Figure 5.1 shows the convergence of states x_1, x_2 and x_3 to zero exponentially fast. Also, Fig. 5.2 shows the smooth evolution of the control inputs.

5.5.2 Reduced Model Order

To test the proposed approach in its ability to control the Lorenz system even if the NF model is of lower order than the real system dynamics and to compare its performance against the use of a simple RHONN scheme we performed a number of simulations, which are analyzed below.

(A) We consider first that (5.129) assumes the following initial condition:

$$x_0 = [0.39, 0.59, 0.25]^T.$$

Fig. 5.2 Evolution of Lorenz system control inputs u_1 (*red dotted line*), u_2 (*blue solid line*) and u_3 (*green dashed line*) using the direct adaptive NF control

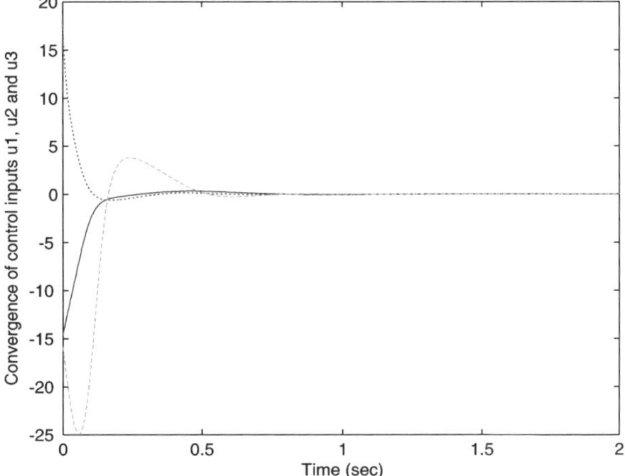

Fig. 5.3 A phase-space trajectory generated by the Lorenz chaotic system with $x_0 = [0.39, 0.59, 0.25]^T$

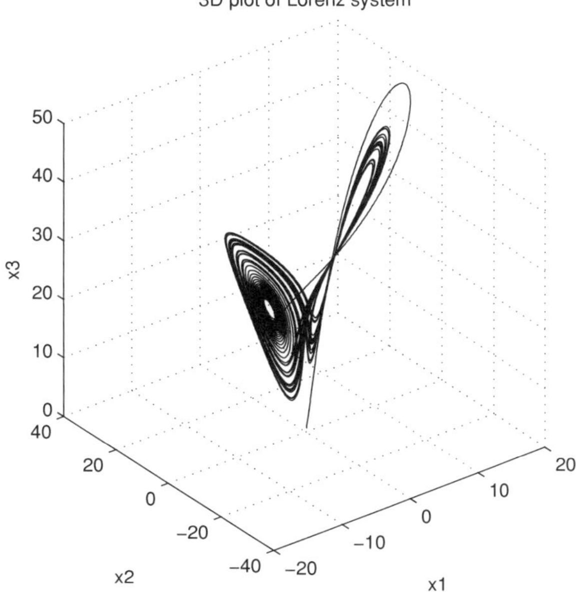

Figures 5.3 and 5.4 shows the evolution of Lorenz chaotic system states without using control signals.

The control objective is to derive appropriate state feedback control law to regulate the system to one of its equilibria, which is $(0, 0, 0)$ when state x_3 and its dynamics are completely omitted in the NF approximator. The main parameters for the control law (5.58), (5.59) and the learning laws (5.68), (5.69) are selected as

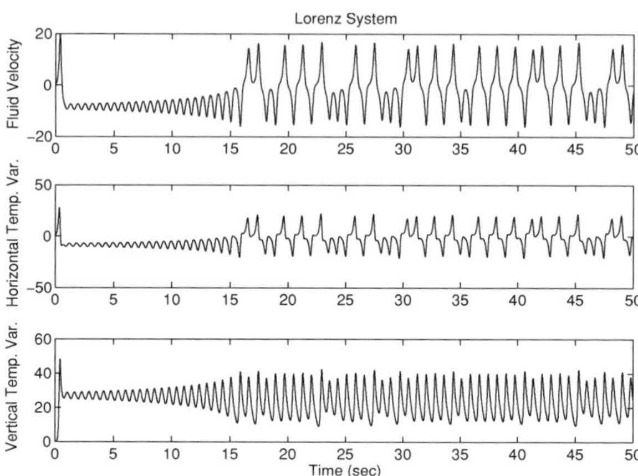

Fig. 5.4 Evolution of Lorenz chaotic system states with $x_0 = [0.39, 0.59, 0.25]^T$

$$A = \mathrm{diag}(-15, -15),$$

$$K = \mathrm{diag}(3, 4).$$

The parameters of the sigmoidals that have been used are $\alpha_1 = \alpha_2 = 1$, $\beta_1 = \beta_2 = 1$ and $\gamma_1 = \gamma_2 = 0$. Also, the learning rates are selected appropriately as $d_{f_l} = 0.1$ and $d_{g_i} = 1$.

In the sequel, we first assume that the system is described, within a degree of accuracy, by an NF system of the form (5.2) with the number of states being $n = 2$, the number of fuzzy partitions being $p = 5$, and the depth of high-order sigmoid terms $k = 5$. In this case $s_i(x)$ assume high-order connection up to the second order. Figures 5.5 and 5.6 shows the convergence of states x_1, x_2 and x_3 to zero exponentially fast when we assume the existence of the states x_1, x_2 only.

(B) In a second case, we assume that (5.129) has the following random initial condition:

$$x_0 = [-0.5, 0.8, 2]^T.$$

The main parameters for the control law (5.58), (5.59) and the learning laws (5.69), (5.68) are selected as

$$A = \mathrm{diag}(-15, -15),$$

$$K = \mathrm{diag}(4, 5).$$

The parameters of the sigmoidals that have been used in both methods are $\alpha_1 = \alpha_2 = 1$, $\beta_1 = \beta_2 = 1$ and $\gamma_1 = \gamma_2 = 0$. Also, the learning rates are selected appropriately as $d_{f_l} = 0.1$ and $d_{g_i} = 1$. Figure 5.7 shows the convergence of states x_1, x_2 and x_3 to zero exponentially fast when in the NF approximator (F-RHONN model) we assume the existence of only the states x_1, x_2, for our approach (solid line) in comparison with the simple RHONNs (dotted line) which denotes the

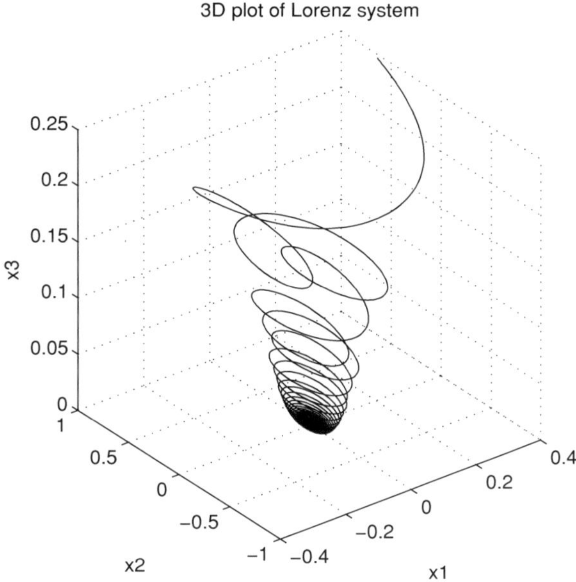

Fig. 5.5 Regulation of the Lorenz chaotic system when we assume only the existence of states x_1, x_2 and the initial condition $x_0 = [0.39, 0.59, 0.25]^T$

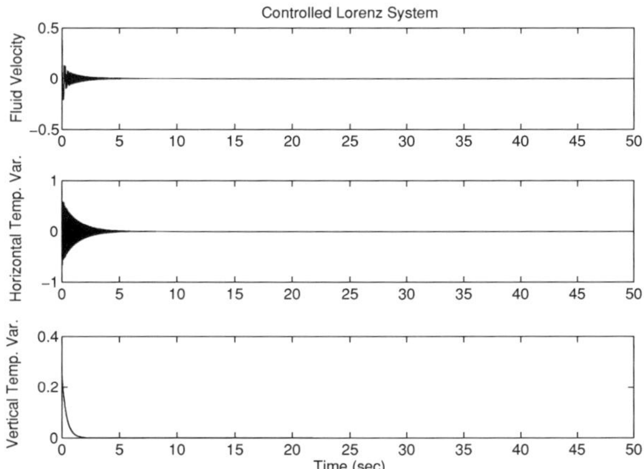

Fig. 5.6 Regulation of Lorenz chaotic system states

superiority of the proposed NF approach against the simple RHONN scheme. Regarding the parameters of RHONNs [which are extensively explained in Rovithakis and Christodoulou (2000)] that have been used in the above simulations, have the same values as F-RHONNs concerning the adaptive gains, the sigmoidals, and the matrices A, K.

Furthermore, as shown in Fig. 5.8, we have similar results when we assume the existence of states x_2, x_3 and ignore state x_1. In case x_2 is omitted (only x_1, x_3 are considered), Fig. 5.9 shows that the proposed NF approach performs equally well, but not the simple RHONN approach as shown in Fig. 5.10, which becomes unstable

Fig. 5.7 Evolution of Lorenz states to zero, for RHONNs (*dotted line*) and NF (*solid line*) when only x_1, x_2 dynamics are considered

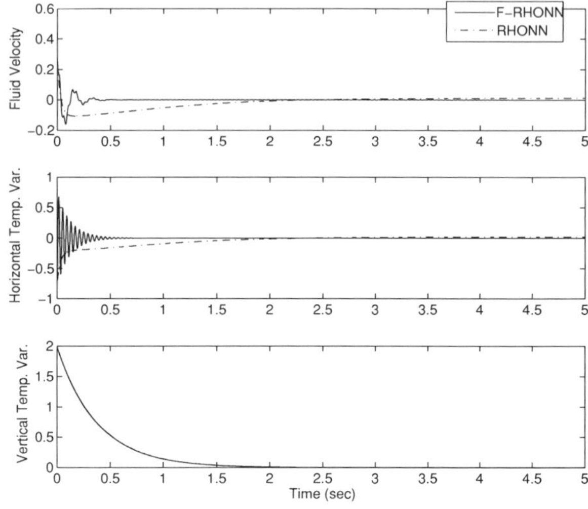

Fig. 5.8 Evolution of Lorenz states to zero, for RHONNs (*dotted line*) and F-RHONNs (*solid line*) when only x_2, x_3 dynamics are considered

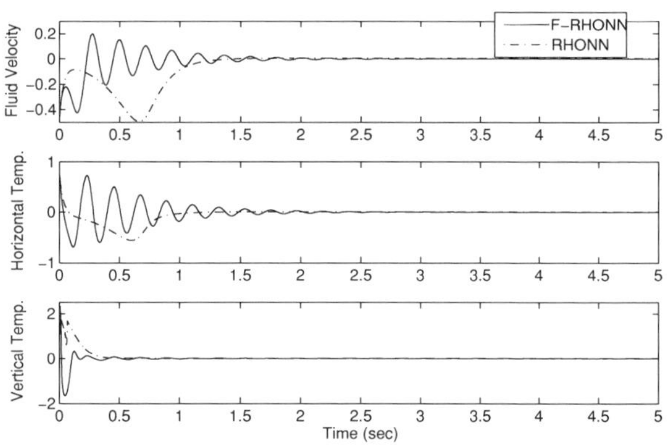

and can not be included in the figure. This is due to the faster convergence of the proposed approach, which prevents inherent marginal instabilities of the dynamics of x_2 from occurring. Conclusively, the proposed approach seems to be much more superior against RHONNs working quite well even in small values of k_i, which is desirable in order to avoid big oscillations during the convergence.

5.5.3 Trajectory Tracking

To test the proposed approach in its ability to control the system and track a reference signal, we compare its performance against the use of a simple RHONN scheme (Rovithakis and Christodoulou 2000).

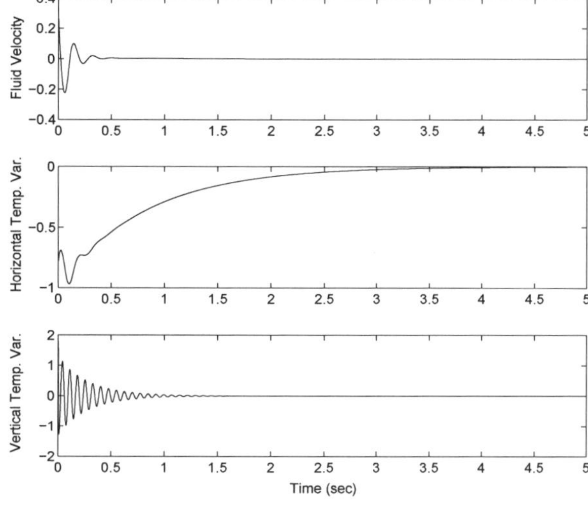

Fig. 5.9 Evolution of Lorenz states when we use the F-RHONNs model and assume only the existence of states x_1, x_3

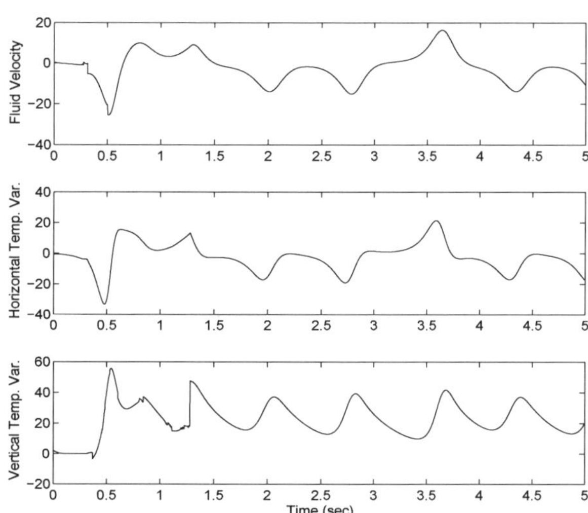

Fig. 5.10 Evolution of Lorenz states when we use the RHONN model and the knowledge of states x_1, x_3

Chaos synchronization of two different dynamical systems are discussed with the aid of the tracking control. We consider the states of a dynamical system from Elabbasy et al. (2006) to be the reference signals $z_d = (z_1, z_2, z_3)^T$, with dynamical equations:

$$\dot{z}_1 = \alpha z_1 - z_2 z_3$$
$$\dot{z}_2 = -\beta z_2 + z_1 z_3$$
$$\dot{z}_3 = -\gamma z_3 + z_1 z_2 \tag{5.130}$$

which become chaotic in the presence of $(\alpha, \beta, \gamma)^T = (0.4, 12, 5)^T$.

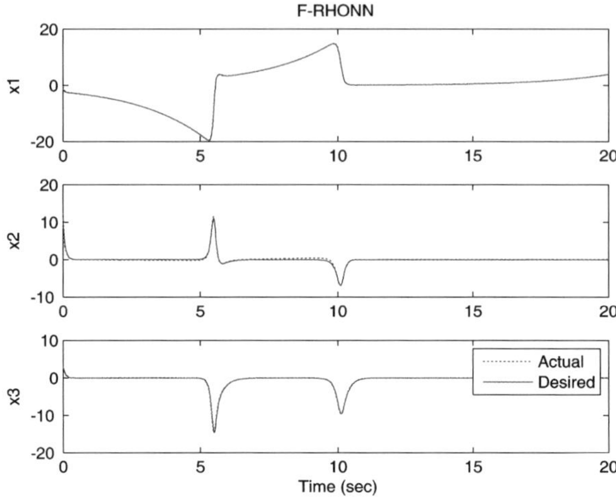

Fig. 5.11 The states of the controlled Lorenz system with F-RHONNs model against the desired trajectory when $z_r = (z_1, z_2, z_3)^T$

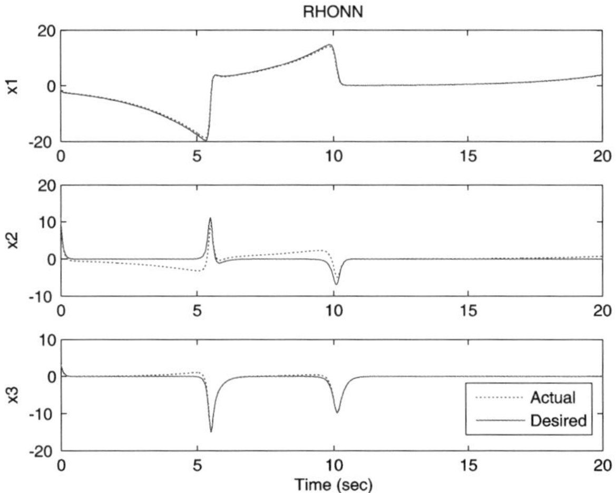

Fig. 5.12 The states of the controlled Lorenz system with RHONN model against the desired trajectory when $z_r = (z_1, z_2, z_3)^T$

We also consider the following initial conditions for Eqs. (5.129), (5.130):

$$x_0 = [-1.5, 1, 4]^T,$$

and

$$z_{d_0} = [-1, 11, 3]^T.$$

Disturbance terms are also added in Eq. 5.129, selected to be random values in the interval $[-1, 1]$. The main parameters for the control law (5.109) and the learning laws (5.124), (5.126) are selected as

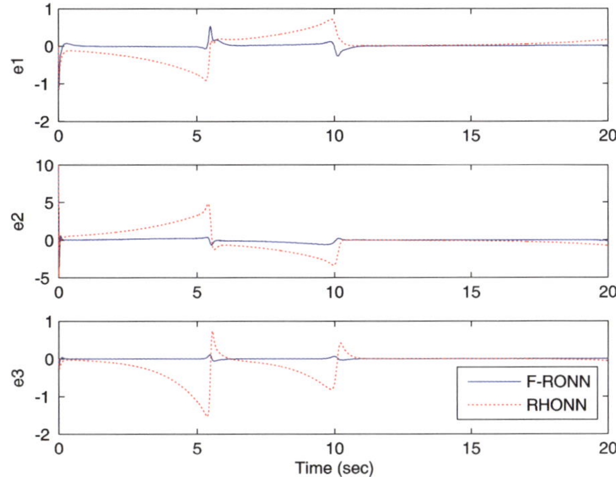

Fig. 5.13 Time evolution of error signals for F-RHONNs (*blue line*) and RHONN (*red line*) model

$$A = \mathrm{diag}(-3.31, -0.22, -0.04),$$
$$K = \mathrm{diag}(66.97, 130.31, 145.65).$$

The parameters of the sigmoidals that have been used are $\alpha_1 = 0.68$, $\beta_1 = 0.35$ and $\gamma_1 = 0.91$. Also, the learning rate is selected appropriately as $d_{f_l} = 0.058$ and the outer boundary for the adaptation of weights as $\varepsilon_i^l = 250$ with $\kappa_{f_i}^{l,\mathrm{outer}} = 0.116$.

The main parameters for the control law and the learning laws for the RHONN model are selected as:

$$A = \mathrm{diag}(-1.93, -0.28, -0.38),$$
$$K = \mathrm{diag}(45.44, 138.62, 148.04).$$

The parameters of the sigmoidals that have been used are $\alpha_2 = 2.75$, $\beta_2 = 1.97$ and $\gamma_2 = 4.23$. Also, the learning rate is selected appropriately as $d_f = 1.582$.

In the sequel, we first assume that the system is described, within a degree of accuracy, by an NF system of the form (5.2) with the number of states being $n = 3$, the number of fuzzy partitions being $p = 3$, and the depth of high-order sigmoid terms $k = 3$. In this case $s_i(x)$ assume high-order connection up to the first order. Figures 5.11 and 5.12 shows the evolution of Lorenz states for F-RHONNs and RHONN models, respectively, while Fig. 5.13 shows the convergence of errors demonstrating the superiority of F-RHONNs (blue line) against RHONNs (red line).

5.6 Summary

Direct adaptive regulation and tracking schemes were considered in this chapter to control nonlinear unknown affine in the control plants. The control schemes are based on the NF modeling, which interweaves the concepts of fuzzy systems (FS) with recurrent multiple high-order neural networks (F-RHONNs). Since the plant is considered unknown, we propose its approximation by the special form of an NF model, which, however, may assume modeling errors or a smaller number of states than the original unknown model.

This practically transforms the original unknown system into an NF model of known structure, which contains a disturbance term to account for the effect of the modeling errors or omitting states, but contains a number of unknown synaptic weights that have to be estimated. Also, a number of different cases were analyzed in regard to the modeling errors (depending on system states, control inputs, or a nonzero constant value) and the reduced model order. Weight updating laws for the synaptic weights of the involved HONNs were provided, which ensure that the system states reach zero or a small region around zero exponentially fast, keeping at the same time all signals in the closed loop bounded. The state trajectory problem was also considered following the same NF modeling assuming the existence of modeling errors.

From the analysis carried out, we observed that the controllers work better when the values of feedback matrix K are the maximum allowable and the NF model matches closer the true unknown nonlinear dynamical plant. Also, when the modeling errors are uniformly bounded by small constant values, in conjunction with the above statement for matrix K we concluded that $\|x(t)\|$ can be arbitrarily close to zero or to a residual set around zero.

Several simulations on the chaotic benchmark Lorenz system demonstrated the performance of the proposed control schemes verifying the theoretical conclusions.

References

Boutalis, Y. S., Theodoridis, D. C., & Christodoulou, M. A. (2009). A new neuro fds definition for indirect adaptive control of unknown nonlinear systems using a method of parameter hopping. *IEEE Transactions on Neural Networks, 20,* 609–625.

Elabbasy, E., Agiza, H., & El-Dessoky, M. (2006). Adaptive synchronization for four-scroll attractor with fully unknown parameters. *Physics Letters A, 34,* 187–191.

Michel, A., & Miller, R. (1977). *Qualitative Analysis of Large Scale Dynamic Systems.* New York: Academic Press.

Moore, E. H. (1920). On the reciprocal of the general algebraic matrix. *Bulletin of the American Mathematical Society, 26,* 394–395.

Penrose, R. (1955). A generalized inverse for matrices. *Proceedings of the Cambridge Philosophical Society, 51,* 406–413.

Popov, V. M. (1973). *Hyperstability of Control Systems.* New York: Springer.

Rovithakis, G., & Christodoulou, M. A. (2000). *Adaptive control with recurrent high order neural networks (theory and industrial applications), in advances in industrial control.* London: Springer Verlag London Limited.

Strogatz, S.H. (1994). Nonlinear dynamics and Chaos. Westview Press (Perseus Books Group), Cambridge

Theodoridis, D. C., Boutalis, Y. S., & Christodoulou, M. A. (2009). Direct adaptive control of unknown nonlinear systems using a new neuro-fuzzy method together with a novel approach of parameter hopping. *Kybernetica, 45,* 349–386.

Theodoridis, D., Boutalis, Y., & Christodoulou, M. (2011a). Neuro-fuzzy direct adaptive control of unknown nonlinear systems with analysis on the model order problem. *Journal of Zhejiang University-SCIENCE C (Computers and Electronics), 12*, 1–16.

Theodoridis, D., Boutalis, Y., & Christodoulou, M. (2011b). Robustifying analysis of the direct adaptive control of unknown multivariable nonlinear systems based on a new neuro-fuzzy method. *Journal of Artificial Intelligence and Soft Computing Research, 1*, 59–79.

Theodoridis, D., Boutalis, Y., & Christodoulou, M. (2012). Direct adaptive neuro-fuzzy trajectory tracking of uncertain nonlinear systems. *International Journal of Adaptive Control and Signal Processing, 26*, 660–688.

Yeap, T. H., & Ahmed, N. U. (1994). Feedback control of chaotic systems. *Dynamic and Control, 4*, 97–114.

Chapter 6
Selected Applications

6.1 Trajectory Tracking of Robot Manipulators

Trajectory tracking control is one of the fundamental tasks in robotic control applications (Braganza et al. 2008; Craig 1985; EL-Hawwary and Elshafei 2006; Lewis et al. 1993; Najim et al. 2008; Yu and Li 2006). Traditional control approaches, like computed torque control, rely on an exact knowledge of the robot dynamics and may lead to poor results under parametric and dynamic uncertainties. Therefore, adaptive control of robotic manipulators has gained significant interest being an active area of research for many decades (Craig 1985; Fateh and Azarfar 2009; Lai et al. 2009; Lewis et al. 1999; Li et al. 2009).

In this section, an adaptive control method for trajectory tracking of robot manipulators, based on the new NF modeling is presented (Theodoridis et al. 2011). The proposed control scheme uses a NF model to estimate system uncertainties. The function of robot system dynamics is first modeled by a fuzzy system, which in the sequel is approximated by a combination of high order neural networks (HONNs). The overall representation is linear in respect to the unknown NN weights leading to weight adaptation laws that ensure stability and convergence to unique global minimum of the error functional.

Due to the adaptive NF modeling, the proposed controller is independent of robot dynamics, since the free parameters of the NF controller are adaptively updated to cope with changes in the system and the environment. Adaptation laws for the network parameters are derived, which ensure network convergence and stable control. A weight hopping technique is also introduced to ensure that the estimated weights stay within prespecified bounds. The simulation results show very good tracking abilities under disturbance torque compared to conventional computed torque PD control.

Y. Boutalis et al., *System Identification and Adaptive Control*,
Advances in Industrial Control, DOI: 10.1007/978-3-319-06364-5_6,
© Springer International Publishing Switzerland 2014

6.1.1 Robot Controller Based on Filtered Tracking Error

Consider an n-DOF robot manipulator, which can be freely moved in its working space and its end-effector is following a trajectory. The manipulator dynamics are described by the following nonlinear differential equation (Lewis et al. 1999; Spong and Vidyasagar 1989):

$$\tau(t) = M\left(\theta\left(t\right)\right)\ddot{\theta}(t) + V_m\left(\theta\left(t\right),\dot{\theta}\left(t\right)\right)\dot{\theta}\left(t\right) + G\left(\theta\left(t\right)\right) + F\left(\dot{\theta}\left(t\right)\right) + \tau_d, \quad (6.1)$$

where $\tau,\theta,\dot{\theta},\ddot{\theta}$ are vectors of joint torques, positions, velocities and accelerations, respectively. Also, M, V_m, G and F are the inertia matrix, the coriolis/centripental matrix, the gravity vector and the friction vector, respectively, (see Lewis et al. (1993), Spong and Vidyasagar (1989) for the physical meaning of these matrices) and τ_d is a disturbance term. Furthermore, we can define the following skew symmetry property (Lewis et al. 1993).

Definition 6.1 The Coriolis/Centripetal matrix can always be selected so that the matrix:

$$S(\theta,\dot{\theta}) \equiv \dot{M}(\theta) - 2V_m(\theta,\dot{\theta}), \quad (6.2)$$

is skew symmetric. Therefore, $x^T S x = 0$ for all vectors x. This is a statement of the fact that the fictitious forces in the robot system do not work.

Consider now the following tracking error e given by:

$$e(t) = \theta_d(t) - \theta(t), \quad (6.3)$$

where θ_d is the desired joint variables trajectories. The filtered version r of the tracking error is given by:

$$r = \dot{e} + \Lambda e, \quad (6.4)$$

where Λ is a positive definite design parameter matrix. The robot dynamics are expressed in terms of this error as (Lewis et al. 1999):

$$M\dot{r} = -V_m r + f(x) + \tau_d - \tau, \quad (6.5)$$

where taking into consideration (6.3), (6.4), the unknown nonlinear function $f(x)$ is defined as:

$$f(x) = M\left(\theta\right)\left(\ddot{\theta}_d + \Lambda\dot{e}\right) + V_m\left(\theta,\dot{\theta}\right)\left(\dot{\theta}_d + \Lambda e\right) + G\left(\theta\right) + F\left(\dot{\theta}\right). \quad (6.6)$$

One may also define $z_1 = \left(\ddot{\theta}_d + \Lambda\dot{e}\right)$, $z_2 = \left(\dot{\theta}_d + \Lambda e\right)$ and

$$x \equiv \left[z_1^T \; z_2^T \; \sin\left(\theta^T\right) \; \cos\left(\theta^T\right) \; \dot{\theta}^T\right]^T. \quad (6.7)$$

A general sort of approximation-based controller is derived by setting:

$$\tau = \hat{f} + K_v r - \upsilon(t), \tag{6.8}$$

where $\hat{f}$ is an estimate of f, $K_v r = K_v \dot{e} + K_v \Lambda e$ an outer PD tracking loop, and $\upsilon(t)$ an auxiliary signal to provide robustness in the face of disturbance and modeling errors. The filtered error approach relies on PD compensation as well as on proper estimation of f. Simple PD controllers have been shown to be able to control the motion of a robot under partial or zero knowledge of its dynamics (Arimoto and Miyazaki 1984; Lewis et al. 1999). However, PD controllers rely on the proper selection of the PD gains, which may vary according to the robot manipulator used. Arbitrarily increasing the gains may result in good performance, producing however large control signals, which for implementation reasons are sometimes undesirable. Moreover, simple PD schemes usually lack stability proofs. Therefore, proper estimation of f using a short of adaptive technique may result in good performance even in the presence of unknown dynamics and disturbances, providing at the same time mathematical proofs for stability and performance. Moreover, such advanced techniques may be directly extended to more complicated control objectives such as force control for grinding, polishing, etc, where straight PD methods are inadequate (Lewis et al. 1999).

6.1.2 Neurofuzzy Model for the Unknown Nonlinear Robot Dynamics

In this section, we present the NF model, for a general serial-link robot arm.

Let $\left(\tau, \theta, \dot{\theta}, \ddot{\theta}\right) \in \Omega$, where Ω is a bounded simple connected set of R^n and $f(\cdot) : \Omega \to R^p$. Also, we assume that the desired trajectory is inside the workspace of the robot manipulator and moreover all of the points of the trajectory are reachable by the robot's end-effector. Define $C^p(\Omega)$ as the space of continuous functions $f(\cdot)$. Since f is a continuous, smooth function it can be approximated by the NF representation given in Sect. 2.1. Therefore, according to (2.11) we can approximate the unknown robot dynamics by several HONNs with L_F high order terms (Theodoridis et al. 2009a):

$$\hat{f}(x) = X_f W_f s_f(x), \tag{6.9}$$

where X_f is a $n \times n \cdot q$ block diagonal matrix, $X_f = \text{diag}\left\{X_{f_1}, \ldots, X_{f_n}\right\}$ with the ith ($i = 1, \ldots, n$) diagonal block element being a row vector containing the q fuzzy output partition centers of f_i. Here, for notational simplicity, it is assumed that each f_i has the same number p of partitions. W_f is a $n \cdot q \times L_F$ matrix of synaptic weights, which can be considered as a block matrix $W_f = \left[W_{f_i}^l\right]$, $i = 1, \ldots, n$, $l = 1, \ldots, L_f$, where each $W_{f_i}^l$ is a q-dimensional column vector. Finally, s_f is a

L_F-dimensional vector with each element $s_{f_l}(x)$ being a high order combination of sigmoid functions of the state. Define now the optimal weight vector $W_f^\star$ as follows:

$$W_f^\star \stackrel{\Delta}{=} \arg \min_{|X_f W_f| \le M_f} \left\{ \sup_{x \in \Omega} |f(x) - X_f W_f s_f| \right\},$$

where M_f is a sufficiently large positive constant. Let us also define the modeling error μ as follows:

$$\mu(x) \stackrel{\Delta}{=} f(x) - X_f W_f^\star s_f(x).$$

with the modeling error bounded by $\|\mu\| < \mu_N$. From the above two equations, we have that the robot dynamics satisfies:

$$f(x) = X_f W_f^\star s_f(x) + \mu. \tag{6.10}$$

Note that from Lemma 2.1, we have that the term μ can be made arbitrarily small, and therefore the high order NF network (6.9) can approximate any smooth function arbitrarily close by appropriately selecting the number of high order collections L_F and the number and values of the centers of fuzzy output partitions. Then, the control law (6.8) becomes:

$$\tau = X_f W_f s_f(x) + K_v r - \upsilon(t). \tag{6.11}$$

It is now necessary to show how to tune the NF weights W_f on line in order to guarantee stable tracking. To this end, define the weight estimation error as $\tilde{W}_f = W_f - W_f^\star$. Then, the closed-loop filtered error dynamics (6.5) becomes:

$$M\dot{r} = -(K_v + V_m)r - X_f \tilde{W}_f s_f(x) + (\mu + \tau_d) + \upsilon(t). \tag{6.12}$$

6.1.3 Neurofuzzy Controller and Error System Dynamics

Suppose that a NF network is used to approximate the nonlinear robot function (6.6) according to (6.10) with W_f^* the optimum approximating weights. The optimum weights are unknown and may even be nonunique. Let us also assume, that the desired trajectory $\theta_d(t)$ is bounded by $\bar{\theta}_B$ a known scalar bound and the initial tracking error is bounded as well. We can now distinguish two possible cases.

Case 1: NF controller in an ideal case.
Suppose the NF network functional reconstruction error μ and the unmodeled disturbances $\tau_d(t)$ are both zero. In the next theorem, weight updating laws are given, which can serve the control objectives. It has to be mentioned that according to Boutalis et al. (2009), Theodoridis et al. (2009a) gradient-type updating laws of the

form given in the next theorem have been found sufficient for the approximation of general nonlinear functions using the NF representation of (6.9).

Theorem 6.1 *The control scheme for (6.1) given by (6.11) (with $\upsilon(t) = 0$), the filtered error system:*

$$M\dot{r} = -(K_v + V_m)r - X_f \tilde{W}_f s_f(x), \tag{6.13}$$

and the weight updating law:

$$\dot{W}_{f_i}^l = (\bar{x}_{f_i})^T d_{f_i} r_i s_l(x), \tag{6.14}$$

guarantees the following properties:

- $r, \tilde{W}_f \in L_\infty, \quad r \in L_2,$
- $\lim_{t \to \infty} r(t) = 0, \ \lim_{t \to \infty} \dot{W}_f(t) = 0.$

Proof Consider the Lyapunov function candidate:

$$V = \frac{1}{2} r^T M r + \frac{1}{2} tr\{\tilde{W}_f^T D_f^{-1} \tilde{W}_f\}, \tag{6.15}$$

where D_f is a positive definite gain matrix which is selected by the designer. Taking the time derivatives of the Lyapunov function candidate (6.15) we obtain:

$$\dot{V} = \frac{1}{2} r^T \dot{M} r + r^T M \dot{r} + tr\left\{\dot{\tilde{W}}_f^T D_f^{-1} \tilde{W}_f\right\}, \tag{6.16}$$

and after substituting Eq. (6.13) we have:

$$\dot{V} = -r^T K_v r + \frac{1}{2} r^T (\dot{M} - 2V_m) r - r^T X_f \tilde{W}_f s_f(x) + tr\left\{\dot{\tilde{W}}_f^T D_f^{-1} \tilde{W}_f\right\}. \tag{6.17}$$

But according to Definition 6.1, $(\dot{M} - 2V_m)$ is skew symmetric and therefore $r^T(\dot{M} - 2V_m)r = 0, \forall r \in \Re^n$. Let us consider that:

$$tr\{\dot{\tilde{W}}_f^T D_f^{-1} \tilde{W}_f\} = r^T X_f \tilde{W}_f s_f(x),$$

then, the above equation using matrix trace properties, results in the following learning law:

$$\dot{W}_f = D_f X_f^T r s_f^T. \tag{6.18}$$

Taking into account the block representation of W_f given in (6.9) the above law can also be written in respect to the updating of each $W_{f_i}^l$ according to (6.14), where d_{f_i} is a $m \times m$ gain matrix, which for simplicity could be taken to be diagonal with equal elements.

Then, $\dot{V}$ assumes the form:

$$\dot{V} = -r^T K_v r \leq -\lambda_{min} (K_v) \|r\|^2 . \tag{6.19}$$

Since $V > 0$ and $\dot{V} \leq 0$, we conclude that $V \in L_\infty$, which implies that $r, \tilde{W}_f \in L_\infty$. Furthermore, $W_f = \tilde{W}_f + W_f^*$ are also bounded. Since V is a nonincreasing function of time and bounded from below, $\lim_{t \to \infty} V = V_\infty$ exists; therefore, by integrating $\dot{V}$ from 0 to ∞ we have:

$$\lambda_{\min} (K_v) \int_0^\infty \|r\|^2 \, dt \leq [V(0) - V_\infty] < \infty,$$

which implies that $r \in L_2$.

Since $X_f, \tilde{W}_f, s_f$ are bounded, $r \in L_\infty$. Since $r \in L_2 \cap L_\infty$, using Barbalat's Lemma we conclude that $\lim_{t \to \infty} r(t) = 0$. Hence, given that Λ in (6.4) is positive definite, we also have that:

$$\lim_{t \to \infty} e(t) = 0.$$

Now, using the boundedness of X_f, s_f and the convergence of $r(t)$ to zero, we have that $\dot{W}_f$ also converges to zero (Ioannou and Fidan 2006).

Case 2: NF controller in non-Ideal case.
As shown in Case 1 above, under the ideal situation of no modeling errors or unmodeled disturbances, the controller of Theorem 6.1 suffices to make the tracking error to go to zero. However, in actual systems there are disturbances. Moreover, the approximation error increases as the number of high order terms or fuzzy output centers decreases. In Non-ideal case, it can be proved that, if the approximation error and system disturbance are not zero but bounded, then our NF controller still works under an additional assumption of persistence of excitation (PE). In this case, the tracking error does not converge to zero, but is bounded by small enough values to guarantee good tracking performance. The high order sigmoid terms could be chosen to be persistently exciting by appropriately selecting parameters α, γ in (2.8). However, if the vector s_f in (6.9) is not persistently exciting, it is well known from adaptive control theory (Ioannou and Fidan 2006) that the phenomenon of *parameter drifting* may occur. To correct this problem, one may use techniques from adaptive control, including dead-zone, σ or switching σ-modification and e-modification (Ioannou and Fidan 2006). Alternatively, one could use a modified weight updating algorithm, that includes a weight hopping (Theodoridis et al. 2009b) when, during the adaptation, term $X_f W_f$ in (6.9) exceeds a prespecified bound.

Theorem 6.2 *The control scheme for* (6.1) *given by* (6.11) *(with* $v(t) = 0$, $\tau_d \neq 0$, $\mu \neq 0$), *the filtered error system:*

$$M\dot{r} = -(K_v + V_m)\, r - X_f \tilde{W}_f s_f(x) + \tau_d + \mu, \tag{6.20}$$

and the modified learning law:

$$\dot{W}^l_{f_i} = H_{P_f}\left[\left(\bar{x}_{f_i}\right)^T d_{f_i} r_i s_l(x); \quad \left(\bar{x}_{f_i}\right)^T d_{f_i} r_i s_l(x) - \frac{\kappa d_{f_i}\, \|r\|\, \left(\bar{x}_{f_i}\, W^l_{f_i}\, \left(\bar{x}_{f_i}\right)^T\right)}{tr\left\{\left(\bar{x}_{f_i}\right)^T \bar{x}_{f_i}\right\}}\right],$$

$$(6.21)$$

where $P_f \overset{\Delta}{=} \left\{\bar{x}_{f_i} \cdot W^l_{f_i} : \left|\bar{x}_{f_i} \cdot W^l_{f_i}\right| < M_{f_i}\right\}$, κ a positive constant selected from the designer and H_{P_f} represents the switching criterion between the two updating laws inside the square brackets, guarantees the uniform ultimate boundedness of r with respect to the set:

$$\Re = \left\{r(t) : \|r(t)\| \le \frac{(\tau_B + \bar{\mu}_N)}{\lambda_{\min}(K_v)}, \quad \lambda_{\min}(K_v) \ne 0\right\}.$$

Proof Consider the Lyapunov function candidate (6.15). Taking its time derivatives and taking into account (6.12) with the disturbance τ_d and the NF reconstruction error μ different than zero we obtain:

$$\dot{V} = -r^T K_v r + \frac{1}{2}r^T(\dot{M} - 2V_m)r - r^T X_f \tilde{W}_f s_f(x) + tr\left\{\dot{\tilde{W}}^T_f D^{-1}_f \tilde{W}_f\right\} + r^T(\tau_d + \mu).$$

$$(6.22)$$

The skew symmetric property of $(\dot{M} - 2V_m)$ together with the learning law (6.14) (which is equivalent to (6.21) if no hopping occurs) results to the following form of $\dot{V}$:

$$\dot{V} \le -\lambda_{\min}(K_v)\|r\|^2 + \|r\|(\tau_B + \bar{\mu}_N)$$
$$\le -(\lambda_{\min}(K_v)\|r\| - (\tau_B + \bar{\mu}_N))\|r\| \le 0, \qquad (6.23)$$

provided that:

$$\|r\| > \frac{(\tau_B + \bar{\mu}_N)}{\lambda_{\min}(K_v)}, \qquad (6.24)$$

and $\lambda_{\min}(K_v) \ne 0$ which is valid from definition.

Inequality (6.24) demonstrates that the trajectory of $r(t)$ is uniformly bounded with respect to the arbitrarily small, (since K_v can be chosen sufficiently large), set $\Re$ shown below:

$$\Re = \left\{r(t) : \|r(t)\| \le \frac{(\tau_B + \bar{\mu}_N)}{\lambda_{\min}(K_v)}, \quad \lambda_{\min}(K_v) \ne 0\right\}.$$

In case the complete modified weight updating law (6.21) is applied, that is when the switching condition is occasionally met, then the second updating law inside the square brackets of (6.21) is applied. In this case, and since the weight jump
$-\frac{\kappa d_{f_i}\|r\|(\bar{x}_{f_i} W^l_{f_i}(\bar{x}_{f_i})^T)}{tr\{(\bar{x}_{f_i})^T \bar{x}_{f_i}\}} = -\kappa d_{f_i}\|r\|h_b$ can also be expressed in respect to the error

weight vector $\tilde{W}_{f_i}^l$ as $-\kappa d_{f_i}\|r\|h_b\dfrac{\tilde{W}_{f_i}^l}{\|\tilde{W}_{f_i}^l\|}$ according to Sect. 3.2.2, $\dot{V}$ satisfies the following inequality:

$$\dot{V} \le -\lambda_{\min}(K_v)\|r\|^2 + \|r\|(\tau_B + \bar{\mu}_N) - \|r\|\,\Theta, \tag{6.25}$$

with Θ being a positive constant expressed as:

$$\Theta = \sum \kappa d_{f_i} h_b \frac{\tilde{W}_{f_i}^T \tilde{W}_{f_i}}{\|\tilde{W}_{f_i}\|},$$

where the summation includes all weight vectors which require hopping. Therefore, the negativity of $\dot{V}$ is actually enhanced by the presence of Θ, with $\Theta_l \le \Theta \le \Theta_u$, where Θ_l stands for the case where only one hopping occurs and Θ_u when all possible hoppings occur. After substituting with the worst case of $\Theta = \Theta_l$, $\dot{V}$ becomes:

$$\begin{aligned}\dot{V} &\le -\lambda_{\min}(K_v)\|r\|^2 + \|r\|(\tau_B + \bar{\mu}_N) - \Theta_l\|r\| \\ &\le -(\lambda_{\min}(K_v)\|r\| + \Theta_l - (\tau_B + \bar{\mu}_N))\|r\| \le 0,\end{aligned} \tag{6.26}$$

provided that:

$$\|r\| > \frac{(\tau_B + \bar{\mu}_N) - \Theta_l}{\lambda_{\min}(K_v)}. \tag{6.27}$$

Finally, inequality (6.27) demonstrates that the trajectory of $r(t)$ is uniformly bounded with respect to the arbitrarily small, (since K_v can be chosen sufficiently large, and the presence of Θ_l can make the ratio even smaller), set $\Re$ shown below

$$\Re = \left\{r(t) : \|r(t)\| \le \frac{(\tau_B + \bar{\mu}_N) - \Theta_l}{\lambda_{\min}(K_v)}\right\}.$$

Thus, $\theta(t)$ is also bounded with respect to $r(t)$. Furthermore, we have:

$$M\dot{r} = -(K_v + V_m)r - X_f\tilde{W}_f s_f(x) + \tau_d + \mu.$$

Hence and since the boundedness of $X_f\tilde{W}_f$ is ensured by the use of the modified weight hopping algorithm and $\tau_d + \mu$ is bounded by definition, we conclude that $\dot{r} \in L_\infty$.

Remark 6.1 The previous analysis reveals that in the case where we have a modeling error and disturbance different from zero at $\|\theta\| = 0$, our adaptive algorithm can guarantee at least uniform ultimate boundedness of all signals in the closed-loop. In particular, Theorem 6.2 shows that if $\tau_B + \bar{\mu}_N$ is sufficiently small, or if the design constant K_v is chosen appropriately, then $\|r(t)\|$ can be arbitrarily close to zero and in the limit as $K_v \to \infty$, actually becomes zero but implementation issues constrain the maximum allowable value of K_v.

Remark 6.2 The switching condition H_{P_f} in (6.21) could be simply a threshold function determining whether $X_{f_i} \cdot W_{f_i}^l \in P_f \triangleq \left\{ \bar{x}_{f_i} \cdot W_{f_i}^l : \left| \bar{x}_{f_i} \cdot W_{f_i}^l \right| < M_{f_i} \right\}$ or not. In a more elegant version one could also take into account the direction of the weight updating. In case the threshold function is activated but the weight updating is in the direction that moves $\bar{x}_{f_i} \cdot W_{f_i}^l$ toward P_f, the switching is finally not activated. Details for this more elegant version can be found in Boutalis et al. (2009), Theodoridis et al. (2009b) and in Sect. 2.4 of this book.

Remark 6.3 The parameter hopping condition arrives from the desire to keep (constraint) $\left| X_f W_f s_f \right| < M_f$, where M_f is an upper bound. Since $\bar{x}_{f_i} W_{f_i} s_f = \sum_{l=1}^{k} \bar{x}_{f_i} W_{f_i}^l s_{fl}(x)$ and $|\bar{x}_{f_i} W_{f_i} s_f(x)| \le \sum_{l=1}^{k} |\bar{x}_{f_i} W_{f_i}^l s_{fl}(x)|$, one is allowed to have a better insight in the algorithm if puts specialized bounds, each one for each component of $X_f W_f s_f(x)$. Each specialized bound may be expressed by $\left| \bar{x}_{f_i} \cdot W_{f_i}^l \right| < M_{f_i}$, with M_{f_i} being again a design parameter determining an external limit for $\bar{x}_{f_i} \cdot W_{f_i}^l$. In the sequel, we consider the forbidden hyperplanes being defined by the equation $\left| \bar{x}_{f_i} \cdot W_{f_i}^l \right| = M_{f_i}$. Note that the direction of the weight updating (6.14) is perpendicular to the forbidden hyperplane. When the weight vector reaches one of the forbidden hyperplanes $\left| \bar{x}_{f_i} \cdot W_{f_i}^l \right| = M_{f_i}$ and the direction of updating is toward the forbidden hyperplane, a new *hopping* is introduced which moves the weights back to the desired hyperspace P_f, allowing thus the algorithm to search the entire space for a better weight solution. This procedure is depicted in Fig. 2.3, in a simplified two-dimensional representation. The magnitude of hopping is $-\dfrac{\kappa d_{f_i} \|r\| (\bar{x}_{f_i} W_{f_i}^l (\bar{x}_{f_i})^T)}{tr\{(\bar{x}_{f_i})^T \bar{x}_{f_i}\}}$ being determined by following the vectorial proof in Sect. 3.2.2 and Fig. 3.5 and d_{f_i} is the i-th element of the gain matrix D_f.

6.1.4 Simulation Results

The planar two-link revolute arm shown in Fig. 6.1 is used extensively in the literature for easy simulation of robotic controllers, with results that can be easily illustrated in 2-D plots.

Its dynamics are given as (Lewis et al. 1993):

$$\dot{x}_1 = x_3$$
$$\dot{x}_2 = x_4$$
$$[\dot{x}_3 \ \dot{x}_4]^T = -M^{-1}(N + \tau)$$

where $x_1 = \theta_1$ is the angular position of joint 1, $x_2 = \theta_2$ is the angular position of joint 2, $x_3 = \dot{\theta}_1$ is the angular velocity of joint 1 and $x_4 = \dot{\theta}_2$ is the angular velocity of joint 2. Also, the matrices M, N have the following form:

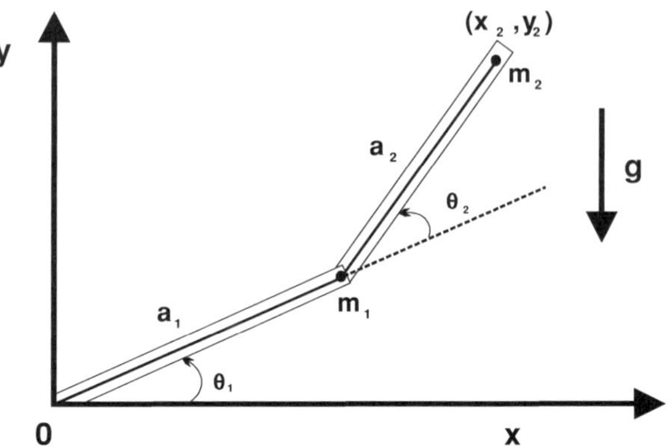

Fig. 6.1 Two-link planar elbow arm

$$
M = \begin{bmatrix} (m_1 + m_2)\, a_1^2 + m_2 a_2^2 + 2m_2 a_1 a_2 \cos(x_2) & m_2 a_2^2 + m_2 a_1 a_2 \cos(x_2) \\ m_2 a_2^2 + m_2 a_1 a_2 \cos(x_2) & m_2 a_2^2 \end{bmatrix},
$$

$$
N = \begin{bmatrix} -m_2 a_1 a_2 \left(2x_3 x_4 + x_4^2\right) \sin(x_2) + (m_1 + m_2)\, g a_1 \cos(x_1) + m_2 g a_2 \cos(x_1 + x_2) \\ m_2 a_1 a_2 x_3^2 \sin(x_2) + m_2 g a_2 \cos(x_1 + x_2) \end{bmatrix}.
$$

We took the arm parameters as $a_1 = a_2 = 1\,m$, $m_1 = m_2 = 1\,kg$, and the desired trajectory $x_{1d} = sin(t)$, $x_{2d} = cos(t)$.

To test the applicability and the performance of the NF technique, we present simulations and comparison with conventional PD Computed Torque (CT) controller in trajectory tracking under unknown disturbance. The interesting reader may find more simulations in Theodoridis et al. (2011), including comparisons with other well established NN-based control techniques (Lewis et al. 1995a, b, 1996, 1999).

The weight updating law for the NF approximator is given by (6.18) or equivalently by (6.14). In this simulations only first-order sigmoid neural functions were used. The number of output membership partitions required by the NF approach was arbitrarily selected to be $p = 5$ and the centers of these partitions were again arbitrarily selected to be:

$$
\bar{x}_{f_1} = \begin{bmatrix} -35 & -15 & 20 & 75 & 100 \end{bmatrix},
$$
$$
\bar{x}_{f_2} = \begin{bmatrix} 10 & 30 & 75 & 110 & 170 \end{bmatrix}.
$$

That is, no actual expert's knowledge has been used. No other fine tuning has been followed regarding the functions of the regression vector apart from the selection of sigmoid parameters and adaptation gains. The controller parameters were taken as: $K_v = \mathrm{diag}\{20, 20\}$, $\Lambda = \mathrm{diag}\{5, 5\}$, $D_f = \mathrm{diag}\{1.8, 1.5\}$ and $F = \mathrm{diag}\{50, 50\}$. The sigmoid values for the NF approach were $\alpha = 0.5$, $\beta = 0.4$ and $\gamma = 0.3$. Moreover, in the control law of (6.11) a simple robustifying term $\upsilon(t) = 300 e(t)$ was added to account for modeling errors. The performance of the proposed NF approach was compared to conventional computed torque PD (CT-PD) control (Lewis et al. 1993),

Fig. 6.2 Tracking error of the first joint for the NF and CT-PD approaches when a sudden disturbance is applied between 10 and 20 s

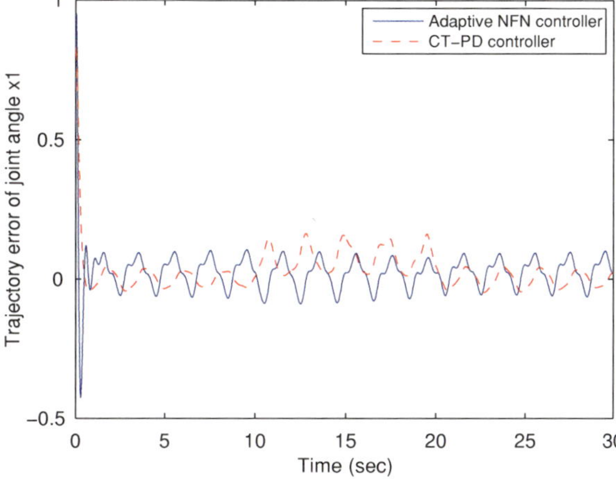

Fig. 6.3 Tracking error of the second joint for the NF and CT-PD approaches when a sudden disturbance is applied between 10 and 20 s

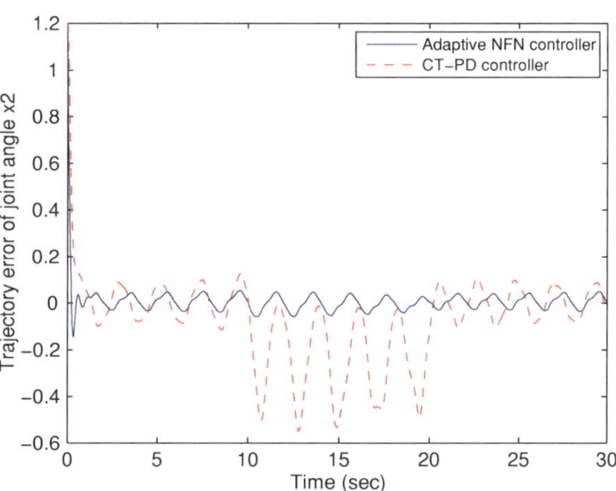

where the PD gains were selected to be $K_p = 100$, $K_v = 20$. It has to be mentioned that the conventional computed torque controller assumes the exact knowledge of the robot model and its parameters.

In the testing scenario, we assume that the robot is tracking the desired trajectory and a sudden disturbance is applied. This happened between 10 and 20 s of simulation time and the disturbance had the following form:

$$\tau_d = 10 + 5 \cdot \sin(18 \cdot \pi \cdot t). \tag{6.28}$$

Figures 6.2 and 6.3 shows the respective tracking errors. It can be seen that the performance of the CT-PD is seriously affected especially the joint angle x_2, while the performance of the NF approach is only slightly affected.

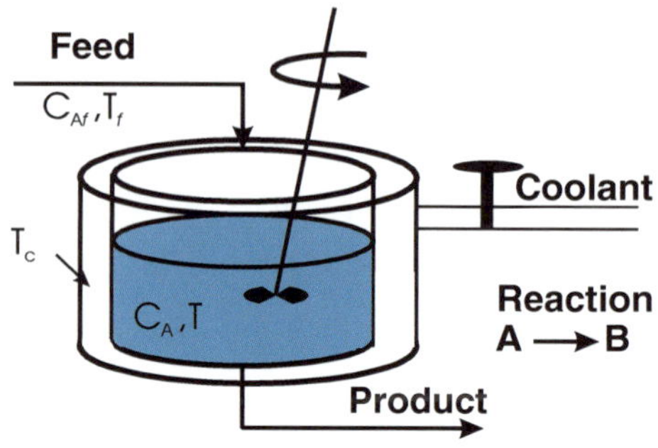

Fig. 6.4 Chemical stirred tank reactor (CSTR) process

6.2 Bioreactor Application

Bioprocesses, even those that have few variables, are in general difficult to be controlled, because of their strong nonlinearities which are difficult to model. Therefore, they offer a good application for NF control.

Such processes usually involve a liquid substrate in a tank containing cells that consume substrates and yield desired products and undesired byproducts. There are many different models for such bioreactors; as an example, we take the continuously stirred tank reactor (CSTR), which is described as a benchmark problem in adaptive control (Ungar 1990), as it is adapted in Lewis et al. (1999).

6.2.1 Model Description and Control Objective

A more complete and realistic model using a cooler around the tank is shown in Fig. 6.4. Let the reactor vessel volume be V. It is assumed that the concentration C_A inside the tank is uniform having also a uniform temperature T. The inlet reactant is supplied with concentration C_{Af}, feed rate F, and temperature T_f. In our example, the control input is the coolant temperature T_c.

A dynamic model for the CSTR temperature control is then given by Ray (1975), Lewis et al. (1999) and used in many real applications for design purposes:

$$V\frac{dC_A}{dt} = F\left(C_{Af} - C_A\right) - V k_o e^{-\frac{E}{RT}} C_A$$

$$V\rho C_p \frac{dT}{dt'} = \rho C_p F\left(T_f - T\right) - \Delta H V k_o e^{-\frac{E}{RT}} C_A - hA\left(T - T_c\right), \quad (6.29)$$

where the system parameters ρC_p, h, A, E, k_o, and ΔH are assumed to be constant. Let $t = \frac{F}{V} t'$ and define a transformation as:

Table 6.1 Different cases of bioreactor constant values b, β, D_a

Case	B	β	D_a
Case 1	7	0.5	0.110
Case 2	8	0.3	0.072
Case 3	11	1.5	0.135

$$\begin{bmatrix} x_1 \\ x_2 \end{bmatrix} = \begin{bmatrix} \frac{C_{Af}-C_A}{C_{Af}} \\ \frac{T-T_{fd}}{T_{fd}} \end{bmatrix}, \tag{6.30}$$

where T_{fd} is the nominal inlet temperature. It has to be noted that the normalized $x_1(t)$ is the difference between inlet and reactor concentrations and it is always less than or equal to 1 (since $C_{Af} \geq C_A \geq 0$).

Our objective is to control the temperature T of the reactor, because this is crucial for the growth of cells. Then, (6.29) can be rewritten in terms of dimensionless variables as:

$$\dot{x}_1 = -x_1 + D_a\,(1-x_1)\,e^{\frac{\gamma x_2}{\gamma + x_2}}$$
$$\dot{x}_2 = -x_2 + BD_a\,(1-x_1)\,e^{\frac{\gamma x_2}{\gamma + x_2}} - \beta\,(x_2 - x_{2c}) + d + \beta u$$
$$y = x_2 - x_{2d} \tag{6.31}$$

with constants $D_a = \frac{k_o V}{F}e^{-\gamma}, \gamma = \frac{E}{RT_{fd}}, B = \frac{-\Delta H C_{Af}\gamma}{\rho C_p T_{fd}}, \beta = \frac{hA}{\rho C_p F}, d = \gamma\frac{T_f - T_{fd}}{T_{fd}}$, $x_{2c} = \gamma\frac{T_{cd}-T_{fd}}{T_{fd}}$ and $x_{2d} = \gamma\frac{T_d - T_{fd}}{T_{fd}}$. Where T_{cd} and T_d are the nominal design values of coolant and reactor temperatures, respectively. Although it is assumed that D_a, γ, B, and β are constant, there are some uncertainties in them. The Damkohler constant D_a is, in fact, a function of the reactor catalyst, and the dimensionless heat transfer constant β is a slowly varying parameter. Note that this is a feedback linearizable model (Lewis et al. 1999).

Table 6.1 shows values of b, β, D_a that have already been studied in the literature (Lewis et al. 1999; Liu and Lewis 1994; Ray 1975). Here, we will use the values $B = 7, \beta = 0.5$ and $D_a = 0.110$, with $\gamma = 20$ (case 1) in order to study the behavior of the bioreactor and the performance of the proposed controller. The desired inlet reactant temperature, T_{fd} is here selected as $300\,°K$, and the control objective is to keep the reactor temperature T at this level. Under this temperature, the tank concentration will converge to some constant steady-state value. In order to show the sensitivity to variations in the inlet temperature, at time $t = 50\,s$, we added a disturbance effect of a $7\,°K$ increase in the inlet reactant temperature. Observe the response of the reactor for Case 1, in Fig. 6.5. Clearly, the temperature T fluctuates widely, something very undesirable for the health of the cell population. The tank temperature T increases by about $50\,°K$.

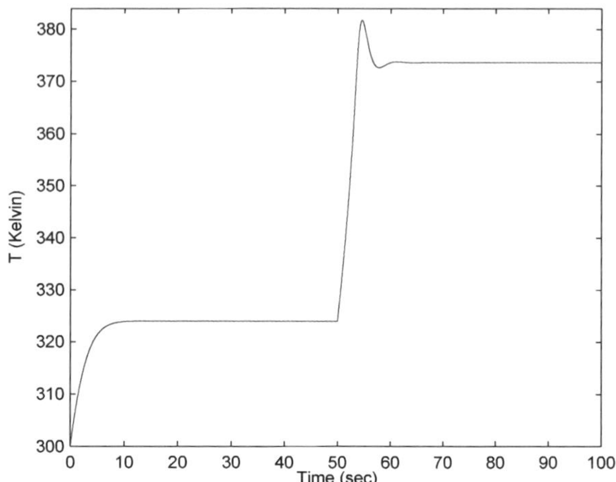

Fig. 6.5 CSTR open-loop response to a disturbance

6.2.2 Simulation Results

The NF control structure introduced in Sect. 4.1 along with the weight updating laws of Theorem 4.1, was now used to regulate the CSTR state x_2 in order the temperature T to capture the desired value $T_{fd} = 300\,°\text{K}$ (Theodoridis et al. 2010). We use 2 output partitions of f and 2 output partitions of g, which means that we have 12 neurons in the hidden layers for both $\hat{f}$ and $\hat{g}$. The design parameters was selected as $k = 4$, the initial weights as $W_f(0) = 0$, $W_g(0) = 1$, the number of high order sigmoidal terms (HOST) used in HONNs were chosen to be 3 $(s(x_1), s(x_2), s(x_1) \cdot s(x_2))$, the updating learning rates $\gamma_1 = 0.18$, $\gamma_2 = 1.64$ and the parameters of the sigmoidal terms such as: $a_1 = 0.81$, $a_2 = 0.55$, $\beta_1 = 0.5$, $\beta_2 = 1.9$, and $\gamma_1 = 2.22$, $\gamma_2 = 1.45$. The performance of the proposed scheme is shown in Figs. 6.6 and 6.7. The results demonstrate the fast on-line convergence to the desired temperature and the robustness against disturbances that can be achieved by using the NF-based controller.

6.3 Experimental Setup and NF Regulation of a DC Motor

Industrial motors are mostly used in the industry but they also find applications to several daily human activities. For the engineers working with the electric traction, the control of these motors has become one of the most significant problems to solve, since their control can permit the motor to follow a prespecified task. The development of computers and specifically of the microcontrollers has given the opportunity to the engineers to control the motors digitally. However, in order to successfully operate the digital control it is necessary to create appropriate circuit

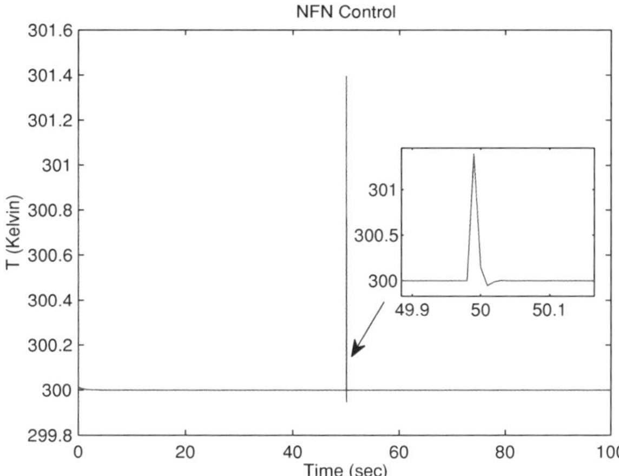

Fig. 6.6 Response with NFN controller. Reactant temperature, T

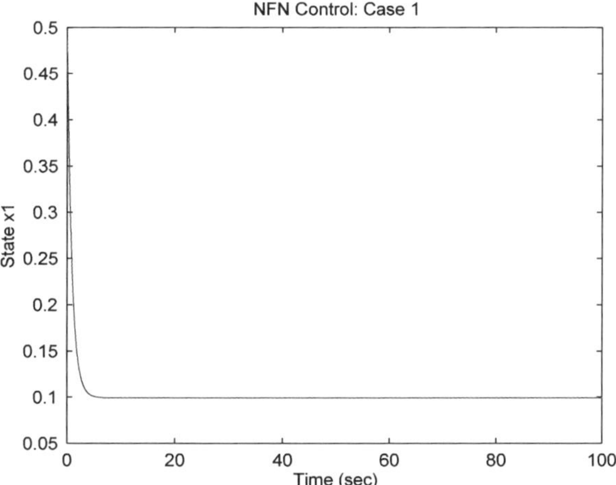

Fig. 6.7 Response with NFN controller. The state $x_1(t)$

arrays (electronic measurements and power signals), which come between sensor measurements and the microcontroller as well as between the digital signals of the microcontroller and the analog inputs of the motor.

By the cooperation of the Laboratory of Automatic Control Systems and the Electrical Machines Laboratory,[1] an experimental device has been setup in the framework of a diploma thesis (Sotiriadis 2010). In this setup the NF algorithm presented in Chap. 3 was applied and displayed an expected good performance. Simulations with a DC motor have been already presented in Chap. 3. In the following subsection, we proceed to a brief description of the most significant parts that

---

[1] The authors wish to acknowledge the contribution of Dr. A. Karlis from the Electrical Machines Laboratory in setting up the experiment and thank Mr. Sotiriadis for his help and the permission to include some of the experimental results in this book.

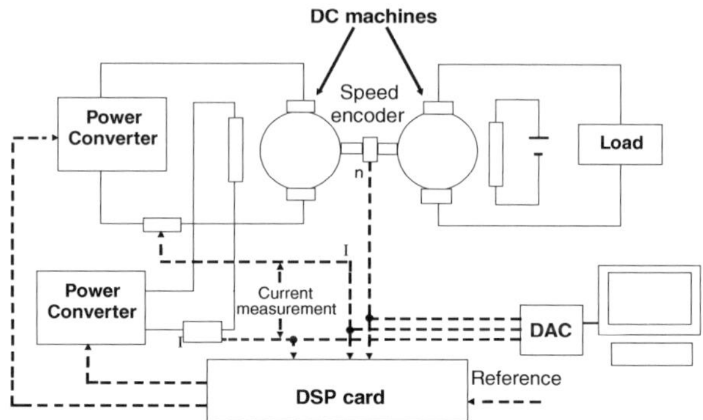

Fig. 6.8 Schematic diagram of the experimental setup

have been used in the experimental setup and we report on the performance of the algorithm.

6.3.1 Description of the Experiment

The general schematic diagram of the experiment is shown in Fig. 6.8. The most important parts of the experiment are the following:

Two separately excited dc machines: One is operating as a motor providing the second one with mechanical power, which in turn is operating as a generator supplying a resistor bank. In the actual experimental setup the motor is accompanied by a power supply and by a dynamometer.

A Digital Signal Processor (DSP) card: The PIC32 integrated circuits were used, which are produced by the Microchip Technology Inc., and belong to the category of microcontrollers. Pic embeds a central processing unit and a memory. Although the size of the memory for the code storage is not big, it is sufficiently enough for most of the applications. The Pic architecture is Harvard-type and is called Risc controller (Reduced instruction set controller) by the manufacturing company. The model of the micro controller that has been used in this experiment is PIC32MX360F512L which is embedded to PIC32 Starter Kit.

The applications of the Pic microcontroller that are of interest for our experimental setup are the following:

- A/D converter
- An output comparator for the production of PWM signal
- A serial output UART for recording data.
  The programming of the Pic takes place in a MPLAB environment of microchip with a language C Compiler, reinforced with many libraries for the programming of microcontroller.

H-bridge or full bridge: The operation of the driving circuit for the excitation of the motor is similar to a driving circuit of the motor armature. Thus, in this case two subcircuits are distinguished, the one is driving the motor and the other the control. Once again for the driving, a H-bridge or otherwise full bridge, is used as well as a control circuit similar to that of the motor armature. The motor excitation has a nominal current value 0.3 A and a nominal voltage value 220 V, with ohmic resistance 550 Ω. The transistors which are used for the implementation of the bridge that drives the armature of the motor (IRF740), are also used for the implementation of the H-bridge that drives the excitation of the motor. Schematically, the H-bridge which has been implemented for the excitation of the motor is the same with that of the armature.

Measurement circuit for sensor measurements with the use of optocouplers: For the electrical isolation of PIC32 from the circuit measurements of motor armature current I_a and field current I_f and the analog output voltage we use optocoupler IL300. The linear optocoupler IL300 consists of an infrared led, an isolated feedback, and an output optodiode. The measurement circuit measures the armature current I_a, the field current I_f, and motor rotations.

Shunt resistors: For the measurements of the armature or field current have been used watt resistances of 1.2 Ω in order to measure the voltage which depends on the current leakage.

A personal computer: A serial port (RS 232) is interfaced between the computer and the PIC microcontroller in order to acquire the speed and the current information.

6.3.2 Experiment

We present a simple regulation task, aiming solely to test the performance of the indirect regulation scheme given in Chap. 3. The controller is applied to the motor when it has some velocity and it drives it to zero. The experimental circuit has been developed in such a way as to drive the motor to work in four quadrants. The programming of the microcontroller is made in order to accomplish simple actions such as digitize, sampling and sensors' data recording as well as to implement our algorithm given in Sect. 3.2.1. The data acquisition in the PC, the data acquisition and the control in the DSP card is accomplished using C++ programming.

The setup can be used for armature and field weakening control of the separately excited DC motor but it is also used to test the proposed control algorithm, which assumes control of both armature and field excitation. Motor data are

Table 6.2 Motor data used in the setup

Parameters (nominal)	Value
Armature voltage	220 V
Field voltage	220 V
Armature current	1.3 A
Field current	0.3 A
Power	175 W
Velocity	1450 rpm

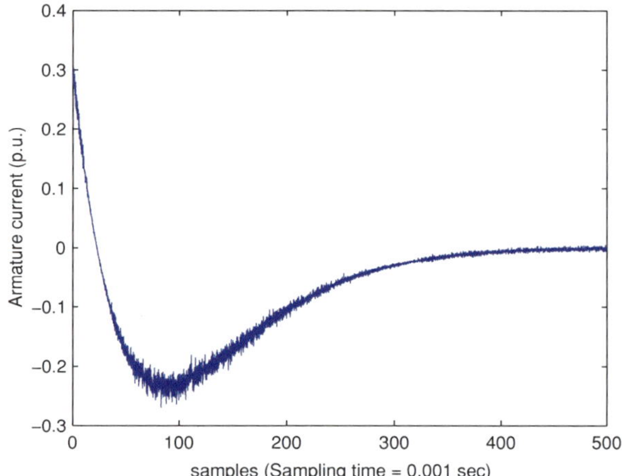

Fig. 6.9 Convergence of current to zero from 0.3 (p.u) initially

given in Table 6.2. The control algorithm performs weight updating and control input calculation according to Eqs. (3.43), (3.46), (3.48).

Similarly to the simulations given in Chap. 3, we assume a second order affine in the control F-RHONNs modeling of the system. The number of fuzzy partitions in X_f was chosen to be $p = 5$ and the depth of high order terms was $k = 5$ (up to second order sigmoidal terms $s(x_i)$, were used). The number of fuzzy partitions of each g_{ii} in X_g is $p = 3$. The control objective was to drive both angular velocity and armature current to zero starting from 0.3 of their nominal values. The code of the algorithm was programmed in C and compiled to run in the Pic. The results of the experiment are shown in Figs. 6.9 and 6.10, verifying the effectiveness of the proposed technique.

In Figs. 6.9 and 6.10, we notice that the motor velocity as well as the armature current are both driven to zero from an initial value as it has been proved by our algorithm. The motor velocity is normalized and is displayed in the vertical axis of Fig. 6.10 as a percentage of its nominal value. Thus, when our algorithm is applied the motor has $0.3 \cdot 1450 = 435$ rotations per minute and after the controller is applied the velocity goes to zero.

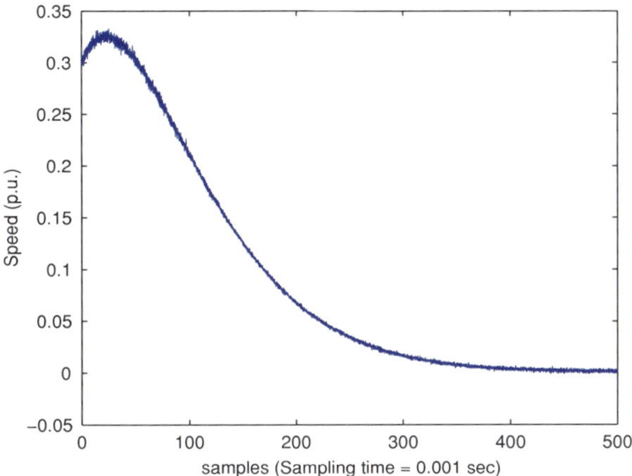

Fig. 6.10 Convergence of speed to zero from 0.3 (p.u) initially

6.4 Summary

In this chapter, we presented a number of selected applications of the NF modeling approach on the adaptive control of unknown nonlinear plants. In the first place, we presented a NF controller for robot manipulators along with the respective theoretical background and a chemical bioreactor both at a simulation level. Also, we gave a brief description of an experimental setup of a DC motor control, built in the Laboratory of Automatic Control Systems, where our NF algorithm for indirect adaptive control was applied. The presented applications demonstrate the applicability of proven theoretical results in the engineering field.

As concerning the robot manipulator, a NF representation of unknown robotic functions f, which is used in a filtered-error approximation-based robotic controller was presented. It was shown that in the ideal case, where the modeling error and the disturbances are zero, the tracking error goes asymptotically to zero. In case that the modeling error and the unmodeled disturbances are not zero, a modified weight updating law was used, based on a weight hopping technique. It was shown that in this case the tracking error remains bounded. Simulation results demonstrate that the new controller exhibits better behavior in comparison with the performance of a conventional computed torque PD control under the presence of unknown disturbance. The bioreactor application illustrates the potency of the NF method, applying the results of Chap. 4, where it is shown that despite the presence of modeling errors, one can obtain once again asymptotic regulation quite well. Finally, the applicability of the method developed in Chap. 3 is tested on a real DC Motor system where it is shown that by following the proposed procedure one can obtain asymptotic regulation.

References

Arimoto, S. & Miyazaki, F. (1984). Stability and robustness of pid feedback control for robotics manipulators of sensory capabilities. In *1st International Symposium of Robotics Research* (pp. 385–407).

Boutalis, Y. S., Theodoridis, D. C., & Christodoulou, M. A. (2009). A new neuro fds definition for indirect adaptive control of unknown nonlinear systems using a method of parameter hopping. *IEEE Transactions on Neural Networks, 20,* 609–625.

Braganza, D., Dixon, W., Dawson, D., & Xian, B. (2008). Tracking control for robot manipulators with kinematic and dynamic uncertainty. *International Journal of Robotics and Automation, 23,* 117–126.

Craig, J. (1985). *Adaptive control of mechanical manipulators.* Reading, MA: Addison-Wesley.

EL-Hawwary, M. & Elshafei, A. (2006). Robust adaptive fuzzy control of a two-link robot arm. *International Journal of Robotics and Automation, 21,* 266–272.

Fateh, M. M., & Azarfar, A. (2009). Improving fuzzy control of robot manipulators by increasing scaling factors. *ICIC Express Letters, 3,* 513–518.

Ioannou, P., & Fidan, B. (2006). Advances in design and control series. In *Adaptive control tutorial.* Philadelphia: SIAM.

Lai, C. W., Chen, P. K., Chung, Y. N., & Hsu, C. H. (2009). Applying adaptive estimator to maneuvering tracking system. *ICIC Express Letters, 3,* 427–432.

Lewis, F. L., Abdallah, C. T., & Dawson, D. M. (1993). *Control of robot manipulators.* New York: Macmillan.

Lewis, F. L., Liu, K., & Yesildirek, A. (1995a). Neural net robot controller with guaranteed tracking performance. *IEEE Transactions on Neural Networks, 6,* 703–715.

Lewis, F.L., Yesildirek, A. & Liu, K. (1995b). Neural net robot controller: Structure and stability proofs. *Journal of Intelligent and Robotic Systems, 12,* 277–299.

Lewis, F. L., Yesildirek, A., & Liu, K. (1996). Multilayer neural net robot controller: Structure and stability proofs. *IEEE Transactions on Neural Networks, 7,* 1–12.

Lewis, F. L., Jaganathan, S. & Yesildirek, A. (1999). *Neural network control of robot manipulators and nonlinear systems.* Philadelphia: Taylor and Francis.

Li, B., Yang, X., Zhao, J., & Yan, P. (2009). Minimum time trajectory generation for a novel robotic manipulator. *International Journal of Innovative Computing, Information and Control, 5,* 369–378.

Liu, K. & Lewis, F.L. (1994). Robust control of a continuous stirred-tank reactor. In *Proceedings of the American Control Conference, Baltimore, Maryland.* (pp. 2350–2354).

Najim, K., Ikonen, E., & Ramirez, E. G. (2008). Trajectory tracking control based on a genealogical decision tree controller for robot manipulators. *International Journal of Innovative Computing, Information and Control, 4,* 53–62.

Ray, W. H. (1975). New approaches to the dynamics of nonlinear systems with implications of process and control system design. *Chemical Process Control, 2,* 245–267.

Sotiriadis, P. (2010). *Hardware implementation of a motor control unit for the application of nonlinear control algorithms.* Laboratory of Automatic Control Systems—Diploma Thesis (supervisor Y. Boutalis), Democritus University of Thrace (DUTH).

Spong, M. W., & Vidyasagar, M. (1989). *Robot dynamics and control.* New York: Wiley.

Theodoridis, D. C., Boutalis, Y. S. & Christodoulou, M. A. (2009a). A new neuro-fuzzy dynamical system definition based on high order neural network function approximators. In *European Control Conference ECC-09, Budapest, Hungary.*

Theodoridis, D. C., Boutalis, Y. S. & Christodoulou, M. A. (2009b). Direct adaptive control of unknown nonlinear systems using a new neuro-fuzzy method together with a novel approach of parameter hopping. *Kybernetica, 45,* 349–386.

Theodoridis, D. C., Boutalis, Y. S., & Christodoulou, M. A. (2010). A new neuro-fuzzy approach with robustness analysis for direct adaptive regulation of systems in brunovsky form. *International Journal of Neural Systems, 20,* 319–339.

Theodoridis, D. C., Boutalis, Y. S., & Christodoulou, M. A. (2011). A new adaptive neuro-fuzzy controller for trajectory tracking of robot manipulators. *International Journal of Robotics and Automation, 26*, 64–75.

Ungar, L.H. (1990). A bioreactor benchmark for adaptive network-based process control. In *Neural Networks for Control* (Chap. 16). Boston: MIT press.

Yu, W., & Li, X. (2006). PD control of robot with velocity estimation and uncertainties compensation. *International Journal of Robotics and Automation, 21*, 1–9.

Part II
The FCN Model

Chapter 7
Introduction and Outline of Part II

7.1 Introduction

Fuzzy Cognitive Maps (FCM) have been introduced by Kosko (1986b) based on Axelrod's work on cognitive maps (Axelrod 1976). They are inference networks using cyclic directed graphs that represent the causal relationships between concepts. They use a symbolic representation for the description and modeling of the system. In order to illustrate different aspects in the behavior of the system, a FCM consists of nodes where each one represents a system characteristic feature. The node interactions represent system dynamics. An FCM integrates the accumulated experience and knowledge on the system operation, as a result of the method by which it is constructed, i.e., by using human experts who know the operation of the system and its behavior. Different methodologies to develop FCM and extract knowledge from experts have been proposed in Stylios and Groumpos (1999, 2000), Stylios et al. (1999), Schneider et al. (1995).

Kosko enhanced the power of cognitive maps considering fuzzy values for their nodes and fuzzy degrees of interrelationships between nodes (Kosko 1986a, b). He also proposed the differential Hebian rule (Kosko 1986a) to estimate the FCM weights expressing the fuzzy interrelationships between nodes based on acquired data. After this pioneering work, fuzzy cognitive maps attracted the attention of scientists in many fields and have been used to model behavioral systems in many different scientific areas. Application examples can be found in political science (Craiger and Coovert 1994; Tsadiras and Kouskouvelis 2005), in economic field (Koulouriotis et al. 2001; Carvalho and Tome 2004; Xirogiannis and Glykas 2004; Glykas et al. 2005), in representing social scientific knowledge (Coban and Secme 2005) and describing decision-making methods (Kottas et al. 2004; Georgopoulos et al. 2003; Zhang et al. 1989). Other applications include failure modes and effects analysis (Pelaez and Bowles 1996), geographical information systems (Satur and Liu 1999a; Liu and Satur 1999; Satur and Liu 1999b), cellular automata (Carvalho et al. 2006), pattern recognition applications (Papakostas et al. 2006, 2008), and numerical and linguistic prediction of time series functions (Stach et al. 2008a). Fuzzy cognitive

Y. Boutalis et al., *System Identification and Adaptive Control*,
Advances in Industrial Control, DOI: 10.1007/978-3-319-06364-5_7,
© Springer International Publishing Switzerland 2014

maps have also been used to model the behavior and reactions of virtual worlds (Silva 1995; Dickerson and Kosko 1993; Xin et al. 2003), as a generic system for decision analysis (Zhang et al. 1989) and as coordinator of distributed cooperative agents (Zhang et al. 1992). The field of FCM applications is continuously growing over time. The interested reader may find a collection of selected applications in Glykas (2010), while a complete, updated list of applications can be found in the survey paper by Papageorgiou and Salmeron (2013).

Regarding FCM weight estimation and updating, a group of publications (Huerga 2002; Papageorgiou and Groumpos 2004; Papageorgiou et al. 2004; Aguilar 2002; Konar and Chakraborty 2005; Stach et al. 2008b) extend the initially proposed differential Hebian rule (Kosko 1986a) to achieve better weight estimation. Another group of methods for training FCM structure involves genetic algorithms and other exhaustive search techniques (Koulouriotis et al. 2001; Papageorgiou 2005; khan et al. 2004; Stach et al. 2005a, b), where the training is based on a collection of particular values of input output historical examples and on the definition of appropriate fitness function to incorporate design restrictions.

Various extensions of FCMs have been proposed in the literature (Hagiwara 1992; Zhang et al. 2003; Miao et al. 2001; Zhang et al. 2006; Miao and Liu 2000; Liu and Miao 1999; Zhou et al. 2006; Smarandache 2001; Kandasamy and Smarandache 2003; Kottos et al. 2007). Characteristic examples are the Dynamic Cognitive Networks (DCN) appearing in (Miao et al. 2001), the Fuzzy Causal Networks in (Zhang et al. 2006; Miao and Liu 2000; Zhang and Liu 2002; Zhou et al. 2006), while the neutrosophic cognitive maps appear in (Smarandache 2001; Kandasamy and Smarandache 2003). A complete list of the various extensions can be found in Papageorgiou and Salmeron (2013). Recently, Fuzzy Cognitive Networks (FCN) (Kottos et al. 2007), which is the topic of this book, have been proposed as a complete computational and storage framework to facilitate the use of FCM in cooperation with the physical system they describe.

FCNs and their storage mechanism assume that they reach equilibrium points, each one associated with a specific operation condition of the underlying physical system. However, the conditions under which FCMs and consequently FCNs reach an equilibrium point and whether this point is unique had not been adequately studied until recently. In the following sections we give the basic notions regarding the definition and operation of FCNs (and equivalently FCMS) and point out specific peculiarities that may appear in their convergence. This will be the stimulus for developing the material of the following chapters, which is devoted to deriving convergence conditions and weight updating laws that comply with them.

7.2 Basic Notions

The basic notions of FCNs are quite similar to those of FCMs. Therefore, the representation and notation presented in this section may apply to both of them. A graphical representation of FCN is depicted in Fig. 7.1. Each concept represents a characteristic of the system; in general it represents events, actions, goals, values, and trends of the

Fig. 7.1 FCN with 8 nodes

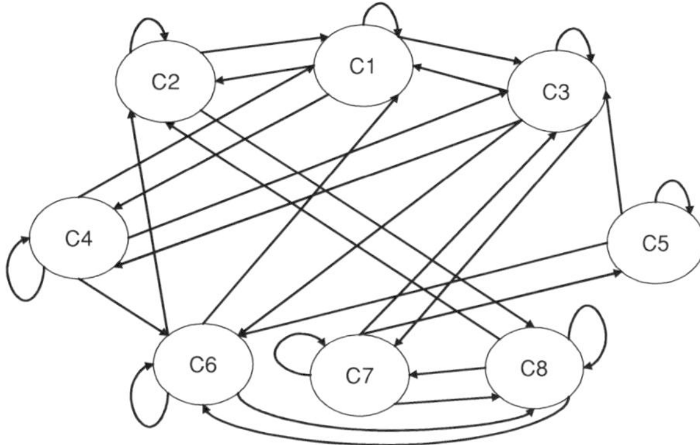

system. Each concept is characterized by a number A_i that represents its value and it results from the transformation of the real-value of the systems variable, represented by this concept, either in the interval [0, 1] or in the interval [−1, 1]. All concept values form vector A as follows:

$$A = \begin{bmatrix} A_1 & A_2 & \dots & A_n \end{bmatrix}^T$$

with n being the number of the nodes (in Fig. 7.1 $n = 8$) . Causality between concepts allows degrees of causality and not the usual binary logic, so the weights of the interconnections may range in the interval [−1,1].

The existing knowledge on the behavior of the system is stored in the structure of nodes and interconnections of the map. The value of w_{ij} indicates how strongly concept C_j influences concept C_i. The sign of w_{ij} indicates whether the relationship between concepts C_j and C_i is direct or inverse.

For the FCN of Fig. 7.1 matrix W is equal to

$$W = \begin{bmatrix} d_{11} & w_{12} & w_{13} & w_{14} & 0 & w_{16} & 0 & 0 \\ w_{21} & d_{22} & 0 & 0 & 0 & w_{26} & 0 & w_{28} \\ w_{31} & 0 & d_{33} & w_{34} & w_{35} & 0 & w_{37} & 0 \\ w_{41} & 0 & w_{43} & d_{44} & 0 & 0 & 0 & 0 \\ 0 & 0 & 0 & 0 & d_{55} & 0 & w_{57} & 0 \\ 0 & 0 & w_{63} & w_{64} & w_{65} & d_{66} & 0 & w_{68} \\ 0 & 0 & w_{73} & 0 & 0 & 0 & d_{77} & w_{78} \\ 0 & w_{82} & 0 & 0 & 0 & w_{86} & w_{87} & d_{88} \end{bmatrix}$$

The equation that calculates the values of concepts of Fuzzy Cognitive Networks, may or may not include self-feedback. In its general form it can be written as:

$$A_i(k) = f\left(\sum_{\substack{j=1 \\ j \neq i}}^{n} w_{ij} A_j(k-1) + d_{ii} A_i(k-1)\right) \tag{7.1}$$

where $A_i(k)$ is the value of concept C_i at discrete time k, $A_i(k-1)$ the value of concept C_i at discrete time $k-1$ and $A_j(k-1)$ is the value of concept C_j at discrete time $k-1$. w_{ij} is the weight of the interconnection from concept C_j to concept C_i and d_{ii} is a variable that takes on values in the interval $[0, 1]$, depending upon the existence of "strong" or "weak" self-feedback to node i.

Regarding the functions f used in FCMs, the following functions are usually found in the literature, allowing also different interpretation of their results:

The bivalent function (Dickerson and Kosko 2006; Tsadiras 2008)

$$f(x) = \begin{cases} 1 & x > 0 \\ 0 & x \leq 0 \end{cases}$$

The use of this function allows the activation of each concept to either 0 or 1, leading to the development of binary FCMs, where each concept is either activated or not.

The trivalent function (Dickerson and Kosko 2006; Tsadiras 2008)

$$f(x) = \begin{cases} 1 & x > 0 \\ 0 & x = 0 \\ -1 & x < 0 \end{cases}$$

In this case, when the concept takes the value 1, it means that this concept increases, when the activation level equals -1, it means that the concept decreases and when level equals to 0, it means that the concept is remaining stable.

The hyperbolic tangent (tansig) sigmoid function (Tsadiras 2008) with saturation level -1 and 1 is used both in FCMs and FCNs. In a general form it can be written as

$$f(x) = \tanh(c_l x) \text{ or } f(x) = \frac{e^{2c_l x} - 1}{e^{2c_l x} + 1}.$$

where $c_l > 0$ is the inclination parameter, used to adjust the inclination of the sigmoid function.

This function squashes the result in the interval $[-1, 1]$.

The sigmoid function with saturation level 0 and 1 (log-sigmoid) is also used both in FCMs and FCNs. f is a sigmoid function commonly used in the Fuzzy Cognitive Maps, which squashes the result in the interval $[0, 1]$ and is expressed as,

$$f = \frac{1}{1 + e^{-c_l x}}.$$

where $c_l > 0$ is used to adjust its inclination.

7.3 Convergence Peculiarities

Equation (7.1) can be rewritten as:

$$A(k) = f(W A(k-1)), \qquad (7.2)$$

where A is a vector containing the nodes values and W contains the weights of the FCN that have to be determined either by human experts or, most desirably, by parameter estimation based on sampled data. In case f are sigmoid, their inclination parameters could be also adjusted based on sampled data.

The application of Eq. (7.1) for each node A_i will calculate its value for the discrete time k based on the values of its influencing nodes at time $k-1$. Repetitive application of Eq. (7.1) or equivalently Eq. (7.2) will probably lead the FCN in an equilibrium point but this is not guaranteed. Alternatively, it may present a limit cycle or a chaotic behavior. Actually, the conditions under which FCMs and consequently FCNs reach an equilibrium point and whether this point is unique had not been adequately studied until recently. Simple FCMs have bivalent node values and trivalent edges (weights) and are equipped with binary threshold functions or sigmoid functions with very large inclination (Dickerson and Kosko 2006; Tsadiras and Margaritis 1997). According to Kosko (Kosko 1997; Dickerson and Kosko 2006), starting from an initial state, simple FCMs follow a path, which ends in a fixed point or limit cycle, while more complex ones may end in an aperiodic or "chaotic" attractor. These fixed points and attractors could represent *meta rules* of the form "If initial state then attractor or fixed point." A more detailed study on the performance of FCMs has been presented in Tsadiras (2008), where the inference capabilities of FCMs equipped with binary, trivalent, or sigmoid functions are compared. Similarly, in Bueno and Salmeron (2009) authors present comparisons stressing the significant advantage of using the sigmoid as activation function in FCMs regarding other functions such as hyberbolic tangent, step, and threshold.

It has been experimentally observed that small cognitive graphs having a small number of nodes and interconnections, equipped with squashing functions like the sigmoid ones, are more likely to reach an equilibrium point after a number of repetitions. Figure 7.3 shows the cognitive graph of a simplified representation of a hydroelectric power station, given in Fig. 7.2, presented in Kottos et al. (2007). It has two tanks (two dams) with two Hydro generators and the whole system is fed by a river, which is represented as an input (or steady) node, that influences but is not influenced by the other nodes. In this type of graphs, it is very likely that the FCM equipped with sigmoid node functions will converge to a fixed (or equilibrium) point that does not depend on the initial state of the nodes but on the weight values and on the input node(s) values.

The relation of the existence of fixed points to the weight interconnections of the FCM and FCN had not been fully investigated until recently. This is, however, of paramount importance if one wants to use FCNs with learning capabilities in

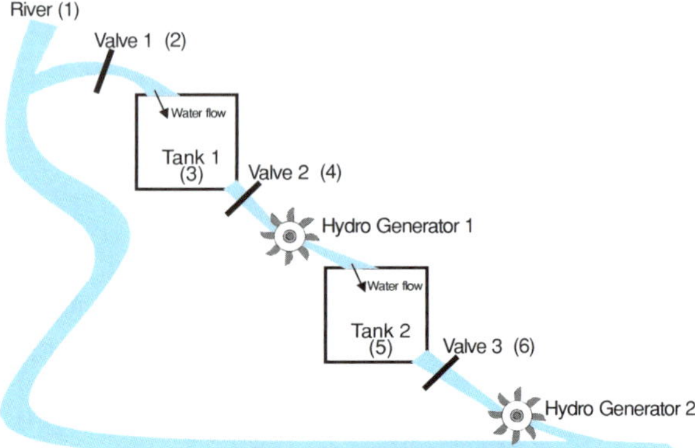

Fig. 7.2 Simplified diagram of a 2-dump hydroelectric power station

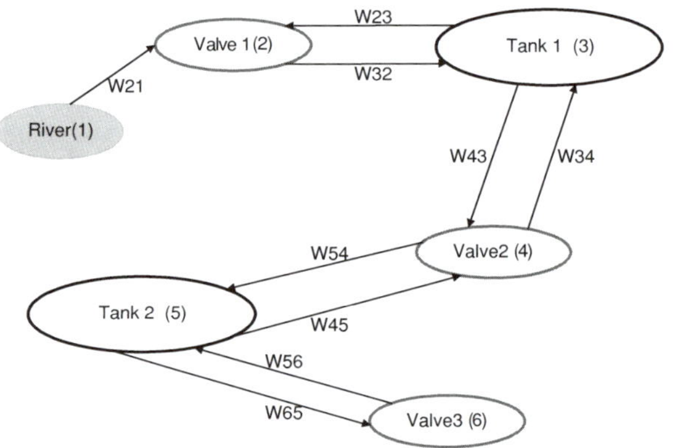

Fig. 7.3 The FCN for the hydroelectric plant

reliable adaptive system identification and control schemes. This is our motivation for devoting Chap. 8 of the book in this topic.

The study of existence of the above fixed points of FCNs equipped with continuous differentiable sigmoid functions has been for the first time introduced by the authors. This is done by using an appropriately defined contraction mapping theorem and the nonexpansive mapping theorem. A first study on this subject has already been presented by the authors in Boutalis et al. (2008, 2009), where however, only the weights of the graph were considered. This work has been extended in Kottos et al. (2012) to include both weights and inclination parameters. Chapter 8 presents a detailed analysis of this topic. It is proved that, when the weight interconnections fulfill certain conditions related to the size of the FCN and *the inclination of the sigmoid functions*, the concept values will converge to a unique solution regardless of their initial states, or in some cases a solution exists that may not necessarily be unique. Otherwise, the existence or the uniqueness of equilibria may or may not exist, it may depend on the initial states, but it cannot be assured. In case the FCN has also input nodes (that is nodes that influence but are not influenced by other nodes), the unique equilibrium does not depend solely on the weight and inclination sets, as in the case of FCNs with no input nodes; it depends also on the values of the input nodes. This in turn gives rise to *meta rules* of the form "If initial weights and input nodes then fixed point," which motivates the definition of the mode of operation of FCN, that is briefly introduced in the next section.

7.4 Fuzzy Cognitive Networks and Its Mode of Operation

FCN were first introduced in Kottos et al. (2007) as an alternative operational framework, allowing FCMs to work in close interaction with the system they describe and consequently become appropriate for control applications and adaptive decision making. The framework consists of

- The representation level. That is a cognitive graph with nodes equipped exclusively with sigmoid functions,
- The updating mechanism which receives feedback from the real system and
- The storage of the acquired knowledge throughout the operation.

The structure of the FCN graph is initially designed using experts knowledge regarding the various concepts or variables of the system and their interdependencies. Also, in this framework it is assumed that the FCN always reaches an equilibrium point, which is associated with an operation condition of the physical system and with a set of FCN parameters (weights and probably sigmoid inclination parameters). The weight updating mechanism serves the purpose of updating the network's weights so that they can represent the system's operating condition. Since different operating conditions might be associated with different weight sets, a fuzzy rule-based scheme was proposed in Kottos et al. (2007) in order to store the different associations. Figure 7.4 shows the interactive operation of the FCN with the system it describes. The input and output of the system consist of the desired FCN node values A^{des}. Given a set of FCN parameters, the FCN will converge to an equilibrium point A^{eq}. The error between A^{des} and A^{eq} is used by the FCN parameters estimator. Once the parameters are adequately adjusted, they contribute to the formation of the information related with the current operational condition, which is stored in the form of fuzzy meta rules. Details of the framework will be given later, in Chap. 10. It has, however, to be pointed out that, this kind of operation, associating operating conditions (equilibria) with different parameter sets, was intuitively conceived by the authors in (Kottos et al. 2005, 2007) without adequate theoretical justification. Moreover, the weight updating algorithms developed in Kottos et al. (2007) are based on a heuristic modification of a delta rule, which attempts to update the weights so that they are appropriate to drive the FCN in unique equilibria. However, this algorithm is not associated with conditions for the FCN parameters nor it theoretically guaranties that the weight updating procedure will always result in weights complying with these conditions. In Chapter 8, the conditions which determine the existence of a unique solution of Eq. (7.2) are derived and in Chapter 9 weight updating algorithm

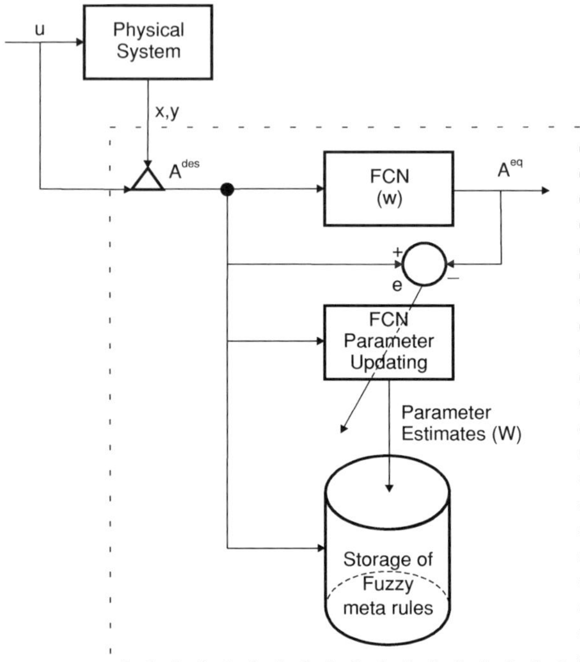

Fig. 7.4 Interactive mode of FCN operation

based on the linear and bilinear parametric models are presented, which guarantee that the parameters always fulfill them.

7.5 Goals and Outline of Part II

The main goals of Part II of this book are the following:

1. The presentation of basic theoretical results related to the existence and uniqueness of equilibrium points in FCN and their association with conditions on the size of the FCN and its parameters.
2. The derivation of adaptive parameter estimation algorithms which will estimate the parameters of the FCN related to specific operation conditions and will respect the conditions of existence of the equilibrium points.
3. The description of an operational framework of the FCN, which is in continuous interaction with the system it describes, learns from past operational conditions and can be used to control the physical system.
4. The presentation of a number of selected applications demonstrating various aspects of the application of the FCN framework in real-life engineering problems.

The material of Part II is deployed in several chapters as follows: In Chap. 8 , the existence and uniqueness of solutions of Eq. (7.2) is analyzed and conditions for the

FCN parameters are derived that guarantee this existence. The solution is actually the equilibrium point of a FCN equipped with sigmoidal nodes, when Eq. (7.2) is repetitively applied. Next, in Chap. 9, adaptive FCN parameter estimation algorithms are derived based on linear and bilinear parametric modeling of the FCN. The proposed algorithms are designed so that the updating procedure does not compromise the conditions of existence of solutions. Finally, Chap. 10, presents the operational framework of FCN and explains its learning mechanism. Moreover, it presents a number of selected applications that use the FCN framework to control benchmark or real-life plants.

More specifically, in Chap. 8 a study is presented for the existence of equilibrium points of FCNs equipped with continuous differentiable sigmoid functions. It is proved that when the weight interconnections fulfill certain conditions related to the size of the FCN and the inclination of the sigmoid functions, the concept values will converge to a unique solution regardless of their initial states, or in some cases a solution exists that may not necessarily be unique. Otherwise, the existence or the uniqueness of equilibria may or may not exist, it may depend on the initial states, but it cannot be assured. In case the FCN has also input nodes (that is nodes that influence but are not influenced by other nodes), the unique equilibrium does not depend solely on the weight set, as in the case of FCNs with no input nodes; it depends also on the values of the input nodes.

Next, in Chap. 9, these conditions form the basis for the derivation of adaptive estimation algorithms of the FCN parameters that fulfill the FCN equilibrium equation. Two algorithms are derived. The first one is based on a linear parametric modeling to describe the FCN equilibrium equation. The algorithm estimates only the weights of the FCN graph. The second algorithm uses a bilinear parametric model and estimates both the weights and the inclination parameters of the nodes' sigmoid functions. Moreover, appropriate modifications of these algorithms are derived in the form of parameter *projection* methods, which prevent the estimated parameters to deviate from values that violate the existence and uniqueness of equilibrium points conditions. The chapter presents simulations that demonstrate the effectiveness of both approaches and shows off their differences.

Finally, Chap. 10 is devoted to the explanation of the framework of operation of FCN and the presentation of a number of selected applications. Based on the conditions derived from Chap. 8 and the adaptive estimation algorithms deployed in Chaps. 9 and 10 introduces a fuzzy rule-based mechanism for storing acquired knowledge during FCN's operation and training. The selected applications demonstrate the different aspects of the FCN framework's use, which can be employed either in decision-making tasks or in the control of a plant based on its FCN equivalent modeling and an inverse control approach. The applications include a conventional benchmark plant and two real-life potential projects related to energy production from renewable power sources.

References

Aguilar, J. (2002). Adaptive random fuzzy cognitive maps, lecture notes in artificial intelligence. In F. J. Garijio, J. C. Riquelme & M. Toro (Eds.), *IBERAMIA 2002* (Vol. 2527, pp. 402–410). Berlin: Springer.

Axelrod, R. (1976). *Structure of decision: The cognitive maps of political elites.* New Jersey: Princeton University Press.

Boutalis, Y., Kottas, T., & Christodoulou, M. (2008). On the existence and uniqueness of solutions for the concept values in fuzzy cognitive maps. In *Proceedings of 47th IEEE Conference on Decision and Control—CDC'08* (pp. 98–104). December 9–11, 2008. Cancun, Mexico.

Boutalis, Y., Kottas, T., & Christodoulou, M. (2009). Adaptive estimation of fuzzy cognitive maps with proven stability and parameter convergence. *IEEE Transactions on Fuzzy Systems, 17,* 874–889.

Bueno, S., & Salmeron, J. L. (2009). Benchmarking main activation functions in fuzzy cognitive maps. *Expert Systems with Applications, 36,* 5221–5229.

Carvalho, J., & Tome, J. (2004). Qualitative modelling of an economic system using rule-based fuzzy cognitive maps. *IEEE International Conference on Fuzzy Systems, 2,* 659–664.

Carvalho, J., Carola, M., & Tome, J. (2006). Using rule-based fuzzy cognitive maps to model dynamic cell behavior in voronoi based cellular automata. In *IEEE International Conference on Fuzzy Systems* (pp. 1687–1694).

Coban, O., & Secme, G. (2005). Prediction of socio-economical consequences of privatization at the firm level with fuzzy cognitive mapping. *Information Sciences, 169,* 131–154.

Craiger, P., & Coovert, M. D. (1994). Modeling dynamic social and psychological processes with fuzzy cognitive maps. *IEEE World Congress on Computational Intelligence and Fuzzy Systems, 3,* 1873–1877.

Dickerson, J., & Kosko, B. (1993). Virtual worlds as fuzzy cognitive maps. In *Virtual Reality Annual International Symposium* (pp. 471–477).

Dickerson, J., & Kosko, B. (2006). Virtual worlds as fuzzy cognitive maps. *Presence, 3,* 173–189.

Koulouriotis, D., Diakoulakis, I., & Emiris, D. (2001). Learning fuzzy cognitive maps using evolution strategies: A novel schema for modeling a simulating high-level behavior. *Proceedings of IEEE Congress on Evolutionary Computation, 1,* 364–371.

Georgopoulos, V., Malandraki, G., & Stylios, C. (2003). A fuzzy cognitive map approach to differential diagnosis of specific language impairment. *Artificial Intelligence in Medicine, 29,* 261–278.

Glykas, M. (Ed.). (2010). Studies in fuzziness and soft computing. *Fuzzy cognitive maps: Advances in theory, methodologies, tools and applications.* Berlin: Springer.

Glykas, M., & Xirogiannis, G. (2005). A soft knowledge modeling approach for geographically dispersed financial organizations. *Soft Computing, 9,* 579–593.

Hagiwara, M. (1992). Extended fuzzy cognitive maps. In *IEEE International Conference on Fuzzy Systems* (pp. 795–801).

Huerga, A. (2002). A balanced differential learning algorithm in fuzzy cognitive maps. In *Proceedings of the 16th International Workshop on Qualitative Reasoning, poster.*

Kandasamy, V., & Smarandache, F. (2003). Fuzzy cognitive maps and neutrosophic cognitive maps. In *ProQuest information and learning.* University of Microfilm International.

Khan, M., Khor, S., & Chong, A. (2004). Fuzzy cognitive maps with genetic algorithm for goal-oriented decision support. *International Journal of Uncertainty, Fuzziness and Knowledge-Based Systems, 12,* 31–42.

Konar, A., & Chakraborty, U. K. (2005). Reasoning and unsupervised learning in a fuzzy cognitive map. *Information Sciences, 170,* 419–441.

Kosko, B. (1986a). Differential hebbian learning. In *American Institute of Physics Conference Proceedings 151 on Neural Networks for, Computing* (pp. 277–282).

Kosko, B. (1986b). Fuzzy cognitive maps. *International Journal of Man-Machine Studies, 24,* 65–75.

Kosko, B. (1997). *Fuzzy engineering.* Boca Raton: Prentice-Hall.

Kottas, T., Boutalis, Y., Devedzic, G., & Mertzios, B. (2004). A new method for reaching equilibrium points in fuzzy cognitive maps. In *Proceedings of 2nd International IEEE Conference of Intelligent Systems* (pp. 53–60). Varna, Burgaria.

Kottas, T., Boutalis, Y., & Christodoulou, M. (2005). A new method for weight updating in fuzzy cognitive maps using system feedback. In *2nd International Conference on Informatics in Control, Automation and Robotics (ICINCO05)* (pp. 202–209). Barcelona, Spain.

Kottas, T., Boutalis, Y., & Christodoulou, M. (2007). Fuzzy cognitive networks: A general framework. *Inteligent Desicion Technologies, 1,* 183–196.

Kottas, T., Boutalis, Y., & Christodoulou, M. (2012). Bi-linear adaptive estimation of fuzzy cognitive networks. *Applied Soft Computing, 21,* doi:10.1016/j.asoc.2012.01.025

Koulouriotis, D., Diakoulakis, I., & Emiris, D. (2001). A fuzzy cognitive map-based stock market model: Synthesis, analysis and experimental results. In *10th IEEE International Conference on Fuzzy Systems* (pp. 465–468).

Liu, Z. Q., & Miao, Y. (1999). Fuzzy cognitive map and its causal inferences. *IEEE International Conference on Fuzzy Systems, 3,* 1540–1545.

Liu, Z. Q., & Satur, R. (1999). Contextual fuzzy cognitive map for decision support in geographic information systems. *IEEE Transactions on Fuzzy Systems, 7,* 495–507.

Miao, Y., & Liu, Z. Q. (2000). On causal inference in fuzzy cognitive maps. *IEEE Transactions on Fuzzy Systems, 8,* 107–119.

Miao, Y., Liu, Z., Siew, C., & Miao, C. (2001). Dynamical cogntive network-an extension of fuzzy cognitive map. *IEEE Transactions on Fuzzy Systems, 9,* 760–770.

Papageorgiou, E., & Groumpos, P. (2004). A weight adaptation method for fuzzy cognitive maps to a process control problem. In *International Conference on Computational Science (ICCS 2004).* June 6–9, 2004 (Vol. 2, pp. 515–522). Krakow, Poland. (Lecture notes in computer science 3037, Berlin: Springer)

Papageorgiou, E., & Salmeron, J. L. (2013). A review of fuzzy cognitive maps research during the last decade. *IEEE Transactions on Fuzzy Systems, 21,* 66–79.

Papageorgiou, E., Stylios, C., & Groumpos, P. (2004). Active hebbian learning algorithm to train fuzzy cognitive maps. *International Journal of Approximate Reasoning, 37,* 219–247.

Papageorgiou, E., Parsopoulos, K., Stylios, C., Groumpos, P., & Vrahatis, M. (2005). Fuzzy cognitive maps learning using particle swarm optimization. *International Journal of Intelligent Information Systems, 25,* 95–121.

Papakostas, G., Boutalis, Y., Koulouriotis, D., & Mertzios, B. (2006). A first study of pattern classification using fuzzy cognitive maps. In *International Conference on Systems, Signals and Image Processing—IWSSIP'06* (pp. 369–374).

Papakostas, G., Boutalis, Y., Koulouriotis, D., & Mertzios, B. (2008). Fuzzy cognitive maps for pattern recognition applications. *International Journal at Pattern Recognition and Artificial Intelligence, 22,* 1461–1486.

Pelaez, C. E., & Bowles, J. B. (1996). Using fuzzy cognitive maps as a system model for failure modes and effects analysis. *Information Sciences, 88,* 177–199.

Satur, R., & Liu, Z. Q. (1999a). A contextual fuzzy cognitive map framework for geographic information systems. *IEEE Transactions on Fuzzy Systems, 7,* 481–494.

Satur, R., & Liu, Z. Q. (1999b). Contextual fuzzy cognitive maps for geographic information systems. In *IEEE International Conference on Fuzzy Systems* (Vol. 2, pp. 1165–1169).

Schneider, M., Shnaider, E., Kandel, A., & Chew, G. (1995). Constructing fuzzy cognitive maps. In *International Joint Conference of the Fourth IEEE International Conference on Fuzzy Systems and the 2nd International Fuzzy Engineering Symposium* (Vol. 4, pp. 2281–2288).

Silva, P. (1995). Fuzzy cognitive maps over possible worlds. In *International Joint Conference of the 4th IEEE International Conference on Fuzzy Systems and the 2nd International Fuzzy Engineering Symposium* (Vol. 2, pp. 555–560).

Smarandache, F. (2001). An introduction to neutrosophy, neutrosophic logic, neutrosophic set, and neutrosophic probability and statistics. In *Proceedings of the 1st International Conference* on Neutrosophy, Neutrosophic Logic, Neutrosophic Set, Neutrosophic Probability and Statistics University of New Mexico—Gallup (Vol. 1–3, pp. 5–22).

Stach, W., Kurgan, L., Pedrycz, W., & Reformat, M. (2005a). Evolutionary development of fuzzy cognitive maps. In *14th IEEE International Conference on Fuzzy Systems* (pp. 619–624).

Stach, W., Kurgan, L., Pedrycz, W., & Reformat, M. (2005b). Genetic learning of fuzzy cognitive maps. *Fuzzy Sets and Systems, 153*, 371–401.

Stach, W., Kurgan, L., & Pedrycz, W. (2008a). Numerical and linguistic prediction of time series with the use of fuzzy cognitive maps. *IEEE Transactions on Fuzzy Systems, 16*, 61–72.

Stach, W., Kurgan, L. A., & Pedrycz, W. (2008b). Data-driven nonlinear hebbian learning method for fuzzy cognitive maps. In *2008 World Congress on Computational Intelligence WCCI'08*.

Stylios, C., & Groumpos, P. (1999). A soft computing approach for modelling the supervisor of manufacturing systems. *Journal of Intelligent and Robotics Systems, 26*, 389–403.

Stylios, C., & Groumpos, P. (2000). Fuzzy cognitive maps in modelling supervisory control systems. *Journal of Intelligent and Fuzzy Systems, 8*, 83–98.

Stylios, C., Groumpos, P., & Georgopoulos, V. (1999). Fuzzy cognitive map approach to process control systems. *Journal of Advanced Computational Intelligence, 3*, 409–417.

Tsadiras, A. (2008). Comparing the inference capabilities of binary, trivalent and sigmoid fuzzy cognitive maps. *Information Science, 178*, 3880–3894.

Tsadiras, A., & Kouskouvelis, I. (2005). Using fuzzy cognitive maps as a decision support system for political decisions: The case of Turkey's integration into the European union. *Lecture Notes in Computer Science, 3746*, 371–381.

Tsadiras, A., & Margaritis, K. (1997). Cognitive mapping and certainty neuron fuzzy cognitive maps. *Information Sciences, 101*, 109–130.

Xin, J., Dickerson, J., & Dickerson, J. (2003). Fuzzy feature extraction and visualization for intrusion detection. In *12th IEEE International Conference on Fuzzy Systems* (pp. 1249–1254).

Xirogiannis, G., & Glykas, M. (2004). Fuzzy cognitive maps in business analysis and performance driven change. *IEEE Transactions on Engineering Management, 51*, 334–351.

Zhang, J., Liu, Z. Q., & Zhou, S. (2006). Dynamic domination in fuzzy causal networks. *IEEE Transactions on Fuzzy Systems, 14*, 42–57.

Zhang, J. Y., & Liu, Z. Q. (2002). Dynamic domination for fuzzy cognitive maps. *IEEE International Conference on Fuzzy Systems, 1*, 1145–1149.

Zhang, J. Y., Liu, Z. Q., & Zhou, S. (2003). Quotient fcms—a decomposition theory for fuzzy cognitive maps. *IEEE Transactions on Fuzzy Systems, 11*, 593–604.

Zhang, W., Chen, S., & Bezdek, J. (1989). A generic system for cognitive map development and decision analysis. *IEEE Transactions on Systems, Man, and Cybernetics, 19*, 31–39.

Zhang, W., Chen, S., Wang, W., & King, R. (1992). A cognitive map based approach to the coordination of distributed cooperative agents. *IEEE Transactions on Systems, Man, and Cybernetics, 22*, 103–114.

Zhou, S., Liu, Z. Q., & Zhang, J. Y. (2006). Fuzzy causal networks: General model, inference and convergence. *IEEE Transactions on Fuzzy Systems, 14*, 412–420.

Chapter 8
Existence and Uniqueness of Solutions in FCN

8.1 Introduction

We recall from Sect. 7.3, that the repetitive equation that computes the new node values based on past ones is given by (7.1), or equivalently by (7.2), which is repeated here as

$$A(k) = f(WA(k-1)),\qquad(8.1)$$

with W being the weight matrix of the FCN and A is the vector containing the node values.

We will now check the existence of solutions in Eq. (8.1), when a continuous and differentiable node function is used, such as sigmoid functions are. We know that the allowable values of the elements of FCN node vectors A lie either in the closed interval $[0, 1]$ or in the closed interval $[-1, 1]$. This is a subset of $\Re$ and is a complete metric space with the usual L_2 metric. Following the analysis given in (Boutalis et al. 2008, 2009; Kottas et al. 2012), we will define the regions where the FCN has a unique solution, which does not depend on the initial condition since it is the unique equilibrium point. The analysis will be based on the contraction mapping theorem and the conditions required so that the sigmoid functions of the FCN nodes retain their contractive properties.

8.2 The Contraction Mapping Principle in Sigmoidal Nodes

We start by introducing the Contraction Mapping Theorem given in Rudin (1964).

Definition 8.1 Let X be a metric space, with metric d. If φ maps X into X and there is a number $0 < c < 1$ such that

$$d(\varphi(x), \varphi(y)) \le cd(x, y)\qquad(8.2)$$

Y. Boutalis et al., *System Identification and Adaptive Control*,
Advances in Industrial Control, DOI: 10.1007/978-3-319-06364-5_8,
© Springer International Publishing Switzerland 2014

for all $x, y \in X$, then φ is said to be a contraction of X into X.

Theorem 8.1 *Rudin* (1964) *If X is a complete metric space, and if φ is a contraction of X into X, then there exists one and only one $x \in X$ such that $\varphi(x) = x$.*

In other words, φ has a unique fixed point. The uniqueness follows from the fact that if $\varphi(x) = x$ and $\varphi(y) = y$, then (8.2) gives $d(x, y) \leq cd(x, y)$, which can only happen when $d(x, y) = 0$ (See Rudin (1964)).

Equation (8.1) can be written as:

$$A(k) = G(A(k-1)) \tag{8.3}$$

where $G(A(k-1))$ is equal to $f(WA(k-1))$.

In FCN $A \in [0, 1]^n$ or $A \in [-1, 1]^n$ and it is also clear according to (8.1) that $G(A(k-1)) \in [0, 1]^n$ or $G(A(k-1)) \in [-1, 1]^n$ depending on which squashing sigmoid function is used. If the following inequality is true:

$$d(G(A), G(A')) \leq cd(A, A')$$

where A and A' are different vectors of concept values and G is defined in (8.3), then G has a unique fixed point A such that:

$$G(A) = A.$$

Before presenting the main theorem we need to explore the role of f as a contraction function.

Theorem 8.2 *The scalar sigmoid function f, ($f = \frac{1}{1+e^{-x}}$) is a contraction of the metric space X into X, were $X = [a, b], a, b,$ finite, according to Definition 8.1, where:*

$$d(f(x), f(y)) \leq cd(x, y). \tag{8.4}$$

Proof Here f is the sigmoid function, $x, y \in X$, X is as defined above and c is a real number such that $0 < c < 1$.

The inclination ℓ_x of a sigmoid function $f(x) = \frac{1}{1+e^{-c_l x}}$ is equal to:

$$\ell_x = \frac{\partial f}{\partial x} = \frac{c_l e^{-c_l x}}{\left(1 + e^{-c_l x}\right)^2} = \frac{c_l}{e^{c_l x}} \left(\frac{1}{1 + e^{-c_l x}}\right)^2 = \frac{c_l}{e^{c_l x}} f^2 \tag{8.5}$$

for $x \in X$. and for ℓ also holds that:

$$\frac{d(f(x_1), f(x_2))}{d(x_1, x_2)} \leq \ell_z \tag{8.6}$$

where ℓ_z is equal to:

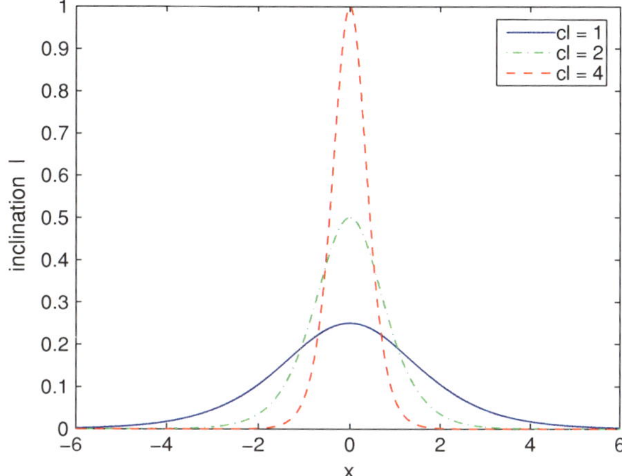

Fig. 8.1 Inclination of sigmoid function $f = \frac{1}{1+e^{-c_l x}}$ when $c_l = 1, 2, 4$

$$\ell_z = \begin{cases} \ell_{x_1}, & \text{if } |x_1| < |x_2| \\ \ell_{x_2}, & \text{if } |x_2| < |x_1|. \end{cases}$$

From (8.5) and (8.6) we get:

$$\frac{\mathrm{d}(f(x_1), f(x_2))}{\mathrm{d}(x_1, x_2)} \leq \ell_z = \frac{c_l}{e^{c_l z}} f^2. \tag{8.7}$$

Equation (8.5) for $c_l = 1, 2, 4$ is plotted in Fig. 8.1. According to Eq. (8.5) one can see that the inclination ℓ of $f(x)$ in the bounded set X depends on c_l and x. In particular, taking into account Fig. 8.1 one can conclude that when $c_l < 4$ the inclination is always smaller than 1 regardless the value of x. Consequently, for the sigmoid with $c_l \geq 4$ the contraction mapping is not valid for every x. There is an interval, for which it is not valid. Figure 8.2 shows the inclination of the sigmoid function when $c_l = 5$. It can be seen that when $-0.1925 < x < 0.1925$ the inclination exceeds 1. In this interval, if one wants to keep the contraction property of the sigmoid used he should probably consider lowering c_l.

We assume for simplicity that $c_l = 1$. More general results, for any values of c_l, will be derived by corollary 8.1 that follows. According to Fig. 8.1 when $c_l = 1$ it always holds that:

$$\frac{1}{4} \geq \ell \tag{8.8}$$

and for any x, y

$$\frac{\mathrm{d}(f(x), f(y))}{\mathrm{d}(x, y)} \leq 1/4. \tag{8.9}$$

From (8.8) and (8.9) we get:

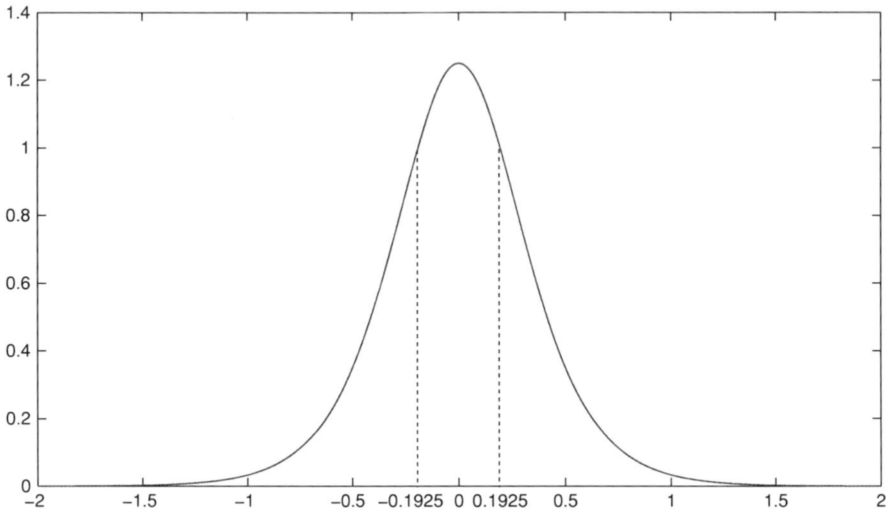

Fig. 8.2 Inclination of sigmoid function $f = \frac{1}{1+e^{-c_l x}}$ when $c_l = 5$

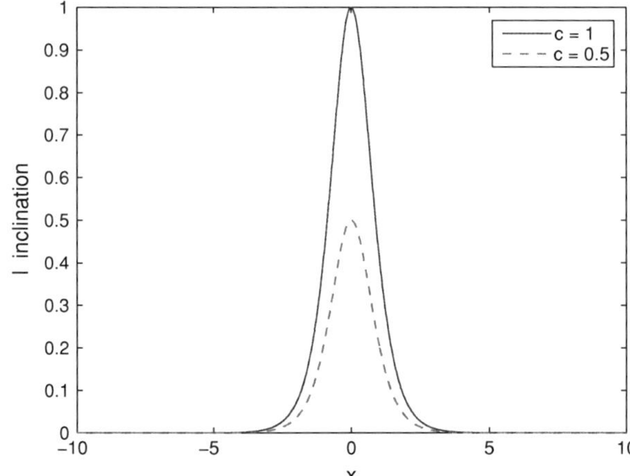

Fig. 8.3 Inclination of sigmoid function $f = \tanh(c_l x)$ when $c_l = 1$ and $c_l = 0.5$ respectively

$$\frac{d(f(x), f(y))}{d(x, y)} < 1. \tag{8.10}$$

Thus, there is always a number c for which $0 \le c < 1$, such that (8.10) is:

$$\frac{d(f(x), f(y))}{d(x, y)} < c < 1. \tag{8.11}$$

Theorem 8.2 can be easily expanded for the continuous and differentiable sigmoid function $f = \tanh(c_l x)$. The inclination ℓ of $f = \tanh(c_{li} x)$ is equal to:

$$\ell = c_l(1 - f^2) \tag{8.12}$$

and its plot for $c_l = 1$ and $c_l = 0.5$ is given in Fig. 8.3. According to Fig. 8.3 one can see that for $c_l = 1$ the inclination l of $f(x)$ in the bounded set X is always smaller than 1. Thus for the hyperbolic tangent function we get:

$$\frac{d(f(x), f(y))}{d(x, y)} \leq \ell_z$$

and there is always a number c, such that:

$$\frac{d(f(x), f(y))}{d(x, y)} \leq c \leq 1$$

In case where $0 < c \leq 1$, that is we allow $c = 1$, the map f becomes nonexpansive and the following theorem holds:

Theorem 8.3 *The scalar sigmoid function f, ($f = \frac{1}{1+e^{-x}}$) is a nonexpansive map of the metric space X into X, were $X = [a, b]$, a, b finite, where:*

$$d(f(x), f(y)) \leq c\,d(x, y) \tag{8.13}$$

and $0 < c \leq 1$.

Proof Use the results in the previous theorem together with the Browder-Gohde-Kirk theorem Kranas and Dugundji (2003, p. 52) for nonexpansive maps in Hilbert spaces. Here, it should be noted that a solution exists but it **is not unique**.

The above theorems were derived assuming that $c_l = 1$. Actually, the analysis is valid provided that $c_l < 4$ in log-sigmoid function and $c_l \leq 1$ in the hyperbolic tangent. In the following, corollary we expand the results to include any value of c_l.

Corollary 8.1 *The scalar sigmoid function f, ($f = \frac{1}{1+e^{-c_l x}}$) is a contraction of the metric space X into X, were $X = [a, b]$, a, b, finite, according to Definition 8.1, if the following inequality is true :*

$$\frac{c_l}{e^{c_l x}} f^2 < 1. \tag{8.14}$$

Similarly, the scalar sigmoid function f, ($f = \tanh(c_l x)$) is a contraction of the metric space X into X, were $X = [a, b]$, a, b, finite, according to Definition 8.1, if

$$c_l(1 - f^2) < 1. \tag{8.15}$$

Proof The proof follows immediately from Theorem 8.2 and the Eqs. (8.5), (8.12), by requiring $\ell < 1$.

Remark 8.1 Similar to Theorem 8.3, if the left hand part of (8.14), (8.15) is allowed to become equal to 1 the solutions exist but they are not unique.

8.3 Conditions for the Existence and Uniqueness of Solutions in FCN

Using the contraction properties of the sigmoid functions and the theorems of the previous section, we can proceed to the next theorem which determines the conditions for the parameters of the FCN so that its unique equilibrium always exists. It is evident that the parameters of the FCN are its weights and probably the inclination parameters of the involved sigmoid functions.

Theorem 8.4 *There is one and only one solution for any concept value A_i of any FCN where the sigmoid function $f = \frac{1}{1+e^{-c_l x}}$ is used, if:*

$$\left(\sum_{i=1}^{n} (c_{l_i} \ell_i \, \|w_i\|)^2 \right)^{1/2} < 1 \tag{8.16}$$

where w_i is the ith row of matrix W, $\|w_i\|$ is the L_2 norm of w_i, ℓ_i is the inclination of function f, equal to $\ell_i = \frac{c_{l_i}}{e^{c_{l_i} w_i \cdot A}} f^2(c_{l_i} w_i \cdot A)$, and c_{l_i} is the inclination parameter c_l of function f corresponding to concept A_i.
If

$$\left(\sum_{i=1}^{n} (c_{l_i} \ell_i \, \|w_i\|)^2 \right)^{1/2} = 1 \tag{8.17}$$

*there exists at least one solution for each concept value A_i of any FCN, which is **not necessarily unique**.*

Proof Let X be the complete metric space $[a, b]^n$ and $G : X \to X$ be a map such that:

$$d(G(A), G(A')) \le c\, d(A, A') \tag{8.18}$$

for some $0 < c < 1$.
   Vector G is equal to:

$$G = \left[\, f(c_{l_1}(w_1 \cdot A)) \; f(c_{l_2}(w_2 \cdot A)) \; \cdots \; f(c_{l_n}(w_n \cdot A)) \,\right]^{T} \tag{8.19}$$

where n is the number of concepts of the FCM, f is the sigmoid function $f = \frac{1}{1+e^{-x}}$, w_i is the ith row of matrix W of the FCM, where $i = 1, 2, \ldots, n$, and by $\cdot$ we denote the inner product between two equidimensional vectors which both belong to $\Re^n$.

Assume that A and A' are two different concept values for the FCM. Then we want to prove the following inequality:

$$\|G(A) - G(A')\| \leq c \|A - A'\|. \tag{8.20}$$

But $\|G(A) - G(A')\|$ according to (8.19) equals to:

$$\|G(A) - G(A')\| = \left(\sum_{i=1}^{n} (f(c_{l_i}(w_i \cdot A)) - f(c_{l_i}(w_i \cdot A')))^2 \right)^{1/2}.$$

According to Theorem 8.2 and Corollary 8.1, for the scalar argument of $f(.)$, which is $c_{l_i}(w_i \cdot A)$ in the bounded and closed interval $[a, b]$ with a and b being finite numbers, it is true that:

$$\left| f(c_{l_i}(w_i \cdot A)) - f(c_{l_i}(w_i \cdot A')) \right| \leq \ell_i \left| (c_{l_i}(w_i \cdot A)) - (c_{l_i}(w_i \cdot A')) \right|$$

for every $i = 1, 2, \ldots, n$,

where $\ell_i = \max \left(\frac{c_{l_i}}{e^{c_{l_i} w_i \cdot A}} f^2(c_{l_i} w_i \cdot A), \frac{c_{l_i}}{e^{c_{l_i} w_i \cdot A'}} f^2(c_{l_i} w_i \cdot A') \right)$.

By using the Cauchy-Schwartz inequality we get:

$$\ell_i |(c_{l_i}(w_i \cdot A)) - (c_{l_i}(w_i \cdot A'))| = \ell_i |c_{l_i}(w_i \cdot (A - A'))| \leq \ell_i c_{l_i} \|w_i\| \|A - A'\|.$$

Subsequently, we get:

$$\|G(A) - G(A')\| = \left(\sum_{i=1}^{n} (f(c_{l_i}(w_i \cdot A)) - f(c_{l_i}(w_i \cdot A')))^2 \right)^{1/2}$$

$$\leq \left(\sum_{i=1}^{n} (\ell_i c_{l_i} \|w_i\| \|A - A'\|)^2 \right)^{1/2}.$$

Finally:

$$\|G(A) - G(A')\| \leq \|A - A'\| \left(\sum_{i=1}^{n} (c_{l_i} \ell_i \|w_i\|)^2 \right)^{1/2}.$$

A necessary condition for the above to be a contraction is:

$$\left(\sum_{i=1}^{n} (c_{l_i} \ell_i \|w_i\|)^2 \right)^{1/2} < 1. \tag{8.21}$$

The proof of the equality according to (8.17) is similar, using the results of Theorem 8.3.

Remark 8.2 Using the same analysis one can prove that for the hyperbolic tangent sigmoid function $f = \tanh(x)$ Eqs. (8.16) and (8.17) are still valid but now, $\ell_i = c_{l_i}(1 - f_i^2)$.

Remark 8.3 Equation (8.16) is computed using the node values vector A. It can be easily observed that ℓ_i receives its maximum value at the origin, that is when the arguments of the exponential functions being involved in its computation become zero. In this case, we say that (8.16) is computed at the origin, $\ell_i = \frac{c_{l_i}}{4}$ and the left-hand side of (8.16) gets its maximum value: $\left(\sum_{i=1}^{n} (\frac{c_{l_i}^2}{4} \|w_i\|)^2 \right)^{1/2}$. If condition (8.16) is fulfilled at the origin, the FCN will converge to its equilibrium regardless its initial conditions, even if $w_i A$ becomes zero.

8.3.1 Conditions Involving Only Weights

Neglecting c_{l_i} in the analysis of previous section is equivalent to setting $c_{l_i} = 1, \forall i$. This simplified case was presented initially in Boutalis et al. (2009) and results in conditions that involve only the weights of the FCN. The next Corollary of Theorem 8.4 gives these conditions.

Corollary 8.2 *There is one and only one solution for any concept value A_i of any FCN where the sigmoid function $f = \frac{1}{1+e^{-x}}$ is used, if:*

$$\left(\sum_{i=1}^{n} \|w_i\|^2 \right)^{1/2} < 4 \tag{8.22}$$

where w_i is the ith row of matrix W, $\|w_i\|$ is the L_2 norm of w_i.
If

$$\left(\sum_{i=1}^{n} \|w_i\|^2 \right)^{1/2} = 4 \tag{8.23}$$

*there exists at least one solution for each concept value A_i of any FCN, which is **not necessarily unique**.*

Proof The result comes immediately from (8.21) if one considers $c_{l_i} = 1$ and recalls from (8.8) that $\ell \leq \frac{1}{4}$.

Remark 8.4 Using the same analysis one can prove that for the hyperbolic tangent sigmoid function $f = \tanh(x)$, with $c_l = 1$, Eqs. (8.22) and (8.23) become:

$$\left(\sum_{i=1}^{n} \| w_i \|^2 \right)^{1/2} < 1 \tag{8.24}$$

and

$$\left(\sum_{i=1}^{n} \| w_i \|^2 \right)^{1/2} = 1 \tag{8.25}$$

respectively.

8.3.2 The Role of the Inclination Parameters

One obvious benefit of including the inclination parameter in the above analysis, that is, assuming $c_{l_i} \neq 1$, $\forall i$, is a significant increase in the degrees of freedom provided to the designer. One may choose appropriate c_{l_i} so that condition (8.21) is fulfilled even for weight sets that alone (i.e., with $c_{l_i} = 1$) they cannot fulfill it. Another benefit is that the required size of the graph (i.e., the number of nodes and their interconnections) so that it is allowed to reach the upper and lower limit node values is significantly reduced. This is explained below.

For each node $A_i(k)$, Eq. (8.1) can be expressed as:

$$A_i(k) = f(w_i A(k-1))$$

where w_i is the i-th row of matrix W. Taking into account the inclination parameter, the above equation is rewritten as:

$$A_i(k) = f(c_{li} w_i A(k-1))$$

with c_{li} being the inclination of sigmoid function f corresponding to node C_i. We also know that by the FCN definition

$$A_i(k) \in [0 \quad 1], A(k) \in [0 \quad 1]^n, w_i \in [-1 \quad 1]^n \text{ and } c_{li} \in (0 \quad \infty).$$

with n being the number of FCN nodes C.

Since f is a sigmoid function, in order that $A_i(k) \to 1$ then $f(w_i A(k-1)) \to 1$, this in turn means that $w_i A(k-1) \to \infty$. According to $w_i \in [-1 \quad 1]^n$ we have that:

$$w_i A(k-1) \in \left[-\sum_{i=1}^{n} A_i(k-1) \quad \sum_{i=1}^{n} A_i(k-1) \right]$$

and since $A_i(k) \in [0 \quad 1]$ we have that:

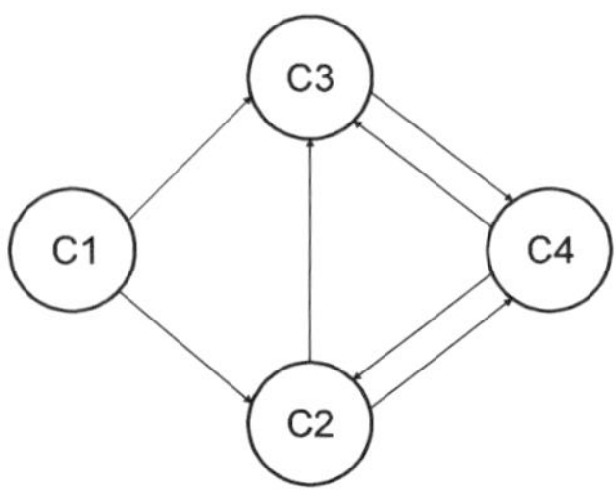
Fig. 8.4 FCN with one input node

$$w_i\, A(k-1) \in \left[-n'\quad n'\right]$$

where n' is the number of nodes C that affect node C_i. This way, $w_i\, A(k-1) \to \infty$ implies $n' \to \infty$ which physically means that an FCN must have a very large number of nodes C affecting node C_i in order to achieve target $A_i(k) \to 1$ or $A_i(k) \to 0$.

Including the inclination parameter we require $f(c_{li}\, w_i\, A(k-1)) \to 1$, which means that $c_{li}\, w_i\, A(k-1) \to \infty$. Applying the same analysis as above, we conclude that:

$$c_{li}\, w_i\, A(k-1) \in \left[-c_{li}n'\quad c_{li}n'\right]$$

taking into account that $c_{li} \in (0\quad \infty)$ we conclude that with a logical number of FCN nodes C affecting node C_i one may achieve target $A_i(k) \to 1$ or $A_i(k) \to 0$ by just using a number for the c_{li} into space $(0\quad \infty)$.

8.4 The Effect of Input Nodes

So far we have not considered the existence of input nodes. This kind of nodes is also called "*steady*" nodes in Kottas et al. (2007) in the sense that they influence but are not influenced by the other nodes of the FCN. In Fig. 8.4, C_1 is such an input (or steady) node.

In a FCN with m input nodes, matrix G assumes the general form:

$$G = \left[\, U_1 \,\ldots\, U_m \;\; f(c_{l_{m+1}} w_{m+1} \cdot A) \,\ldots\, f(c_{l_{m+n}} w_{m+n} \cdot A)\,\right]^T \qquad (8.26)$$

corresponding to vector: $A = \left[\, U_1\, U_2 \ldots U_m\, A_{m+1}\, A_{m+2}\, \ldots\, A_{m+n}\,\right]^T$, where m is the number of inputs and n is the number of the other concept nodes in FCM. Under this definition Eq. (8.3) assumes the same form:

$$A(k) = G(A(k-1)). \qquad (8.27)$$

The next theorem states that for matrix G and vector A defined above the results of Theorem 8.4 are still valid. The proof follows the same lines of the proof of

Theorem 8.4 taking into account that, since input nodes are not influenced by the other nodes, they do not change their values during the iterative solution of (8.27).

Theorem 8.5 *For an FCN with input nodes, with its concept values driven by* (8.27), *where G is described in* (8.26), *there is one and only one solution for any concept value A_i if Eq.* (8.16) *is fullfiled, that is:*

$$\left(\sum_{i=1}^{n} (c_{l_{m+i}} \ell_{m+i} \, \|w_{m+i}\|)^2 \right)^{1/2} < 1$$

where w_{m+i} is the $(m+i)$th row of matrix W, $\|w_{m+i}\|$ is the L_2 norm of w_{m+i}, $c_{l_{m+1}}$ is the inclination parameter c_l of function f corresponding to A_{m+i} concept and ℓ_{m+i} is the inclination of its function f, being equal to $\ell_{m+i} = \dfrac{c_{l_{m+i}}}{e^{c_{l_{m+i}} w_{m+i} \cdot A}} f^2(c_{l_{m+i}} w_{m+i} \cdot A)$.
If

$$\left(\sum_{i=1}^{n} (c_{l_{m+i}} \ell_{m+i} \, \|w_{m+i}\|)^2 \right)^{1/2} = 1$$

a **not always unique** *solution exists.*

Proof Assume A and A' are two different concept values for the FCN having one or more inputs. Then, we want to prove again inequality (8.20), that is:

$$\left\| G(A) - G(A') \right\| \le c \left\| A - A' \right\|.$$

But since input node values are not influenced by the other nodes of the FCN $\left\| G(A) - G(A') \right\|$ according to (8.26) is equal to:

$$\left\| G(A) - G(A') \right\| = \left(\sum_{i=1}^{m} (U_i - U_i)^2 + \sum_{i=1}^{n} (f(c_{l_{m+i}} w_{m+i} \cdot A) - f(c_{l_{m+i}} w_{m+i} \cdot A'))^2 \right)^{1/2}$$

where m is the number of inputs and n is the number of the other nodes in the FCM. The above equation is equivalent to the following:

$$\left\| G(A) - G(A') \right\| = \left(0 + \sum_{i=1}^{n} (f(c_{l_{m+i}} w_{m+i} \cdot A) - f(c_{l_{m+i}} w_{m+i} \cdot A'))^2 \right)^{1/2}.$$

Therefore, $\left\| G(A) - G(A') \right\|$ assumes quite the same form with that appearing in Theorem 8.4 leading to the same condition, that is:

$$\left(\sum_{i=1}^{n} (c_{l_{m+i}} \ell_{m+i} \, \|w_{m+i}\|)^2 \right)^{1/2} < 1.$$

Similarly, for the $=$ case the respective results of Theorem 8.4 still hold and the results obtained for the hyperbolic tangent function are still valid.

Theorem 8.5 says that the existence and uniqueness of equilibrium points in FCN with input nodes depends only on the weights and inclination parameters that influence the non-input nodes. However, as it will be explained in Sect. 8.6, the values of the equilibrium point are determined not only by the FCN parameters but by the input nodes values as well.

Remark 8.5 Taking into account Theorem 8.5, Corollary 8.2 and Remark 8.4 it follows immediately that, for the FCN with input nodes and $c_{l_i} = 1$, the conditions for existence and uniqueness are

$$\left(\sum_{i=1}^{n} \|w_{m+i}\|^2 \right)^{1/2} \leq 4 \tag{8.28}$$

and for FCN involving the hyperbolic tangent function

$$\left(\sum_{i=1}^{n} \|w_{m+i}\|^2 \right)^{1/2} \leq 1, \tag{8.29}$$

where in both formulas the $=$ case signals a non-unique solution.

8.5 Academic Examples

We proceed to present simple academic examples demonstrating the theoretical results of the previous sections and the benefit of using both weight and inclination parameters in the design of the FCN. We start with the hypothesis that we have a FCN that uses log-sigmoid function with $c_l = 1$. In case the FCN is too small, that is, it has two or three nodes, inequality (8.22) is **always true** because $|w_{ij}|_{i \neq j} \leq 1$ and $d_{ii} = w_{ii}$ (the diagonal elements of weight matrix) can at most take the value of 1. FCNs with four or more nodes need more attention and are examined below.

Example 8.1 FCN with four nodes.

Suppose that we have a FCN with four concepts (nodes). The weight matrix W of this FCM is:

$$W = \begin{bmatrix} d_{11} & w_{12} & w_{13} & w_{14} \\ w_{21} & d_{22} & w_{23} & w_{24} \\ w_{31} & w_{32} & d_{33} & w_{34} \\ w_{41} & w_{42} & w_{43} & d_{44} \end{bmatrix}.$$

The square root of the sum of the square L_2 norm of each row of matrix W is equal to:

$$\sqrt{\sum_{i=1}^{4} \|w_i\|^2} = \sqrt{\|w_1\|^2 + \|w_2\|^2 + \|w_3\|^2 + \|w_4\|^2}. \qquad (8.30)$$

The L_2 norm of each row is equal to: $\|w_i\| = \sqrt{\sum_{j=1}^{4} w_{ij}^2}$, where i denotes the ith row of matrix W and j denotes the column index. Equation (8.30) is now:

$$\sqrt{\sum_{i=1}^{4} \|w_i\|^2} = \sqrt{\|w_1\|^2 + \|w_2\|^2 + \|w_3\|^2 + \|w_4\|^2}$$

$$= \sqrt{\left(\sqrt{\sum_{j=1}^{4} w_{1j}^2}\right)^2 + \left(\sqrt{\sum_{j=1}^{4} w_{2j}^2}\right)^2 + \left(\sqrt{\sum_{j=1}^{4} w_{3j}^2}\right)^2 + \left(\sqrt{\sum_{j=1}^{4} w_{4j}^2}\right)^2}$$

$$= \sqrt{\sum_{j=1}^{4} w_{1j}^2 + \sum_{j=1}^{4} w_{2j}^2 + \sum_{j=1}^{4} w_{3j}^2 + \sum_{j=1}^{4} w_{4j}^2}$$

$$= \sqrt{\sum_{i=1}^{4} \left(\sum_{j=1}^{4} w_{ij}^2\right)} = \sqrt{\sum_{j=1}^{4} (d_{jj}) + \sum_{i=1}^{4} \left(\sum_{j=1, i\neq j}^{4} w_{ij}^2\right)}.$$

Since for the nondiagonal elements $|w_{ij}| < 1$, then : $w_{ij}^2 < 1$.

Finally, the above equation concludes to:

$$\sqrt{\sum_{j=1}^{4} (d_{jj}) + \sum_{i=1}^{4} \left(\sum_{j=1, i\neq j}^{4} w_{ij}^2\right)} \leq \sqrt{4 + 12} = 4.$$

Subsequently, we get:

$$\sqrt{\sum_{i=1}^{4} \|w_i\|^2} \leq 4.$$

According to Corollary 8.2, in order that only one solution exists for the concepts of a FCN the following inequality must be true: $\sqrt{\sum_{i=1}^{4} \|w_i\|^2} < 4$.

If $\sqrt{\sum_{i=1}^{4} \|w_i\|^2} = 4$ a solution always exists which **is not necessarily unique**.

We finally get the next conclusion: Since it is a usual practice that, by designing the FCN, in general most of the d_{ii} to be close to zero, "a FCN with four concepts has a unique solution generically."

Example 8.2 FCN with more than four nodes

Suppose that we have a FCN with more than four nodes. The weight matrix W of the FCM is:

$$W = \begin{bmatrix} d_{11} & w_{12} & w_{13} & \ldots & w_{1n} \\ w_{21} & d_{22} & w_{23} & \ldots & w_{2n} \\ w_{31} & w_{32} & d_{33} & \ldots & w_{3n} \\ \ldots & \ldots & \ldots & \ldots\ldots \\ w_{n1} & w_{n2} & w_{n3} & \ldots & d_{nn} \end{bmatrix}$$

where $n > 4$ The square root of the sum of the square L_2 norm of each row of matrix W is given by:

$$\sqrt{\sum_{i=1}^{n} \|w_i\|^2} = \sqrt{\|w_1\|^2 + \|w_2\|^2 + \cdots + \|w_n\|^2}$$

$$\Rightarrow \sqrt{\sum_{i=1}^{n} \|w_i\|^2} = \sqrt{\sum_{i=1}^{n}\left(\sum_{j=1}^{n} w_{ij}^2\right)}$$

$$\Rightarrow \sqrt{\sum_{i=1}^{n} \|w_i\|^2} = \sqrt{\sum_{j=1}^{n}\left(d_{jj}\right) + \sum_{i=1}^{n}\left(\sum_{j=1,i\neq j}^{n} w_{ij}^2\right)}$$

$$\Rightarrow \sqrt{\sum_{i=1}^{n} \|w_i\|^2} \leq \sqrt{\sum_{j=1}^{n}\left(d_{jj}\right)} + \sqrt{\sum_{i=1}^{n}\left(\sum_{j=1,i\neq j}^{n} w_{ij}^2\right)}.$$

Finally, we conclude that for an FCM with $n > 4$ concepts Corollary 8.2 is true when:

$$\sqrt{\sum_{i=1}^{n}\left(\sum_{j=1,i\neq j}^{n} w_{ij}^2\right)} < 4 - \sqrt{\sum_{j=1}^{n}\left(d_{jj}\right)}. \tag{8.31}$$

When

$$\sqrt{\sum_{i=1}^{n}\left(\sum_{j=1,i\neq j}^{n} w_{ij}^2\right)} = 4 - \sqrt{\sum_{j=1}^{n}\left(d_{jj}\right)} \tag{8.32}$$

a **not necessarily unique solution** always exists.

Therefore, when $n > 4$ the condition for the uniqueness of solution of (8.1) depends on the number and the strength of diagonal d_{ii} elements of the FCN that are nonzero and on the strength and the number of its weights, that is, on the **size** of the

FCN. As it is demonstrated in Example 8.4 below, the effect of the size of the FCN may be alleviated if one includes a sigmoid inclination parameter $c_{l_i} \neq 1$.

When the weights are within the bound provided by Eq. (8.22) (or equivalently Eq. (8.32) in the $n > 4$ case of the previous example) the solution of (8.1) is unique and therefore the FCN will converge to one value regardless of its initial concept values. If the condition of uniqueness is not met, then the solution may or may not exist, and if it exists it is probable to depend on the initial concept values. Therefore, the fulfillment of the condition gives rise to a *meta rules* representation of the FCN having the form "If weights then fixed point." This result is also supported by the following illustrative example.

Example 8.3 Convergence to unique equilibrium points.

Let a 5 nodes FCN, having the log-sigmoid function with $c_l = 1$ and matrix W being equal to:

$$W = \begin{bmatrix} 0 & 0 & 0 & 0.2 & 0.3 \\ 0.8 & 0 & 0.7 & 0 & 0.6 \\ 0 & 0.6 & 0 & 0 & 0.7 \\ 0 & 0.4 & 0.2 & 0 & 0 \\ 0.3 & 0 & 0 & 0.8 & 0 \end{bmatrix}.$$

For matrix W Eq. (8.22) is true, that is:

$$\left(\sum_{i=1}^{5} \|w_i\|^2 \right)^{1/2} = 1.8439 < 4.$$

Repetitive application of Eq. (8.1) will drive the FCM to a unique equilibrium point A which is equal to:

$$A = \begin{bmatrix} 0.5795 & 0.7963 & 0.7191 & 0.6136 & 0.6603 \end{bmatrix}.$$

Let us assume that the weight matrix W of the FCN becomes:

$$W = \begin{bmatrix} 0 & 0 & 0 & 0.2 & 0.3 \\ 0.8 & 0 & 0.7 & 0 & 0.6 \\ 0 & 0.5 & 0 & 0 & 0.6 \\ 0 & 0.5 & 0.2 & 0 & 0 \\ 0.3 & 0 & 0 & 0.8 & 0 \end{bmatrix}.$$

Then, Eq. (8.22) is also true

$$\left(\sum_{i=1}^{5} \|w_i\|^2 \right)^{1/2} = 1.8028 < 4$$

and another unique equilibrium point for the FCM nodes emerges as follows:

$$A = \begin{bmatrix} 0.5806 & 0.7933 & 0.6888 & 0.6305 & 0.6634 \end{bmatrix}.$$

It has to be noted that the equilibrium point obtained this way is independent of the initial concept values in Eq. (8.1). It depends solely on W provided that for the elements of W Eq. (8.22) is valid.

Example 8.4 The effect of the inclination parameter.

Suppose that we desire $A_i(k) = f(a) = 0.999$. If we use $c_{li} = 1$, a must be $a = w_i A(k-1) = 6.9068$. This means that we need at least seven (7) nodes C affecting node C_i with their values being equal to one, in order to achieve the target value for concept $A_i(k)$.

Suppose again that $A_i(k) = f(a) = 0.999$. If now we use $c_{li} = 5$, a has to be $a = 5w_i A(k-1) = 6.9068$, then $w_i A(k-1) = 1.3814$. This means that we need at least two (2) nodes C affecting node C_i, in order to achieve the same target value for concept $A_i(k)$. That is, in order for the FCN nodes to achieve the desired values require the influence of fewer nodes. This means that the designer has the ability to design smaller FCN structures with similar performance, by omitting unnecessary nodes (not related to crucial physical quantities) and increasing respectively the inclination parameter values.

8.6 Discussion and Interpretation of the Results

Theorem 8.4 provides conditions that the parameters of the FCN should fulfill so that it reaches an equilibrium condition and this condition be unique. It is clear that the above results hold only for continuous differentiable squashing functions like the log-sigmoid and the hyperbolic tangent functions. Therefore, the results are not valid for FCMs equipped with bivalent or trivalent functions; we cannot conclude anything for this type of FCMs using the above analysis. That's why the FCN framework assumes that the nodes of the cognitive graph are equipped exclusively with sigmoid functions (see Sect. 7.4).

Equations (8.16), (8.17) are valid for FCN having log-sigmoid functions. Equation (8.16) provides with conditions for the existence and uniqueness of an equilibrium condition, while Eq. (8.17) refers to condition that guarantees only the existence. Similar respective results arise from Remark 8.2 for FCNs equipped with hyperbolic tangent functions. In case none of the conditions hold, this does not imply that an equilibrium does not exist. It may or may not exist, just there is not any guarantee for its existence. Equations (8.22), (8.23) and (8.24), (8.25) give the respective conditions when the inclination parameters of all the sigmoid functions are $c_l = 1$.

It can be observed and further verified by the exploration given in the examples of Sect. 8.5 that the conditions provide with upper bounds for the FCN weights, which depend on the size of the FCN. Moreover, taking into account the observations made in the proof of Theorem 8.2, it can be concluded that these upper bounds depend also on the inclination of the sigmoid used. The larger the inclination becomes the smaller these bounds are, so that in the limit no weight set can be found to guarantee the uniqueness or the existence of the equilibrium. This is however expected, because sigmoids with large inclinations tend to reach the behavior of bivalent or trivalent functions and therefore no guarantee can exist based on the above analysis. FCMs that use sigmoids with large inclinations (or in the limit bivalent or trivalent functions) tend to give mainly qualitative results, while FCMs using small inclination sigmoids give both quantitative and qualitative results (see also the relevant conclusions in Tsadiras (2008)). This is again mathematically expected since sigmoids with large inclination have very low discriminative abilities because they produce very similar outputs for inputs that may be quite dissimilar. For FCNs with sigmoids having small inclinations it is much easier to find weight sets that fulfill the existence and uniqueness conditions.

On the other hand, as it was pointed out in Sect. 8.3.2 and Example 8.4, large inclination parameters contribute to lowering the required size of the FCN. Therefore, a trade off between these two factors (easiness of achieving equilibrium points versus FCN size reduction) has to be made by the FCN designer.

Another issue that needs a careful reading is the effect of input nodes. Similar to Theorems 8.4 and 8.5 states that the weights and the parameters c_{l_i} of the sigmoids have to comply with the bounds implied by its conditions in order to assure the existence or the uniqueness of the FCN equilibrium. However, when these conditions are fulfilled the unique equilibrium does not depend solely on the weight and parameter set, as in the case of FCN with no input nodes. It depends also on the values of the input nodes. This is a quite reasonable implication because the input node values are steady and unlike the other node values they are not changing during the repetitive application of Eq. (7.1) until the FCN reaches its equilibrium. Therefore, they act as an external excitation. Different values of the excitation will drive the FCN in different equilibria.

The results obtained so far are quite significant. They are valid for a class of sigmoid functions for which we assure, through the derived conditions, that they present contractive or nonexpansive properties and are used in FCN. Many physical systems can reach an equilibrium point after initial perturbations. If FCNs are used for the representation of such systems, it is important to know, which FCN structure may guarantee the existence and probably the uniqueness of the equilibrium. This gives rise also to the development of estimation algorithms, which can learn the structure (weights and sigmoid parameters) based on real system's data. Such algorithms are given in Chap. 9. Moreover, in FCNs with input nodes this unique equilibrium may be different depending on the value of the exciting input, a behavior quite similar with the behavior of stable nonlinear physical systems. Therefore, FCNs could be used not only for modeling but also for the control of unknown nonlinear systems. This is demonstrated later on, in Chap. 10.

8.7 Summary

The existence and uniqueness of equilibrium points of FCNs, equipped with continuous differentiable sigmoid functions, were examined in this chapter. After exploring the contraction mapping properties of sigmoid functions, conditions involving the FCN weights, and the sigmoid inclination parameters were derived, which guarantee the existence and the uniqueness of equilibrium points in the repetitive operation of the FCN. The effect of the existence of input nodes in the convergence properties of the FCN was also analyzed. Special attention was given to the numerical exploration and the interpretation of the results that were presented in separate sections. The theoretical results derived in this chapter form the basis for the development of FCN parameter estimation algorithms that are given in Chap. 9.

References

Boutalis, Y., Kottas, T. & Christodoulou, M. (2008). On the existence and uniqueness of solutions for the concept values in fuzzy cognitive maps. In *Proceedings of 47th IEEE Conference on Decision and Control—CDC'08, December 9–11, 2008, Cancun, Mexico* (pp. 98–104)

Boutalis, Y., Kottas, T., & Christodoulou, M. (2009). Adaptive estimation of fuzzy cognitive maps with proven stability and parameter convergence. *IEEE Transactions on Fuzzy Systems, 17,* 874–889.

Kottas, T., Boutalis, Y., & Christodoulou, M. (2007). Fuzzy cognitive networks: A general framework. *Inteligent Desicion Technologies, 1,* 183–196.

Kottas, T., Boutalis, Y., & Christodoulou, M. (2012). Bi-linear adaptive estimation of fuzzy cognitive networks. In *Applied Soft Computing,* 21 (2012). doi:10.1016/j.asoc.2012.01.025

Kranas, A., & Dugundji, J. (2003). *Fixed Point Theory.* New York: Springer.

Rudin, W. (1964). *Principles of Mathematical Analysis.* New York: McGraw-Hill

Tsadiras, A. (2008). Comparing the inference capabilities of binary, trivalent and sigmoid fuzzy cognitive maps. *Information Science, 178,* 3880–3894.

Chapter 9
Adaptive Estimation Algorithms of FCN Parameters

9.1 Introduction

Based on the results and observations of Chap. 8 we are now proposing a method of finding appropriate weight sets related to a desired equilibrium point of the FCN. Choosing a desired state A^{des} for the FCN this is equivalent to solving the equilibrium equation

$$A^{\mathrm{des}} = f(W^* A^{\mathrm{des}}) \tag{9.1}$$

in respect to W^*, where $A^{\mathrm{des}} = \left[A_1^{\mathrm{des}}, A_2^{\mathrm{des}}, \ldots A_n^{\mathrm{des}}\right]^T$ and f is a vector-valued function $f : \Re^n \rightarrow \Re$, defined as follows: $f(x) = [f_1(x_1), f_2(x_2), \ldots f_n(x_n)]^T$, where $x \in \Re^n$ and $f_i(x_i) = \frac{1}{1+e^{-c_{l_i} x_i}}$, for $i = 1, 2, \ldots, n$. Then,

$$f^{-1}(A^{\mathrm{des}}) = W^* A^{\mathrm{des}} \tag{9.2}$$

where $f^{-1}(A^{\mathrm{des}}) = \left[f_1^{-1}(A_1^{\mathrm{des}}), f_2^{-1}(A_2^{\mathrm{des}}), \ldots, f_n^{-1}(A_n^{\mathrm{des}})\right]^T$

$$f_i^{-1}(A_i^{\mathrm{des}}) = c_{l_i}^* w_i^* \cdot A^{\mathrm{des}} \tag{9.3}$$

with w_i^* being the ith row of W^* and c_{l_i} is the c_l factor of function f corresponding to concept A_i. The form of f_i^{-1} is straightforwardly computed if one tries to solve the sigmoid function in respect to its argument. That is

$$f_i^{-1}(A_i^{\mathrm{des}}) = \ln\left(\frac{A_i^{\mathrm{des}}}{1 - A_i^{\mathrm{des}}}\right). \tag{9.4}$$

In order to solve Eq. (9.3) we distinguish two cases. In the first one, we assume that the sigmoid inclination parameters are $c_{l_i} = 1$, $\forall i$ and therefore the only parameters to be estimated are the FCN weights. In this case, (9.3) is rewritten simply as

Y. Boutalis et al., *System Identification and Adaptive Control*,
Advances in Industrial Control, DOI: 10.1007/978-3-319-06364-5_9,

Fig. 9.1 Online parameter
estimator designed for FCNs

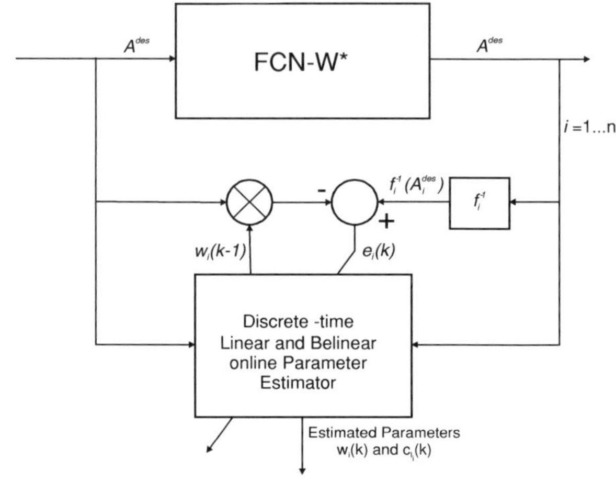

$$f_i^{-1}(A_i^{\mathrm{des}}) = w_i^* \cdot A^{\mathrm{des}}, \tag{9.5}$$

which is clearly a linear parametric model (LPM) because the unknown weights appear linearly in the equation. Moreover, $f_i^{-1}(A_i^{\mathrm{des}})$ is scalar and the unknown parameters are many, forming the vector w_i^*. Therefore, we choose to solve this equation recursively, requiring at each iteration the LPM to track the constant value $f_i^{-1}(A_i^{\mathrm{des}})$. An adaptive estimation algorithm for the LPM is presented in Sect. 9.2, following the development made in Boutalis et al. (2009). Moreover, since the estimated parameters correspond to specific equilibrium points, the existence and uniqueness conditions, derived in Chap. 8, should be always fulfilled. To this end appropriate projection methods are presented in Sect. 9.3, which guarantee that the conditions are not violated during the recursive updating algorithm.

In the second case, both $c_{l_i}^*$ and w_i^* have to be estimated. In Eq. (9.3) the parametric model (right-hand side) is clearly bilinear (BPM), requiring the development of a bilinear estimation algorithm, which is presented in Sect. 9.4, following the development made in Kottas et al. (2012). Similar with the linear case, projection methods are derived in Sect. 9.5, which guarantee that the existence and uniqueness conditions are not violated. The updating scheme is illustrated in Fig. 9.1, demonstrating that the weight learning is performed adaptively based on the desired equilibrium point of the FCN.

9.2 Adaptive Estimation Algorithm Based on Linear Parametric Modeling

In order to solve (9.5), we employ methods from adaptive LPM estimation (Ioannou and Fidan 2006) and proceed according to Boutalis et al. (2009). We take into account that $f_i(x_i)$ is the sigmoid function and its inverse is given by (9.4). The weight updating laws are given as follows:

The error $\varepsilon_i(k)$ of the parametric discrete-time adaptive law is of the form:

$$\varepsilon_i(k) = \frac{f_i^{-1}(A_i^{\mathrm{des}}) - w_i(k-1)A^{\mathrm{des}}}{c + (A^{\mathrm{des}})^T A^{\mathrm{des}}} \tag{9.6}$$

while the updating algorithm is given by:

$$w_i(k) = w_i(k-1) + \alpha \varepsilon_i(k)(A^{\mathrm{des}})^T \tag{9.7}$$

where $w_i(k)$ is the ith row of $W(k)$, which is the estimator of $W^*(k)$. A^{des} is a constant vector and $f_i^{-1}(A_i^{\mathrm{des}})$ is also constant and scalar. $\alpha > 0$ and $c > 0$ are design parameters.

By using the above updating algorithm, we are confident that the estimator converges to W^*. In Chap. 8 we proved that if the elements of W^* fulfill inequality (8.22), then W^* corresponds to the unique solution of (9.5) in respect to the desired equilibrium state A^{des}.

Proof From (9.6), (9.7) and $\tilde{w}_i(k) = w_i(k) - w_i^*$ we obtain the error equation

$$\tilde{w}_i(k) = \left[I - \frac{a A^{\mathrm{des}}(A^{\mathrm{des}})^T}{c + (A^{\mathrm{des}})^T A^{\mathrm{des}}} \right] \tilde{w}_i(k-1) \tag{9.8}$$

$$\varepsilon_i(k) = -\frac{\tilde{w}_i(k-1)A^{\mathrm{des}}}{c + (A^{\mathrm{des}})^T A^{\mathrm{des}}}. \tag{9.9}$$

For each node, consider the Lyapunov function

$$V(k) = \frac{1}{2a}\tilde{w}_i(k)\tilde{w}_i^T(k) \tag{9.10}$$

Then

$$\begin{aligned}
V(k) - V(k-1) =\ & \frac{1}{2a}\tilde{w}_i(k)\tilde{w}_i^T(k) - \frac{1}{2a}\tilde{w}_i(k-1)\tilde{w}_i^T(k-1) \\
=\ & -\frac{\tilde{w}_i(k-1)A^{\mathrm{des}}(A^{\mathrm{des}})^T \tilde{w}_i^T(k-1)}{c + (A^{\mathrm{des}})^T A^{\mathrm{des}}} \\
& + \frac{\tilde{w}_i(k-1)A^{\mathrm{des}}(A^{\mathrm{des}})^T a A^{\mathrm{des}}(A^{\mathrm{des}})^T \tilde{w}_i^T(k-1)}{2(c + (A^{\mathrm{des}})^T A^{\mathrm{des}})^2}
\end{aligned}$$

Using $\varepsilon_i(k)(c + (A^{\text{des}})^T A^{\text{des}}) = -\tilde{w}_i(k-1)A^{\text{des}}$, we obtain

$$V(k) - V(k-1) = -\varepsilon_i^2(k)(c + (A^{\text{des}})^T A^{\text{des})\left[1 - \frac{a(A^{\text{des}})^T A^{\text{des}}}{2(c + (A^{\text{des}})^T A^{\text{des}})}\right].$$

Since $a, c > 0$ we can always choose $0 < a < 2$ such that $\frac{a(A^{\text{des}})^T A^{\text{des}}}{2(c+(A^{\text{des}})^T A^{\text{des}})} < 2$. It then follows that

$$V(k) - V(k-1) \le -c_0\varepsilon_i^2(k)(c + (A^{\text{des}})^T A^{\text{des}}) \le 0 \qquad (9.11)$$

for some constant $c_0 > 0$. From (9.10), (9.11) we have that $V(k)$ and therefore $w_i(k) \in L_\infty$ and $V(k)$ has a limit, i.e., $\lim_{k \to \infty} V(k) = V_\infty$. Consequently, using (9.11) we obtain

$$c_0 \sum_{k=1}^{\infty} (\varepsilon_i^2(k)(c + (A^{\text{des}})^T A^{\text{des}})) \le V(0) - V_\infty < \infty$$

which implies $\varepsilon_i(k)\sqrt{(c + (A^{\text{des}})^T A^{\text{des}})} \in L_2$ and $\varepsilon_i(k)\sqrt{(c + (A^{\text{des}})^T A^{\text{des}})} \to 0$ as $k \to \infty$. Since $\sqrt{(c + (A^{\text{des}})^T A^{\text{des}})} \ge c > 0$, we also have that $\varepsilon_i(k) \in L_2$ and $\varepsilon_i(k) \to 0$ as $k \to \infty$. We have

$$\varepsilon_i(k)A^{\text{des}} = \varepsilon_i(k)\sqrt{(c + (A^{\text{des}})^T A^{\text{des}})}\frac{A^{\text{des}}}{\sqrt{(c+(A^{\text{des}})^T A^{\text{des}})}}. \text{ Since } \frac{A^{\text{des}}}{\sqrt{(c+(A^{\text{des}})^T A^{\text{des}})}} \text{ is}$$

bounded and $\varepsilon_i(k)\sqrt{(c + (A^{\text{des}})^T A^{\text{des}})} \in L_2$, we have that $\varepsilon_i(k)A^{\text{des}} \in L_2$ and $\left\|\varepsilon_i(k)A^{\text{des}}\right\| \to 0$ as $k \to \infty$. This implies (using (9.7)) that $\|w_i(k) - w_i(k-1)\| \in L_2$ and $\|w_i(k) - w_i(k-1)\| \to 0$ as $k \to \infty$. Now

$$w_i(k) - w_i(k-N) = w_i(k) - w_i(k-1) + w_i(k-1) - w_i(k-2)$$
$$+ \ldots + w_i(k-N+1) - w_i(k-N)$$

for any finite N. Using the Schwartz inequality, we have

$$\|w_i(k) - w_i(k-N)\|^2 \le \|w_i(k) - w_i(k-1)\|^2 + \|w_i(k-1) - w_i(k-2)\|^2$$
$$+ \ldots + \|w_i(k-N+1) - w_i(k-N)\|^2.$$

Since each term on the right-hand side of the inequality is in L_2 and goes to zero with $k \to \infty$, it follows that $\|w_i(k) - w_i(k-N)\| \in L_2$ and $\|w_i(k) - w_i(k-N)\| \to 0$ as $k \to \infty$.

Remark 9.1 The n weight updating laws given by (9.7) for each row w_i, $i = 1, \ldots, n$ of W can be written in a compact form as follows

$$\underline{W}(k) = \underline{W}(k-1) + a\underline{\varepsilon}A^{\text{des}} \tag{9.12}$$

where $\underline{W} = [w_1, w_2, \ldots w_n]^T$ is a $n^2 \times 1$ column vector containing all the elements of matrix W arranged in a column, row after row. Also, $\underline{\varepsilon}$ is a $n^2 \times n$ matrix having the form $\underline{\varepsilon} = [\varepsilon_1 I_n, \varepsilon_2 I_n, \ldots \varepsilon_n I_n]^T$, where I_n is a $n \times n$ unit matrix and ε_i is given by (9.6).

9.3 Projection Methods Based on the Existence and Design Conditions

In parameter identification problems, we have some a priori knowledge as to where the unknown parameter W^* is located in $\Re^n$. This knowledge could be the upper and the lower bounds for each element of W^*. In our parameter identification problem inequalities $-1 \le w_{ij} \le 1$ and (8.22) must be true. Employing concepts from adaptive parameter estimation (Ioannou and Fidan (2006)), convex projection methods are next illustrated to modify the adaptive laws of the Sect. 9.2. The projection methods modify the weight updating laws so that, in each iteration, the existence and uniqueness conditions are not violated. It can be proved that, this modification does not alter the sign of ΔV and therefore the estimation error still converges to zero. The proofs are not provided here. Instead, we present the relevant proofs for the more general case when projection laws are applied for the estimation of the bilinear parametric model in Sect. 9.5. A schematic representation of the orthogonal projection method is illustrated in Fig. 9.2.

9.3.1 Projection Method 1

The S_1 set for projection that satisfies $-1 \le w_{ij} \le 1$ is the convex set:

$$S_1 = \{W \in \Re^n | g(w_{ij}) \le 0, \quad g(w_{ij}) = |w_{ij}| - 1, \forall i, j \in \aleph\}.$$

The updating formula of parameters W of FCN is now given by:

$$\begin{aligned}
\overline{W}(k) &= \underline{W}(k-1) + a\underline{\varepsilon}A^{\text{des}} \\
\underline{W}(k) &= \begin{cases} \overline{W}(k), & \text{if } \overline{W}(k) \in S_1 \\ \Pr(\overline{W}(k)), & \text{if } \overline{W}(k) \notin S_1 \end{cases}
\end{aligned} \tag{9.13}$$

where $\Pr(\overline{W}(k)) = \perp\text{proj}$ of $\overline{W}(k)$ on S_1.

Fig. 9.2 Discrete-time
parameter projection

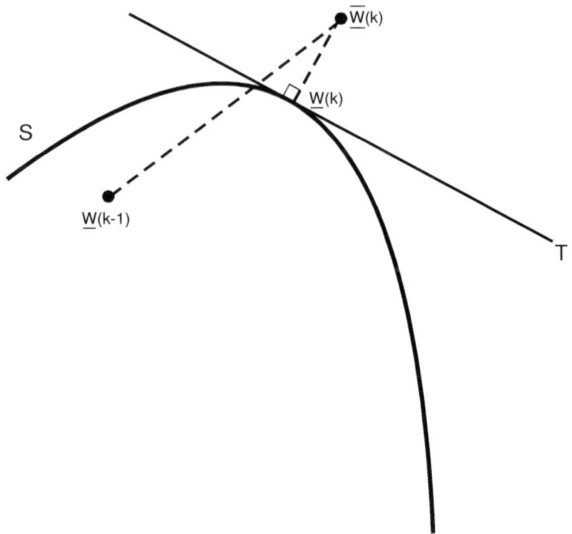

This formula can also be element-wise written as:

$$\overline{w}_{ij}(k) = w_{ij}(k-1) + \alpha \varepsilon_i(k) A_i^{\text{des}}$$
$$w_{ij}(k) = \begin{cases} \overline{w}_{ij}(k), & \text{if } \left|\overline{w}_{ij}(k)\right| \leq 1 \\ -1 & \text{if } \overline{w}_{ij}(k) < -1 \\ 1 & \text{if } \overline{w}_{ij}(k) > 1. \end{cases} \tag{9.14}$$

9.3.2 Projection Method 2

The S_2 set for projection that satisfies (8.22) is the convex set:

$$S_2 = \{\underline{W} \in \mathfrak{R}^{n^2} | g(\underline{W}) \leq 0, \quad g(\underline{W}) = \left\|\underline{W}\right\| - 4\}.$$

The updating equation of parameters W of FCN is now given by:

$$\overline{W}(k) = \underline{W}(k-1) + a\underline{\varepsilon} A^{\text{des}}$$
$$\underline{W}(k) = \begin{cases} \overline{W}(k), & \text{if } \overline{W}(k) \in S_2 \\ \Pr(\overline{W}(k)), & \text{if } \overline{W}(k) \notin S_2 \end{cases} \tag{9.15}$$

where the orthogonal projection $\Pr(\overline{W}(k)) = \perp\text{proj}$ of $\overline{W}(k)$ on S_2 is given by $\dfrac{4}{\left\|\overline{W}(k)\right\|}\overline{W}(k)$. Therefore, Eq. (9.15) can also be written in the following compact form:

$$\overline{W}(k) = \underline{W}(k-1) + a\underline{\varepsilon}A^{\mathrm{des}}$$
$$\underline{W}(k) = \overline{W}(k) \min\left(1, \frac{4}{\|\overline{W}(k)\|}\right).$$

(9.16)

9.3.3 Concurrent Projection Method

In the previous subsections two projection methods were presented. These two methods can be combined if we consider that the set S, that satisfies both inequalities, is the intersection of sets S_1 and S_2.

$$S = S_1 \bigcap S_2.$$

We replace the weight updating equation:

$$\underline{W}(k) = \underline{W}(k-1) + a\underline{\varepsilon}A^{\mathrm{des}}$$

with

$$\overline{W}(k) = \underline{W}(k-1) + a\underline{\varepsilon}A^{\mathrm{des}}$$
$$\underline{W}(k) = \begin{cases} \overline{W}(k), & \text{if } \overline{W}(k) \in S \\ \mathrm{Pr}(\overline{W}(k)), & \text{if } \overline{W}(k) \notin S \end{cases}$$

(9.17)

where $\mathrm{Pr}(\overline{W}(k)) = \perp \mathrm{proj}$ of $\overline{W}(k)$ on S.

9.4 Adaptive Estimation Algorithm Based on the Bilinear Parametric Model

We recall from Sect. 9.1 that, when both $c^*_{l_i}$ and w^*_i are to be estimated, the right-hand side of Eq. (9.3) assumes a bilinear parametric form. In solving repetitively (9.3), we employ methods from adaptive parameter estimation literature (Ioannou and Fidan 2006) and proceed according to the development made in Kottas et al. (2012). The updating scheme is illustrated in Fig. 9.1, demonstrating that the weight and inclination parameters learning is performed adaptively based on the desired equilibrium point of the FCN. The weight updating laws are given as follows:

The error $\varepsilon_i(k)$ of the parametric discrete-time adaptive law is of the form:

$$\varepsilon_i(k) = \frac{f_i^{-1}(A_i^{\mathrm{des}}) - c_{l_i}(k-1)w_i(k-1)A^{\mathrm{des}}}{c + c_{l_i}(0)(A^{\mathrm{des}})^T A^{\mathrm{des}} + \gamma(w_i(k-1)A^{\mathrm{des}})^2}$$

(9.18)

while the updating algorithm is given by:

$$c_{l_i}(k) = c_{l_i}(k-1) + \gamma\varepsilon_i(k)(w_i(k-1)A^{\text{des}}) \tag{9.19}$$

$$w_i(k) = w_i(k-1) + a\,\text{sgn}(c_{l_i}^*)\varepsilon_i(k)\left(A^{\text{des}}\right)^T \tag{9.20}$$

where $0 < \gamma < 1$, $a, c > 0$, $c_{l_i}(0) > 0$ and by the definition of the sigmoid function $c_{l_i}^* > 0$ implying that $\text{sgn}(c_{l_i}^*) = 1$. $w_i(k)$ is the ith row of $W(k)$, which is the estimator of $W^*(k)$. $c_{l_i}(k)$ is the c_l factor of function f corresponding to the ith concept and is the estimator of $c_{l_i}^*(k)$. A^{des} is constant vector and $f_i^{-1}(A_i^{\text{des}})$ is also constant and scalar. α, c and γ are design parameters.

By using the above updating equations $\forall i$, we can now prove that the estimators converge to W^* and c_l^*, respectively. In Chap. 8 we proved that if inequality (8.16) is true then W^* and $c_{l_i}^*$ corresponding to the designed FCN provide a unique solution and satisfies (9.3).

Proof From (9.18), (9.19), (9.20) and $\tilde{w}_i(k) = w_i(k) - w_i^*$, $\tilde{c}_{l_i}(k) = c_{l_i}(k) - c_{l_i}^*$ we obtain the error equation

$$\tilde{w}_i(k) = \left[I - \frac{aA^{\text{des}}(A^{\text{des}})^T}{c + (A^{\text{des}})^T A^{\text{des}}}\right]\tilde{w}_i(k-1) \tag{9.21}$$

$$\varepsilon_i(k) = -\frac{\tilde{c}_{l_i}(k-1)w_i(k-1)A^{\text{des}} + c_{l_i}^*(k)\tilde{w}_i(k-1)A^{\text{des}}}{c + c_{l_i}(0)(A^{\text{des}})^T A^{\text{des}} + \gamma(w_i(k-1)A^{\text{des}})^2}. \tag{9.22}$$

The updating algorithms of (9.19) and (9.20) can also be written as:

$$\tilde{c}_{l_i}(k) = \tilde{c}_{l_i}(k-1) + \gamma\varepsilon_i(k)(w_i(k-1)A^{\text{des}}) \tag{9.23}$$

$$\tilde{w}_i(k) = \tilde{w}_i(k-1) + a\,\text{sgn}(c_{l_i}^*)\varepsilon_i(k)\left(A^{\text{des}}\right)^T. \tag{9.24}$$

For each node i consider the Lyapunov function

$$V_i(k) = \frac{\tilde{c}_{l_i}^2(k)}{2\gamma} + \frac{\left|c_{l_i}^*(k)\right|}{2a}\tilde{w}_i(k)\tilde{w}_i^T(k). \tag{9.25}$$

Then

$$\Delta V_i(k) \overset{\Delta}{=} V_i(k) - V_i(k-1) = \left(\frac{\tilde{c}_{l_i}^2(k)}{2\gamma} - \frac{\tilde{c}_{l_i}^2(k-1)}{2\gamma}\right)$$

$$+ \frac{\left|c_{l_i}^*(k)\right|}{2a}\left(\tilde{w}_i(k)\tilde{w}_i^T(k) - \tilde{w}_i(k-1)\tilde{w}_i^T(k-1)\right)$$

Using (9.23) and (9.24) $\Delta V_i(k)$ is:

$$\Delta V_i(k) = \left(\frac{\gamma \varepsilon_i^2(k) \left(w_i(k-1)A^{\mathrm{des}} \right)^2}{2} + \tilde{c}_{l_i}(k-1)\varepsilon_i(k) \left(w_i(k-1)A^{\mathrm{des}} \right) \right)$$

$$+ \frac{\left| c_{l_i}^* \right|}{2} \operatorname{sgn}\left(c_{l_i}^* \right) \varepsilon_i(k) \left(\tilde{w}_i(k-1)A^{\mathrm{des}} + \left(A^{\mathrm{des}} \right)^T \tilde{w}_i^T(k-1) \right)$$

$$+ \frac{\left| c_{l_i}^* \right|}{2} a \left(\varepsilon_i(k) \right)^2 \left(A^{\mathrm{des}} \right)^T A^{\mathrm{des}}$$

Taking into account that $c_{l_i}^*(k) = \left| c_{l_i}^*(k) \right| \operatorname{sgn}\left(c_{l_i}^*(k) \right)$ and $\tilde{w}_i(k-1)A^{\mathrm{des}} = \left(A^{\mathrm{des}} \right)^T \tilde{w}_i^T(k-1)$, $\Delta V_i(k)$ is now equal to:

$$\Delta V_i(k) = \frac{1}{2}\gamma \varepsilon_i^2(k) \left(w_i(k-1)A^{\mathrm{des}} \right)^2 + \tilde{c}_{l_i}(k-1)\varepsilon_i(k) \left(w_i(k-1)A^{\mathrm{des}} \right)$$

$$+ c_{l_i}^*(k)\varepsilon_i(k) \left(\tilde{w}_i(k-1)A^{\mathrm{des}} \right) + \frac{\left| c_{l_i}^*(k) \right|}{2} a \left(\varepsilon_i(k) \right)^2 \left(A^{\mathrm{des}} \right)^T A^{\mathrm{des}}.$$

Using (9.22), $\Delta V_i(k)$ is:

$$\Delta V_i(k) = \frac{1}{2}\gamma \varepsilon_i^2(k) \left(w_i(k-1)A^{\mathrm{des}} \right)^2$$

$$- \varepsilon_i(k)\varepsilon_i(k) \left(c + c_{l_i}(0) \left(A^{\mathrm{des}} \right)^T A^{\mathrm{des}} + \gamma(w_i(k-1)A^{\mathrm{des}})^2 \right)$$

$$+ \frac{\left| c_{l_i}^* \right|}{2} a \left(\varepsilon_i(k) \right)^2 \left(A^{\mathrm{des}} \right)^T A^{\mathrm{des}}.$$

Finally:

$$\Delta V_i(k) = -\varepsilon_i^2(k)m^2(k) \left[1 - \frac{\gamma \xi^2(k) + \left| c_{l_i}^* \right| a \left(A^{\mathrm{des}} \right)^T A^{\mathrm{des}}}{2m^2(k)} \right]$$

where $m^2(k) = c + c_{l_i}(0) \left(A^{\mathrm{des}} \right)^T A^{\mathrm{des}} + \gamma \xi^2(k)$ and $\xi^2(k) = \left(w_i(k-1)A^{\mathrm{des}} \right)^2$ It is obvious that if $0 < \gamma < 1$ and $a, c > 0$, then

$$\frac{\gamma \xi^2(k) + \left| c_{l_i}^* \right| a \left(A^{\mathrm{des}} \right)^T A^{\mathrm{des}}}{2m^2(k)} < 1$$

which implies that:

$$\Delta V_i(k) \leq -c_0 \varepsilon_i^2(k)m^2(k) \leq 0 \tag{9.26}$$

for some constant $c_0 > 0$.

From (9.25), (9.26) we have that $V_i(k)$ and therefore $w_i(k) \in L_\infty$, $c_{l_i}(k) \in L_\infty$ and $V_i(k)$ has a limit, i.e., $\lim_{k \to \infty} V_i(k) = V_\infty$. Consequently, using (9.26) we obtain

$$c_0 \sum_{k=1}^{\infty} (\varepsilon_i^2(k) m_i^2(k)) \le V_i(0) - V_\infty < \infty$$

which implies $\varepsilon_i(k) m_i(k) \in L_2$ and $\varepsilon_i(k) m_i(k) \to 0$ as $k \to \infty$. Since $m_i(k) = \sqrt{(c + c_{l_i}(0)(A^{\mathrm{des}})^T A^{\mathrm{des}} + \gamma \xi^2(k))} \ge \sqrt{c} > 0$, we also have that $\varepsilon_i(k) \in L_2$ and $\varepsilon_i(k) \to 0$ as $k \to \infty$.

We have $\varepsilon_i(k) A^{\mathrm{des}} = \varepsilon_i(k) m_i(k) \frac{A^{\mathrm{des}}}{m_i(k)}$. Since $\frac{A^{\mathrm{des}}}{m_i(k)}$ is bounded and $\varepsilon_i(k) m_i(k) \in L_2$, we have that $\varepsilon_i(k) A^{\mathrm{des}} \in L_2$ and $\left\| \varepsilon_i(k) A^{\mathrm{des}} \right\| \to 0$ as $k \to \infty$. This implies (using (9.20)) that $\| w_i(k) - w_i(k-1) \| \in L_2$ and $\| w_i(k) - w_i(k-1) \| \to 0$ as $k \to \infty$. Now,

$$w_i(k) - w_i(k-N) = w_i(k) - w_i(k-1) + w_i(k-1) - w_i(k-2)$$
$$+ \ldots + w_i(k-N+1) - w_i(k-N)$$

for any finite N. Using the Schwartz inequality, we have

$$\| w_i(k) - w_i(k-N) \|^2 \le \| w_i(k) - w_i(k-1) \|^2 + \| w_i(k-1) - w_i(k-2) \|^2$$
$$+ \ldots + \| w_i(k-N+1) - w_i(k-N) \|^2 .$$

Since each term on the right-hand side of the inequality is in L_2 and goes to zero with $k \to \infty$, it follows that $\| w_i(k) - w_i(k-N) \| \in L_2$ and $\| w_i(k) - w_i(k-N) \| \to 0$ as $k \to \infty$. Since $\varepsilon_i(k) A^{\mathrm{des}} \in L_2$ and $w_i(k-1) \in L_\infty$ then $w_i(k-1) \varepsilon_i(k) A^{\mathrm{des}} \in L_2 \cap L_\infty$ and $\left| \varepsilon_i(k) w_i(k-1) A^{\mathrm{des}} \right| \to 0$ as $k \to \infty$. This implies (using (9.19)) that $\left| c_{l_i}(k) - c_{l_i}(k-1) \right| \in L_2$ and $\left| c_{l_i}(k) - c_{l_i}(k-1) \right| \to 0$ as $k \to \infty$. Now,

$$c_{l_i}(k) - c_{l_i}(k-N) = c_{l_i}(k) - c_{l_i}(k-1) + c_{l_i}(k-1) - c_{l_i}(k-2)$$
$$+ \cdots + c_{l_i}(k-N+1) - c_{l_i}(k-N)$$

for any finite N. We also have that,

$$\left| c_{l_i}(k) - c_{l_i}(k-N) \right|^2 \le$$
$$\left| c_{l_i}(k) - c_{l_i}(k-1) \right|^2 + \left| c_{l_i}(k-1) - c_{l_i}(k-2) \right|^2$$
$$+ \ldots + \left| c_{l_i}(k-N+1) - c_{l_i}(k-N) \right|^2$$

Since each term of the right-hand side of the inequality is in L_2 and goes to zero with $k \to \infty$, it follows that $\left| c_{l_i}(k) - c_{l_i}(k-N) \right| \in L_2$ and $\left| c_{l_i}(k) - c_{l_i}(k-N) \right| \to 0$ as $k \to \infty$.

Remark 9.2 The n weight updating laws given by (9.20) for each row w_i, $i = 1, \ldots, n$ of W can be written in a compact form as follows

$$\underline{W}(k) = \underline{W}(k-1) + a\underline{\varepsilon}A^{\text{des}} \tag{9.27}$$

where $\underline{W} = [w_1, w_2, \ldots w_n]^T$ is a $n^2 \times 1$ column vector containing all the elements of matrix W arranged in a column, row after row. Also, $\underline{\varepsilon}$ is a $n^2 \times n$ matrix having the form $\underline{\varepsilon} = [\varepsilon_1 I_n, \varepsilon_2 I_n, \ldots \varepsilon_n I_n]^T$, where I_n is a $n \times n$ unit matrix and ε_i is given by (9.18).

9.5 The Projection Methods for the Bilinear Model

In the bilinear model estimation, we have some a priori knowledge as to where the unknown parameters W^* and c_l^* are located in $\mathfrak{R}^n$ and $\mathfrak{R}$, respectively. In our parameter identification problem due to the definition of FCN (see Chap. 7) inequalities $-1 \le w_{ij} \le 1$ and $c_{l_i} \ge 0$ must be true. In order to ensure that the contraction holds then (8.14) and (8.16) must be also true. In the sequel, we are proposing a set of convex projection methods for the bilinear case. They modify the adaptive laws of the previous section so that they produce estimates that fulfill the conditions set. A schematic representation of the orthogonal projection method is illustrated in Fig. 9.3. According to the figure, when, after updating, the estimates (here W and C_l) exceed a prespecified area (say S_2), they are orthogonally projected on the surface of this area. In the following, we are grouping the projection methods in two categories. In projection method 1, we perform projections that are related with the requirements set by the initial definition of FCN, namely $-1 \le w_{ij} \le 1$ and $c_{l_i} \ge 0$. In projection method 2, we perform projections related with the contraction properties, that is, with the fulfillment of (8.14) and (8.16). Next, it will be proved that these projections do not worsen the stability and the convergence properties of the bilinear estimation algorithm.

9.5.1 Projection Method 1: Relation to FCN Definition

(**A**) The S_1' set for projection that satisfies $c_{l_i} > 0$ is the convex set:

$$S_1' = \{c_{l_i} \in \mathfrak{R} | g(c_{l_i}, \theta_i) \ge 0, \quad g(c_{l_i}, \theta_i) = c_{l_i} - \theta_i, \forall i \in \aleph\}$$

where, θ_i is a small positive constant. The updating formula of parameter c_{l_i} of FCN is now given by:

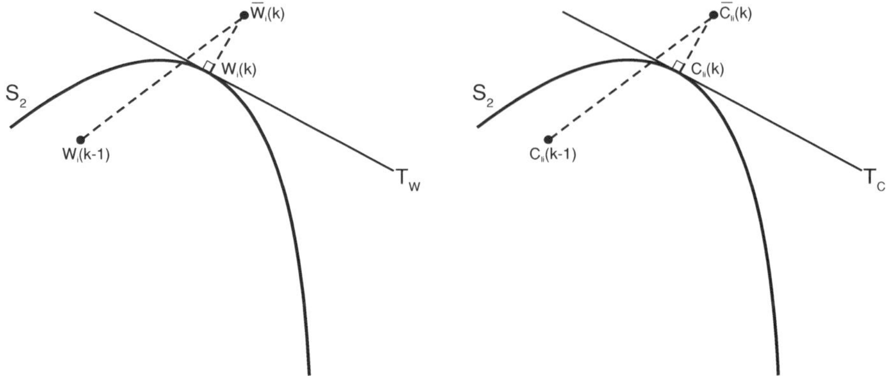

Fig. 9.3 Discrete-time parameter projection. The parameters exceeding the area S_2 are orthogonally projected on its surface

$$\overline{c}_{l_i}(k) = c_{l_i}(k-1) + \gamma \varepsilon_i(k)(w_i(k-1)A_i^{\text{des}})$$
$$c_{l_i}(k) = \begin{cases} \overline{c}_{l_i}(k), & \text{if } g(\overline{c}_{l_i}(k), \theta_i) > 0 \\ \theta_i & \text{if } \overline{g}(\overline{c}_{l_i}(k), \theta_i) \leq 0 \end{cases} \tag{9.28}$$

(B) The S_1 set for projection that satisfies $-1 \leq w_{ij} \leq 1$ is the convex set:

$$S_1 = \{W \in \mathfrak{R}^n | g(w_{ij}) \leq 0, \quad g(w_{ij}) = |w_{ij}| - 1, \forall i, j \in \aleph\}.$$

The updating formula of parameters W of FCN is now given by:

$$\overline{w}_{ij}(k) = w_{ij}(k-1) + \alpha \operatorname{sgn}(c_{l_i}(k))\varepsilon_i(k) A_j^{\text{des}}$$
$$w_{ij}(k) = \begin{cases} \overline{w}_{ij}(k), & \text{if } |\overline{w}_{ij}(k)| \leq 1 \\ \operatorname{sgn}(w_{ij}(k)) & \text{if } |\overline{w}_{ij}(k)| > 1. \end{cases} \tag{9.29}$$

9.5.2 Projection Method 2: Relation to Contraction Conditions

(A) The S_1'' set for projection that satisfies (8.14) is the convex set:

$$S_1'' = \{c_{l_i} \in \mathfrak{R} | g(c_{l_i}) \leq 0, \quad g(c_{l_i}) = c_{l_i} - M_0, \forall i \in \aleph\}$$

where M_0 satisfies the following equation:

$$\frac{M_0}{e^{M_0 w_i(k) A^{\text{des}}}} \left(\frac{1}{1 + e^{-M_0 w_i(k) A^{\text{des}}}} \right)^2 = 1. \tag{9.30}$$

One may estimate M_0 by numerically solving Eq. (9.30).

The updating formula of parameter c_{l_i} of FCN is now given by:

$$
\begin{aligned}
\overline{c}_{l_i}(k) &= c_{l_i}(k-1) + \gamma\varepsilon_i(k)\,(w_i(k-1)A_i^{\mathrm{des}}) \\
c_{l_i}(k) &= \begin{cases} \overline{c}_{l_i}(k), & \text{if } g(\overline{c}_{l_i}(k)) \le 0 \\ M_0 & \text{if } g(\overline{c}_{l_i}(k)) > 0 \end{cases}
\end{aligned}
\tag{9.31}
$$

(**B**) The S_2 set for projection that satisfies (8.16) is the convex set:

$$
S_2 = \left\{ w_i \in \mathfrak{R}^n,\, c_{l_i} \in \mathfrak{R} \,\middle|\, g(w_i, c_{l_i}) \le 0,\ g(w_i, c_{l_i}) \right.
$$

$$
\left. = \left(l_i c_{l_i} \|w_i\|\right)\left(1 - \frac{1}{\sqrt{\sum\limits_{i=1}^{n}\left(l_i \overline{c}_{l_i}(k)\,\|\overline{w}_i(k)\|\right)^2}}\right) \right\}.
$$

The updating equations of parameters w_i and c_{l_i} of FCN are now given by:

$$
\begin{aligned}
\overline{w}_i(k) &= w_i(k-1) + \alpha\,\mathrm{sgn}(c_{l_i}^*)\varepsilon_i(k)(A^{\mathrm{des}})^T \\
w_i(k) &= \overline{w}_i(k)\min\left(1,\left(\frac{1}{\sqrt{\sum\limits_{i=1}^{n}\left(l_i\overline{c}_{l_i}(k)\|\overline{w}_i(k)\|\right)^2}}\right)\right)
\end{aligned}
\tag{9.32}
$$

$$
\begin{aligned}
\overline{c}_{l_i}(k) &= c_{l_i}(k-1) + \gamma\varepsilon_i(k)(w_i(k-1)A^{\mathrm{des}}) \\
c_{l_i}(k) &= \overline{c}_{l_i}(k)\min\left(1,\left(\frac{1}{\sqrt{\sum\limits_{i=1}^{n}\left(l_i\overline{c}_{l_i}(k)\|\overline{w}_i(k)\|\right)^2}}\right)\right).
\end{aligned}
\tag{9.33}
$$

It has to be noted that due to (9.28), $\mathrm{sgn}(c_{l_i}(k))$ in (9.29) and (9.32) is always positive. Also, in method 2.B, after the projection condition is activated, the orthogonal projection affects all w_i and c_{l_i} proportionally to their size according to (9.32) and (9.33).

The above projection methods do not permit the estimates of $w_i(k)$ and $c_{l_i}(k)$ to exceed FCN design bounds or bounds required from the contraction property. We can now prove that they also do not alter the stability and convergence properties of the bilinear algorithm.

Proof To prove the above statement, it suffices to show that, when one projection method is applied the negativity of $\Delta V_i(k)$ is not altered, where $V_i(k)$ was defined in (9.25). We start the proof with projection method 1.A.

Projection method 1.A.

We consider the case where the projection method 1.A is activated. In this case the updating is made according to the second branch of (9.28). From (9.18) and $\tilde{w}_i(k) = w_i(k) - w_i^*$, $\tilde{c}_{l_i}(k) = c_{l_i}(k) - c_{l_i}^*$ we obtain the error equation given by Eq. (9.22), while the updating algorithm is now given by equations.

$$\tilde{c}_{l_i}(k) = \theta_i - c_{l_i}^* \tag{9.34}$$

$$\tilde{w}_i(k) = \tilde{w}_i(k-1) + \alpha \operatorname{sgn}(c_{l_i}^*)\varepsilon_i(k)(A^{\text{des}})^T. \tag{9.35}$$

Equation (9.34) can also be written as:

$$\tilde{c}_{l_i}(k) = A - B \tag{9.36}$$

where $A = \left(\tilde{c}_{l_i}(k-1) + \gamma\varepsilon_i(k)(w_i(k-1)A_i^{\text{des}})\right)$,
$B = \left(\tilde{c}_{l_i}(k-1) + \gamma\varepsilon_i(k)(w_i(k-1)A_i^{\text{des}}) - \theta_i + c_{l_i}^*\right)$
For each node i consider the function (9.25)

$$V_i(k) = \frac{\tilde{c}_{l_i}^2(k)}{2\gamma} + \frac{\left|c_{l_i}^*\right|}{2a}\tilde{w}_i(k)\tilde{w}_i^T(k).$$

Then

$$\Delta V_i(k) \overset{\Delta}{=} V_i(k) - V_i(k-1) = \left(\frac{\tilde{c}_{l_i}^2(k)}{2\gamma} - \frac{\tilde{c}_{l_i}^2(k-1)}{2\gamma}\right)$$
$$+ \frac{\left|c_{l_i}^*\right|}{2a}\left(\tilde{w}_i(k)\tilde{w}_i^T(k) - \tilde{w}_i(k-1)\tilde{w}_i^T(k-1)\right).$$

Using (9.35) and (9.36) $\Delta V_i(k)$ is:

$$\Delta V_i(k) = \left(\frac{\gamma\varepsilon_i^2(k)\left(w_i(k-1)A^{\text{des}}\right)^2}{2} + \tilde{c}_{l_i}(k-1)\varepsilon_i(k)\left(w_i(k-1)A^{\text{des}}\right)\right)$$
$$+ \frac{\left|c_{l_i}^*\right|}{2}\operatorname{sgn}\left(c_{l_i}^*\right)\varepsilon_i(k)\left(\tilde{w}_i(k-1)A^{\text{des}} + \left(A^{\text{des}}\right)^T\tilde{w}_i^T(k-1)\right)$$
$$+ \frac{\left|c_{l_i}^*\right|}{2}a\left(\varepsilon_i(k)\right)^2\left(A^{\text{des}}\right)^T A^{\text{des}}.$$

Taking into account that

$$
\frac{A^2 - \tilde{c}_{l_i}^2(k-1)}{2\gamma} = \frac{1}{2}\gamma\varepsilon_i^2(k)\left(w_i(k-1)A^{\mathrm{des}}\right)^2 + \tilde{c}_{l_i}(k-1)\varepsilon_i(k)\left(w_i(k-1)A^{\mathrm{des}}\right),
$$

$$
c_{l_i}^* = \left|c_{l_i}^*\right|\mathrm{sgn}\left(c_{l_i}^*\right) \quad \text{and} \quad \tilde{w}_i(k-1)A^{\mathrm{des}} = \left(A^{\mathrm{des}}\right)^T \tilde{w}_i^T(k-1).
$$

$\Delta V_i(k)$ is now equal to:

$$
\Delta V_i(k) = \frac{1}{2}\gamma\varepsilon_i^2(k)\left(w_i(k-1)A^{\mathrm{des}}\right)^2 + \tilde{c}_{l_i}(k-1)\varepsilon_i(k)\left(w_i(k-1)A^{\mathrm{des}}\right)
$$

$$
+ c_{l_i}^*\varepsilon_i(k)\left(\tilde{w}_i(k-1)A^{\mathrm{des}}\right) + \frac{\left|c_{l_i}^*\right|}{2}a\left(\varepsilon_i(k)\right)^2\left(A^{\mathrm{des}}\right)^T A^{\mathrm{des}}
$$

$$
+ \frac{1}{2\gamma}\left(B - 2A\right)B.
$$

Using (9.22), $\Delta V_i(k)$ is:

$$
\Delta V_i(k) = \frac{1}{2}\gamma\varepsilon_i^2(k)\left(w_i(k-1)A^{\mathrm{des}}\right)^2
$$

$$
- \varepsilon_i(k)\varepsilon_i(k)\left(c + c_{l_i}(0)\left(A^{\mathrm{des}}\right)^T A^{\mathrm{des}} + \gamma(w_i(k-1)A^{\mathrm{des}})^2\right)
$$

$$
+ \frac{\left|c_{l_i}^*\right|}{2}a\left(\varepsilon_i(k)\right)^2\left(A^{\mathrm{des}}\right)^T A^{\mathrm{des}} + \frac{1}{2\gamma}\left(B - 2A\right)B.
$$

Using $\tilde{c}_{l_i}(k-1) = c_{l_i}(k-1) - c_{l_i}^*$, after straightforward manipulations we conclude to:

$$
B - 2A = -\left(c_{l_i}(k-1) + \gamma\varepsilon_i(k)\left(w_i(k-1)A^{\mathrm{des}}\right) - \theta_i\right) - 2\left(\theta_i - c_{l_i}^*\right)
$$

and $B = \left(c_{l_i}(k-1) + \gamma\varepsilon_i(k)\left(w_i(k-1)A^{\mathrm{des}}\right) - \theta_i\right)$

Finally:

$$
(B - 2A)B = -\left(c_{l_i}(k-1) + \gamma\varepsilon_i(k)\left(w_i(k-1)A^{\mathrm{des}}\right) - \theta_i\right)^2
$$

$$
- 2\left(\theta_i - c_{l_i}^*\right)\left(c_{l_i}(k-1) + \gamma\varepsilon_i(k)\left(w_i(k-1)A^{\mathrm{des}}\right) - \theta_i\right).
$$

When $c_{l_i}(k) < \theta_i$ we have that $c_{l_i}(k) - \theta_i < 0$ and from Eq. (9.28) we have that

$$
c_{l_i}(k-1) + \gamma\varepsilon_i(k)\left(w_i(k-1)A^{\mathrm{des}}\right) - \theta_i < 0,
$$

we also know that $c_{l_i}^* > \theta_i \Rightarrow \theta_i - c_{l_i}^* < 0$.

Finally we have that:

$$\left(\theta_i - c_{l_i}^*\right)\left(c_{l_i}(k-1) + \gamma\varepsilon_i(k)\,(w_i(k-1)A^{\text{des}}) - \theta_i\right) > 0. \tag{9.37}$$

Taking into account (9.37) we conclude that: $(B-2A)B < 0$
   Finally,

$$\Delta V_i(k) = -\varepsilon_i^2(k)m^2(k)\left[1 - \frac{\gamma\xi^2(k) + \left|c_{l_i}^*\right|a\left(A^{\text{des}}\right)^T A^{\text{des}}}{2m^2(k)}\right] + \frac{1}{2\gamma}(B-2A)B$$

where $m^2(k) = c + c_{l_i}(0)\left(A^{\text{des}}\right)^T A^{\text{des}} + \gamma\xi^2(k)$ and $\xi^2(k) = \left(w_i(k-1)A^{\text{des}}\right)^2$
   It is obvious that if $0 < \gamma < 1$ and $a, c > 0$, then

$$\frac{\gamma\xi^2(k) + \left|c_{l_i}^*\right|a\left(A^{\text{des}}\right)^T A^{\text{des}}}{2m^2(k)} < 1$$

which implies that: $\Delta V_i(k) \le -c_0\varepsilon_i^2(k)m^2(k) - c_0'(2A - B)B \le 0$
   for some constant $c_0 > 0$, $c_0' > 0$. Therefore, the negativity of $\Delta V_i(k)$ is not compromised. Actually, it is further strengthened due to the term (B-2A)B, which is negative.

Projection method 1.B.

According to (9.29), the projection method 1.B is activated when $\left|\overline{w}_{ij}(k)\right| > 1$. Here $\overline{w}_{ij}(k)$ denotes w_{ij} after the kth updating operation occurs according to (9.20). In the following, we use $w_{ij}(k)$ and $\overline{w}_{ij}(k)$ equivalently. Obviously, not all elements w_{ij} of w_i require projection. This has to be taken into account in the proof of the maintenance of the negativity of $\Delta V_i(k)$, which follows the same lines as with projection method 1.A. From (9.18) and $\tilde{w}_i(k) = w_i(k) - w_i^*$, $\tilde{c}_{l_i}(k) = c_{l_i}(k) - c_{l_i}^*$ we obtain the error equation given by Eq. (9.22), while the updating algorithm can now be given by eqs.

$$\tilde{c}_{l_i}(k) = \tilde{c}_{l_i}(k-1) + \gamma\varepsilon_i(k)(w_i(k-1)A^{\text{des}}) \tag{9.38}$$

$$\tilde{w}_{ij}(k) = \begin{cases} \tilde{w}_{ij}(k-1) + \alpha\,\text{sgn}(c_{l_i}^*)\varepsilon_i(k)A_j^{\text{des}}, & if\ \left|\overline{w}_{ij}(k)\right| \le 1 \\ \text{sgn}(w_{ij}(k)) - w_{ij}^*, & if\ \left|\overline{w}_{ij}(k)\right| > 1 \end{cases} \tag{9.39}$$

where $\tilde{w}_{ij}(k)$ is the jth element of row $\tilde{w}_i(k)$, A_j^{des} is the jth element of vector A^{des} and w_{ij}^* is the jth element of row w_i^*. We define two sets of indices, $I_p = \left\{j : \left|w_{ij}(k)\right| > 1\right\}$ and $I = \left\{j : \left|w_{ij}(k)\right| \le 1\right\}$, being subsets of $\{1, 2, \ldots, n\}$. That is, I_p is the set of indices denoting weights that require projection, while I contains the remaining indices. Taking into account all w_{ij}, $j = 1 \ldots n$ updatings, equation (9.39) can now be written as:

$$\tilde{w}_i(k) = \mathbf{A} - \mathbf{B} \tag{9.40}$$

where $\mathbf{A} = \tilde{w}_i(k-1) + \alpha \mathrm{sgn}(c_{l_i}^*)\varepsilon_i(k)\left(A^{\mathrm{des}}\right)^T$ is a row vector $\mathbf{A} = \begin{bmatrix} \mathbf{a}_1 & \mathbf{a}_2 & \ldots & \mathbf{a}_n \end{bmatrix}$ with $\mathbf{a}_j = \tilde{w}_{ij}(k-1) + \alpha \mathrm{sgn}(c_{l_i}^*)\varepsilon_i(k)A_j^{\mathrm{des}}$, $j = 1 \ldots n$. Moreover, $\mathbf{B} = \begin{bmatrix} \mathbf{b}_1 & \mathbf{b}_2 & \ldots & \mathbf{b}_n \end{bmatrix}$, is a row vector such that

$$\mathbf{b}_j = \begin{cases} \tilde{w}_{ij}(k-1) + \alpha \mathrm{sgn}(c_{l_i}^*)\varepsilon_i(k)A_j^{\mathrm{des}} - \mathrm{sgn}(w_{ij}(k)) + w_{ij}^*, & \forall j \in I_p \\ 0, & \forall j \in I \end{cases}$$

For each node i consider the Lyapunov function (9.25). Then

$$\Delta V_i(k) \stackrel{\triangle}{=} V_i(k) - V_i(k-1) = \left(\frac{\tilde{c}_{l_i}^2(k)}{2\gamma} - \frac{\tilde{c}_{l_i}^2(k-1)}{2\gamma} \right)$$
$$+ \frac{\left|c_{l_i}^*\right|}{2a}\left(\tilde{w}_i(k)\tilde{w}_i^T(k) - \tilde{w}_i(k-1)\tilde{w}_i^T(k-1) \right).$$

Using (9.38) and (9.40) $\Delta V_i(k)$ is:

$$\Delta V_i(k) = \left(\frac{\gamma \varepsilon_i^2(k)\left(w_i(k-1)A^{\mathrm{des}}\right)^2}{2} + \tilde{c}_{l_i}(k-1)\varepsilon_i(k)\left(w_i(k-1)A^{\mathrm{des}}\right) \right)$$
$$+ \frac{\left|c_{l_i}^*\right|}{2a}\left(\mathbf{A}\mathbf{A}^T - \mathbf{A}\mathbf{B}^T - \mathbf{B}\mathbf{A}^T + \mathbf{B}\mathbf{B}^T - \tilde{w}_i(k-1)\tilde{w}_i^T(k-1) \right).$$

Taking into account that:

$$\frac{\left|c_{l_i}^*\right|}{2a}\left(\mathbf{A}\mathbf{A}^T - \tilde{w}_i(k-1)\tilde{w}_i^T(k-1) \right) = \frac{\left|c_{l_i}^*\right|}{2}\mathrm{sgn}\left(c_{l_i}^*\right)$$
$$\varepsilon_i(k)\left(\tilde{w}_i(k-1)A^{\mathrm{des}} + \left(A^{\mathrm{des}}\right)^T \tilde{w}_i^T(k-1) \right)$$
$$+ \frac{\left|c_{l_i}^*\right|}{2}a\left(\varepsilon_i(k)\right)^2\left(A^{\mathrm{des}}\right)^T A^{\mathrm{des}},$$
$$c_{l_i}^* = \left|c_{l_i}^*\right|\mathrm{sgn}\left(c_{l_i}^*\right), \; \tilde{w}_i(k-1)A^{\mathrm{des}}$$
$$= \left(A^{\mathrm{des}}\right)^T \tilde{w}_i^T(k-1) \text{ and } \mathbf{A}\mathbf{B}^T = \mathbf{B}\mathbf{A}^T.$$

$\Delta V_i(k)$ becomes:

$$\Delta V_i(k) = \frac{1}{2}\gamma\varepsilon_i^2(k)\left(w_i(k-1)A^{\text{des}}\right)^2 + \tilde{c}_{l_i}(k-1)\varepsilon_i(k)\left(w_i(k-1)A^{\text{des}}\right)$$
$$+ c_{l_i}^*(k)\varepsilon_i(k)\left(\tilde{w}_i(k-1)A^{\text{des}}\right) + \frac{\left|c_{l_i}^*\right|}{2}a\left(\varepsilon_i(k)\right)^2\left(A^{\text{des}}\right)^T A^{\text{des}}$$
$$- \frac{\left|c_{l_i}^*(k)\right|}{2a}2\mathbf{AB}^T + \frac{\left|c_{l_i}^*\right|}{2a}\mathbf{BB}^T.$$

Using (9.22), $\Delta V_i(k)$ is:

$$\Delta V_i(k) = \frac{1}{2}\gamma\varepsilon_i^2(k)\left(w_i(k-1)A^{\text{des}}\right)^2$$
$$- \varepsilon_i(k)\varepsilon_i(k)\left(c + c_{l_i}(0)\left(A^{\text{des}}\right)^T A^{\text{des}} + \gamma(w_i(k-1)A^{\text{des}})^2\right)$$
$$+ \frac{\left|c_{l_i}^*\right|}{2}a\left(\varepsilon_i(k)\right)^2\left(A^{\text{des}}\right)^T A^{\text{des}} + \frac{\left|c_{l_i}^*\right|}{2a}\left[\mathbf{B} - 2\mathbf{A}\right]\mathbf{B}^T.$$

Following straightforward manipulation, each element of matrix $\mathbf{B} - 2\mathbf{A}$ is equal to:

$$\mathbf{b}_j - 2\mathbf{a}_j = \begin{cases} -\tilde{w}_{ij}(k-1) - \alpha\text{sgn}(c_{l_i}^*)\varepsilon_i(k)A_j^{\text{des}} + \text{sgn}(w_{ij}(k)) - w_{ij}^* \\ -2\left(\text{sgn}(w_{ij}(k)) - w_{ij}^*\right), & \forall j \in I_p \\ -2\left(\tilde{w}_{ij}(k-1) + \alpha\text{sgn}(c_{l_i}^*)\varepsilon_i(k)A_j^{\text{des}}\right), & \forall j \in I \end{cases}$$

From $\tilde{w}_i(k-1) = w_i(k-1) - w_i^*$ we have that $\mathbf{b}_j - 2\mathbf{a}_j$ is equal to:

$$\mathbf{b}_j - 2\mathbf{a}_j = \begin{cases} -w_{ij}(k-1) - \alpha sgn(c_{l_i}^*)\varepsilon_i(k)A_j^{des} + \text{sgn}(w_{ij}(k)) \\ -2\left(\text{sgn}(w_{ij}(k)) - w_{ij}^*\right), & \forall j \in I_p \\ -2\left(\tilde{w}_{ij}(k-1) + \alpha sgn(c_{l_i}^*)\varepsilon_i(k)A_j^{des}\right), & \forall j \in I \end{cases}$$

Similarly, we have that $\mathbf{b}_j$ is now equal to:

$$\mathbf{b}_j = \begin{cases} w_{ij}(k-1) + \alpha\text{sgn}(c_{l_i}^*)\varepsilon_i(k)A_j^{des} - \text{sgn}(w_{ij}(k)), & \forall j \in I_p \\ 0, & \forall j \in I \end{cases}$$

Finally:

$$
\begin{aligned}
[\mathbf{B} - 2\mathbf{A}]\,\mathbf{B}^T = &- \sum_{j \in I_p} \left(w_{ij}(k-1) + \alpha \mathrm{sgn}(c_{l_i}^*)\varepsilon_i(k)A_j^{\mathrm{des}} - \mathrm{sgn}(w_{ij}(k)) \right)^2 \\
&- 2 \sum_{j \in I_p} \Bigg[\left(w_{ij}(k-1) + \alpha \mathrm{sgn}(c_{l_i}^*)\varepsilon_i(k)A_j^{\mathrm{des}} - \mathrm{sgn}(w_{ij}(k)) \right) \\
&\qquad\qquad \left(\mathrm{sgn}(w_{ij}(k)) - w_{ij}^* \right) \Bigg]
\end{aligned}
$$

When $w_{ij}(k) < -1$ we have that $w_{ij}(k) - \mathrm{sgn}(w_{ij}(k)) < 0$ and consequently

$$
w_{ij}(k-1) + \alpha \mathrm{sgn}(c_{l_i}^*)\varepsilon_i(k)A_j^{\mathrm{des}} - \mathrm{sgn}(w_{ij}(k)) < 0.
$$

By the definition of the FCN, we also know that $w_{ij}^* > -1 \Rightarrow \mathrm{sgn}(w_{ij}(k)) - w_{ij}^* < 0$.
When $w_{ij}(k) > 1$, then $w_{ij}(k) - \mathrm{sgn}(w_{ij}(k)) > 0$ and consequently

$$
w_{ij}(k-1) + \alpha \mathrm{sgn}(c_{l_i}^*)\varepsilon_i(k)A_j^{\mathrm{des}} - \mathrm{sgn}(w_{ij}(k)) > 0
$$

we also know that $w_{ij}^* < 1 \Rightarrow \mathrm{sgn}(w_{ij}(k)) - w_{ij}^* > 0$.
Finally, we conclude to:

$$
\left(w_{ij}(k-1) + \alpha \mathrm{sgn}(c_{l_i}(k))\varepsilon_i(k)A_j^{\mathrm{des}} - \mathrm{sgn}(w_{ij}(k)) \right) \left(\mathrm{sgn}(w_{ij}(k)) - w_{ij}^* \right) > 0
\tag{9.41}
$$

Using (9.41) it is easily concluded that: $[\mathbf{B} - 2\mathbf{A}]\,\mathbf{B}^T < 0$.
Finally

$$
\begin{aligned}
\Delta V_i(k) = &- \varepsilon_i^2(k)m^2(k) \left[1 - \frac{\gamma \xi^2(k) + \left| c_{l_i}^*(k) \right| a \left(A^{\mathrm{des}} \right)^T A^{\mathrm{des}}}{2m^2(k)} \right] \\
&+ \frac{\left| c_{l_i}^*(k) \right|}{2a} [\mathbf{B} - 2\mathbf{A}]\,\mathbf{B}^T
\end{aligned}
$$

where $m^2(k) = c + c_{l_i}(0)\left(A^{\mathrm{des}} \right)^T A^{\mathrm{des}} + \gamma \xi^2(k)$ and $\xi^2(k) = \left(w_i(k-1)A^{\mathrm{des}} \right)^2$.
It is obvious that if $0 < \gamma < 1$ and $a, c > 0$, then

$$
\frac{\gamma \xi^2(k) + \left| c_{l_i}^*(k) \right| a \left(A^{\mathrm{des}} \right)^T A^{\mathrm{des}}}{2m^2(k)} < 1
$$

which implies that:

$$\Delta V_i(k) \le -c_0 \varepsilon_i^2(k) m^2(k) - c_0' [2\mathbf{A} - \mathbf{B}]\mathbf{B}^T \le 0$$

for some constant $c_0 > 0$, $c_0' > 0$. Therefore, the negativity of $\Delta V_i(k)$ is not compromised. Actually, it is further strengthened due to the term $[\mathbf{B} - 2\mathbf{A}]\mathbf{B}^T$, which is negative.

Projection method 2.A.

From (9.18) and $\tilde{w}_i(k) = w_i(k) - w_i^*$, $\tilde{c}_{l_i}(k) = c_{l_i}(k) - c_{l_i}^*$ we obtain the error equation given by Eq. (9.22). When the projection condition is fulfilled the updating equations are given by eqs.

$$\tilde{c}_{l_i}(k) = M_0 - c_{l_i}^* \tag{9.42}$$

$$\tilde{w}_i(k) = \tilde{w}_i(k-1) + \alpha \operatorname{sgn}(c_{l_i}^*)\varepsilon_i(k)(A^{\mathrm{des}})^T \tag{9.43}$$

The rest of the analysis follows the lines of proof given in Projection Method 1.A and concludes to:

$$(B - 2A)B = -\left(c_{l_i}(k-1) + \gamma \varepsilon_i(k)(w_i(k-1)A^{\mathrm{des}}) - M_0\right)^2$$
$$- 2\left(M_0 - c_{l_i}^*\right)\left(c_{l_i}(k-1) + \gamma \varepsilon_i(k)(w_i(k-1)A^{\mathrm{des}}) - M_0\right)$$

Since the projection is activated, $c_{l_i}(k) > M_0$, so we have that $c_{l_i}(k) - M_0 > 0$ and therefore:

$$c_{l_i}(k-1) + \gamma \varepsilon_i(k)(w_i(k-1)A^{\mathrm{des}}) - M_0 > 0$$

we also know that $c_{l_i}^* < M_0 \Rightarrow M_0 - c_{l_i}^* > 0$.
Finally we have that:

$$\left(M_0 - c_{l_i}^*\right)\left(c_{l_i}(k-1) + \gamma \varepsilon_i(k)(w_i(k-1)A^{\mathrm{des}}) - M_0\right) > 0 \tag{9.44}$$

and eventually we conclude that $(B - 2A)B < 0$
Finally

$$\Delta V_i(k) = -\varepsilon_i^2(k) m^2(k)\left[1 - \frac{\gamma \xi^2(k) + \left|c_{l_i}^*\right| a \left(A^{\mathrm{des}}\right)^T A^{\mathrm{des}}}{2m^2(k)}\right] + \frac{1}{2\gamma}(B - 2A)B$$

where $m^2(k) = c + c_{l_i}(0)\left(A^{\mathrm{des}}\right)^T A^{\mathrm{des}} + \gamma \xi^2(k)$ and $\xi^2(k) = \left(w_i(k-1)A^{\mathrm{des}}\right)^2$.

It is obvious that if $0 < \gamma < 1$ and $a, c > 0$, then

$$\frac{\gamma \xi^2(k) + \left| c_{l_i}^* \right| a \left(A^{\mathrm{des}} \right)^T A^{\mathrm{des}}}{2m^2(k)} < 1$$

which implies that:

$$\Delta V_i(k) \leq -c_0 \varepsilon_i^2(k) m^2(k) - c_0' \left(2A - B \right) B \leq 0$$

for some constant $c_0 > 0, c_0' > 0$. Therefore, the negativity of $\Delta V_i(k)$ is not compromised. Actually, it is further strengthened due to the term $(B - 2A)B$, which is negative.

Projection method 2.B.
From (9.18) and $\tilde{w}_i(k) = w_i(k) - w_i^*$, $\tilde{c}_{l_i}(k) = c_{l_i}(k) - c_{l_i}^*$ we obtain the error equation given by Eq. (9.22). Taking into account the projection method 2.B the updating equations can be written as:

$$\tilde{w}_i(k) = \rho \left(\tilde{w}_i(k-1) + \alpha \mathrm{sgn}(c_{l_i}^*) \varepsilon_i(k)(A^{\mathrm{des}})^T \right) - (1 - \rho)w_i^* \tag{9.45}$$

$$\tilde{c}_{l_i}(k) = \rho \left(\tilde{c}_{l_i}(k-1) + \gamma \varepsilon_i(k)(w_i(k-1)A^{\mathrm{des}}) \right) - (1 - \rho)c_{l_i}^* \tag{9.46}$$

where $\rho = \left(\dfrac{1}{\sqrt{\sum\limits_{i=1}^{n} \left(l_i \bar{c}_{l_i}(k) \| \bar{w}_i(k) \| \right)^2}} \right)$.

Equation (9.45) can be written as:

$$\tilde{w}_i(k) = \rho \left(\tilde{w}_i(k-1) + \alpha \mathrm{sgn}(c_{l_i}^*) \varepsilon_i(k)(A^{\mathrm{des}})^T \right)$$
$$- (1 - \rho) \left(\tilde{w}_i(k-1) + \alpha \mathrm{sgn}(c_{l_i}^*) \varepsilon_i(k)(A^{\mathrm{des}})^T \right) + (1 - \rho) \left(\tilde{w}_i(k-1) \right.$$
$$\left. + \alpha \mathrm{sgn}(c_{l_i}^*) \varepsilon_i(k)(A^{\mathrm{des}})^T \right) - (1 - \rho)w_i^* \Rightarrow$$

$$\tilde{w}_i(k) = \left(\tilde{w}_i(k-1) + \alpha \mathrm{sgn}(c_{l_i}^*) \varepsilon_i(k)(A^{\mathrm{des}})^T \right)$$
$$- (1 - \rho) \left(\tilde{w}_i(k-1) + \alpha \mathrm{sgn}(c_{l_i}^*) \varepsilon_i(k)(A^{\mathrm{des}})^T - w_i^* \right) \Rightarrow$$

$$\tilde{w}_i(k) = \left(\tilde{w}_i(k-1) + \alpha \mathrm{sgn}(c_{l_i}^*) \varepsilon_i(k)(A^{\mathrm{des}})^T \right)$$
$$- (1 - \rho) \left(w_i(k-1) + \alpha \mathrm{sgn}(c_{l_i}^*) \varepsilon_i(k)(A^{\mathrm{des}})^T \right)$$

which can also be written as:

$$\tilde{w}_i(k) = \mathbf{A} - \mathbf{B} \tag{9.47}$$

where $\mathbf{A} = \left(\tilde{w}_i(k-1) + \alpha\,\mathrm{sgn}(c_{l_i}^*)\varepsilon_i(k)(A^{\mathrm{des}})^T \right)$

and $\mathbf{B} = (1-\rho)\left(w_i(k-1) + \alpha\,\mathrm{sgn}(c_{l_i}^*)\varepsilon_i(k)(A^{\mathrm{des}})^T \right)$.

Equation (9.46) can be written as:

$$\tilde{c}_{l_i}(k) = \rho\left(\tilde{c}_{l_i}(k-1) + \gamma\varepsilon_i(k)\left(w_i(k-1)A^{\mathrm{des}}\right)\right)$$
$$- (1-\rho)\left(\tilde{c}_{l_i}(k-1) + \gamma\varepsilon_i(k)\left(w_i(k-1)A^{\mathrm{des}}\right)\right)$$
$$+ (1-\rho)\left(\tilde{c}_{l_i}(k-1) + \gamma\varepsilon_i(k)\left(w_i(k-1)A^{\mathrm{des}}\right)\right) - (1-\rho)c_{l_i}^*$$

Using the same steps as above, $\tilde{c}_{l_i}(k)$ concludes to:

$$\tilde{c}_{l_i}(k) = \left(\tilde{c}_{l_i}(k-1) + \gamma\varepsilon_i(k)\left(w_i(k-1)A^{\mathrm{des}}\right)\right)$$
$$- (1-\rho)\left(c_{l_i}(k-1) + \gamma\varepsilon_i(k)\left(w_i(k-1)A^{\mathrm{des}}\right)\right)$$

which can also be written as:

$$\tilde{c}_{l_i}(k) = C - D \tag{9.48}$$

where $C = \left(\tilde{c}_{l_i}(k-1) + \gamma\varepsilon_i(k)\left(w_i(k-1)A^{\mathrm{des}}\right)\right)$

and $D = (1-\rho)\left(c_{l_i}(k-1) + \gamma\varepsilon_i(k)\left(w_i(k-1)A^{\mathrm{des}}\right)\right)$.

For each node i consider the Lyapunov function (9.25). Then,

$$\Delta V_i(k) \overset{\Delta}{=} V_i(k) - V_i(k-1) = \left(\frac{\tilde{c}_{l_i}^2(k)}{2\gamma} - \frac{\tilde{c}_{l_i}^2(k-1)}{2\gamma} \right)$$
$$+ \frac{\left| c_{l_i}^* \right|}{2a}\left(\tilde{w}_i(k)\tilde{w}_i^T(k) - \tilde{w}_i(k-1)\tilde{w}_i^T(k-1) \right)$$

Using (9.47) and (9.48) $\Delta V_i(k)$ is:

$$\Delta V_i(k) = \left(\frac{C^2 + (D-2C)D}{2\gamma} - \frac{\tilde{c}_{l_i}^2(k-1)}{2\gamma} \right)$$
$$+ \frac{\left| c_{l_i}^* \right|}{2a}\left(\mathbf{A}\mathbf{A} + (\mathbf{B}-2\mathbf{A})\,\mathbf{B}^T - \tilde{w}_i(k-1)\tilde{w}_i^T(k-1) \right)$$

Taking into account that

$$\frac{C^2 - \tilde{c}_{l_i}^2(k-1)}{2\gamma} = \frac{1}{2}\gamma\varepsilon_i^2(k)\left(w_i(k-1)A^{\mathrm{des}}\right)^2$$

$$+ \tilde{c}_{l_i}(k-1)\varepsilon_i(k)\left(w_i(k-1)A^{\mathrm{des}}\right), \frac{\left|c_{l_i}^*\right|}{2}$$

$$\left(\mathbf{AA} - \tilde{w}_i(k-1)\tilde{w}_i^T(k-1)\right) = \frac{\left|c_{l_i}^*\right|}{2}\mathrm{sgn}\left(c_{l_i}^*\right)\varepsilon_i(k)$$

$$\left(\tilde{w}_i(k-1)A^{\mathrm{des}} + \left(A^{\mathrm{des}}\right)^T \tilde{w}_i^T(k-1)\right)$$

$$+ \frac{\left|c_{l_i}^*\right|}{2}a\left(\varepsilon_i(k)\right)^2\left(A^{\mathrm{des}}\right)^T A^{\mathrm{des}}, c_{l_i}^* = \left|c_{l_i}^*\right|\mathrm{sgn}\left(c_{l_i}^*\right)$$

$$\text{and } \tilde{w}_i(k-1)A^{\mathrm{des}} = \left(A^{\mathrm{des}}\right)^T \tilde{w}_i^T(k-1)$$

$\Delta V_i(k)$ is now equal to:

$$\Delta V_i(k) = \frac{1}{2}\gamma\varepsilon_i^2(k)\left(w_i(k-1)A^{\mathrm{des}}\right)^2$$

$$- \varepsilon_i(k)\varepsilon_i(k)\left(c + c_{l_i}(0)\left(A^{\mathrm{des}}\right)^T A^{\mathrm{des}} + \gamma(w_i(k-1)A^{\mathrm{des}})^2\right)$$

$$+ \frac{\left|c_{l_i}^*\right|}{2}a\left(\varepsilon_i(k)\right)^2\left(A^{\mathrm{des}}\right)^T A^{\mathrm{des}} + \frac{1}{2\gamma}(D-2C)D + \frac{\left|c_{l_i}^*\right|}{2}(\mathbf{B}-2\mathbf{A})\mathbf{B}^T$$

Regarding $D - 2C$:
$D - 2C = (1-\rho)\left(c_{l_i}(k-1) + \gamma\varepsilon_i(k)(w_i(k-1)A^{\mathrm{des}})\right) - 2\left(\tilde{c}_{l_i}(k-1) + \gamma\varepsilon_i(k)\right.$
$\left.(w_i(k-1)A^{\mathrm{des}})\right)$ Using $\tilde{c}_{l_i}(k-1) = c_{l_i}(k-1) - c_{l_i}^*$ we conclude to:

$$D - 2C = -(1+\rho)\left(c_{l_i}(k-1) + \gamma\varepsilon_i(k)(w_i(k-1)A^{\mathrm{des}})\right).$$

Similarly, $(D-2C)C = -(1-\rho^2)\left(c_{l_i}(k-1) + \gamma\varepsilon_i(k)(w_i(k-1)A_i^{\mathrm{des}})\right)^2$.
Since $\rho < 1$ we conclude that: $(D-2C)C < 0$. Using quite similar analysis we conclude that:

$$(\mathbf{B}-2\mathbf{A})\mathbf{B}^T = -(1-\rho^2)\left(w_{l_i}(k-1) + \alpha\mathrm{sgn}(c_{l_i}^*)\varepsilon_i(k)(A^{\mathrm{des}})^T\right)$$

$$\left(w_{l_i}(k-1) + \alpha\mathrm{sgn}(c_{l_i}^*)\varepsilon_i(k)(A^{\mathrm{des}})^T\right)^T$$

for which also stands that: $(\mathbf{B} - 2\mathbf{A})\,\mathbf{B}^T < 0$

Finally

$$\Delta V_i(k) =$$

$$- \varepsilon_i^2(k) m^2(k) \left[1 - \frac{\gamma \xi^2(k) + \left| c_{l_i}^* \right| a \left(A^{\mathrm{des}} \right)^T A^{\mathrm{des}}}{2 m^2(k)} \right]$$

$$+ \frac{1}{2\gamma}\,(D - 2C)\,D + \frac{\left| c_{l_i}^* \right|}{2}\,(\mathbf{B} - 2\mathbf{A})\,\mathbf{B}^T$$

where $m^2(k) = c + c_{l_i}(0)\left(A^{\mathrm{des}}\right)^T A^{\mathrm{des}} + \gamma \xi^2(k)$ and $\xi^2(k) = \left(w_i(k-1)A^{\mathrm{des}}\right)^2$.

It is obvious that if $0 < \gamma < 1$ and $a, c > 0$, then

$$\frac{\gamma \xi^2(k) + \left| c_{l_i}^* \right| a \left(A^{\mathrm{des}} \right)^T A^{\mathrm{des}}}{2 m^2(k)} < 1$$

which implies that:

$$\Delta V_i(k) \le -c_0 \varepsilon_i^2(k) m^2(k) - c_0'\,(2C - D)\,D - c_0''\,(2\mathbf{A} - \mathbf{B})\,\mathbf{B}^T \le 0$$

for some constant $c_0 > 0$, $c_0' > 0$, $c_0'' > 0$. Therefore, the negativity of $\Delta V_i(k)$ is not compromised. Actually, it is further strengthened due to the terms $(D - 2C)D$ and $(\mathbf{B} - 2\mathbf{A})\,\mathbf{B}^T$, which are negative.

9.6 Simulations and Comparisons Between the Two Parametric Models

In this section, we present simulations, using mostly numerical examples, demonstrating the operation and performance of the linear and bilinear estimation algorithms with and without the respective projection methods. Comparisons are also made between the two algorithms operating on the same examples as well as on an artificial system identification task. The performance of the algorithms on systems arising in real-life engineering applications will be demonstrated in Chap. 10. We start with the algorithm for the LPM. The parameters α, c in Eqs. (9.6) and (9.7) are set to $\alpha = 0.9$ and $c = 1$. The same parameter values are used in (9.18), (9.19) and (9.20) with the additional use of $\gamma = 0.1$.

9.6.1 Simulation Results Based on the LPM

For the numerical examples a FCN with 8 nodes is used. The initial matrix W of FCN is shown in Table 9.1. Except from the diagonal elements, which have the value 1 and the zero elements all other weights were randomly selected. This matrix fulfills Eq. (8.22) because

$$\|\underline{W}\| = 3.856 < 4$$

With this weights matrix the FCN reaches an equilibrium point given by the following vector A.

$$A = \begin{bmatrix} 0.8981 \ 0.8147 \ 0.9037 \ 0.9387 \\ 0.9111 \ 0.8704 \ 0.8551 \ 0.7666 \end{bmatrix}^T$$

In order to test the performance of the method in the extreme case where the self feedback weight connections are all one, we deliberately assumed that d_{ii}, $i = 1, \ldots, n$ are kept constant to $d_{ii} = 1$ and therefore they do not participate in the weight updating. The same holds for the weights with null value, because they represent nonexisting connections.

Example 9.1 **LPM without projection**
Suppose that for the 8 nodes FCN the desired state is equal to:

$$A^{\text{des}} = \begin{bmatrix} 0.76 \ 0.58 \ 0.69 \ 0.69 \ 0.58 \ 0.71 \ 0.63 \ 0.78 \end{bmatrix}^T$$

Applying (9.6) and (9.7) or equivalently (9.12), W concludes to matrix W_1 of Table 9.1.

For the elements of W_1 Eq. (8.22) is true:

$$\left(\sum_{i=1}^{8} \|w_i\|^2 \right)^{1/2} = 3.2421 < 4$$

Example 9.2 **LPM with projection method 1**
Suppose that for the same 8 nodes FCN the desired state is:

$$A^{\text{des}} = \begin{bmatrix} 0.96 \ 0.68 \ 0.97 \ 0.59 \ 0.78 \ 0.75 \ 0.73 \ 0.81 \end{bmatrix}^T$$

Applying (9.6) and (9.14), W concludes to matrix W_2 of Table 9.1. During the updating procedure the value of weights w_{34}, w_{36}, and w_{37} were moving outside the set S_1. By using the projection method given by (9.14) the weights were projected onto the set S_1 and now w_{34}, w_{36}, and w_{37} have the value one, which is desirable.

Table 9.1 Weight matrices of numerical Examples (1–3)

$$
W = \begin{bmatrix}
1 & 0.1 & 0.2 & 0.5 & 0.6 & 0 & 0 & 0 \\
0.4 & 1 & 0 & 0 & 0 & 0 & 0 & 0.4 \\
-0.5 & 0 & 1 & 0.8 & 0 & 0.5 & 0.7 & 0 \\
0.7 & 0 & 0.9 & 1 & 0 & 0.4 & 0 & 0 \\
0 & 0 & 0.7 & 0 & 1 & 0.9 & 0 & 0 \\
0.1 & 0.5 & 0 & 0 & 0 & 1 & 0 & 0.7 \\
0 & 0 & 0.7 & 0 & 0.4 & 0 & 1 & -0.1 \\
0 & 0.2 & 0 & 0 & 0 & 0.2 & 0.1 & 1
\end{bmatrix}
$$

$$
W_1 = \begin{bmatrix}
1 & -0.0771 & -0.0107 & 0.2893 & 0.4229 & 0 & 0 & 0 \\
-0.1596 & 1 & 0 & 0 & 0 & 0 & 0 & -0.1743 \\
-0.8335 & 0 & 1 & 0.4972 & 0 & 0.1884 & 0.4235 & 0 \\
0.0527 & 0 & 0.3123 & 1 & 0 & -0.2048 & 0 & 0 \\
0 & 0 & -0.2709 & 0 & 1 & -0.0990 & 0 & 0 \\
-0.2627 & 0.2232 & 0 & 0 & 0 & 1 & 0 & 0.3277 \\
0 & 0 & 0.3432 & 0 & 0.1001 & 0 & 1 & -0.5034 \\
0 & 0.2772 & 0 & 0 & 0 & 0.2945 & 0.1838 & 1
\end{bmatrix}
$$

$$
W_2 = \begin{bmatrix}
1 & 0.4438 & 0.6904 & 0.7983 & 0.9943 & 0 & 0 & 0 \\
0.0141 & 1 & 0 & 0 & 0 & 0 & 0 & 0.0744 \\
0.4543 & 0 & 1 & 1 & 0 & 1 & 1 & 0 \\
-0.1199 & 0 & 0.0716 & 1 & 0 & -0.2405 & 0 & 0 \\
0 & 0 & 0.1397 & 0 & 1 & 0.4668 & 0 & 0 \\
-0.2079 & 0.2819 & 0 & 0 & 0 & 1 & 0 & 0.4402 \\
0 & 0 & 0.4161 & 0 & 0.1717 & 0 & 1 & -0.3370 \\
0 & 0.3227 & 0 & 0 & 0 & 0.3353 & 0.2317 & 1
\end{bmatrix}
$$

$$
W_3 = \begin{bmatrix}
1 & 0.4217 & 0.6304 & 0.8579 & 1 & 0 & 0 & 0 \\
0.0873 & 1 & 0 & 0 & 0 & 0 & 0 & 0.1131 \\
-0.2034 & 0 & 1 & 1 & 0 & 0.7416 & 0.9599 & 0 \\
0.1830 & 0 & 0.3937 & 1 & 0 & -0.0210 & 0 & 0 \\
0 & 0 & 0.9123 & 0 & 1 & 1 & 0 & 0 \\
-0.1342 & 0.3286 & 0 & 0 & 0 & 1 & 0 & 0.4851 \\
0 & 0 & 0.6765 & 0 & 0.3770 & 0 & 1 & -0.1220 \\
0 & 0.5129 & 0 & 0 & 0 & 0.5482 & 0.4747 & 1
\end{bmatrix}
$$

$$
W_4 = \begin{bmatrix}
1.0000 & 0.4548 & 0.6620 & 0.8201 & 0.9746 & 0 & 0 & 0 \\
0.0905 & 1.0000 & 0 & 0 & 0 & 0 & 0 & 0.1097 \\
-0.0435 & 0 & 1.0000 & 0.9322 & 0 & 0.6975 & 0.8814 & 0 \\
0.1883 & 0 & 0.3576 & 1.0000 & 0 & 0.0159 & 0 & 0 \\
0 & 0 & 0.9122 & 0 & 1.0000 & 1.0000 & 0 & 0 \\
-0.0638 & 0.2981 & 0 & 0 & 0 & 1.0000 & 0 & 0.4328 \\
0 & 0 & 0.6096 & 0 & 0.3663 & 0 & 1.0000 & -0.0395 \\
0 & 0.5029 & 0 & 0 & 0 & 0.5414 & 0.4894 & 1.0000
\end{bmatrix}
$$

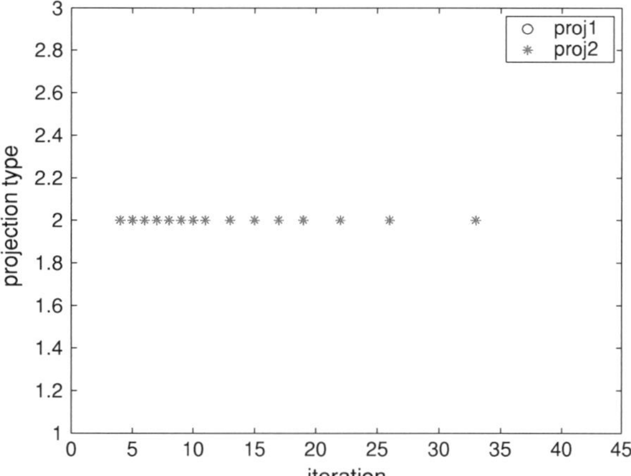

Fig. 9.4 Activated projection methods at updating iteration k (Example 9.3)

For the elements of W_2 Eq. (8.22) is true:

$$\left(\sum_{i=1}^{8}\|w_i\|^2\right)^{1/2} = 3.8379 < 4$$

Example 9.3 **LPM with projection method 2**
Suppose that for the 8 nodes FCN the desired state is:

$$A^{\text{des}} = \left[\,0.97\ 0.71\ 0.95\ 0.79\ 0.93\ 0.79\ 0.85\ 0.89\,\right]^{T}$$

Applying (9.6) and (9.7) or equivalently (9.12), W concludes to matrix W_3 of Table 9.1.

For the elements of W_3 Eq. (8.22) is not true:

$$\left(\sum_{i=1}^{8}\|w_i\|^2\right)^{1/2} = 4.0779 > 4$$

Therefore, $g(\underline{W}) = \|\underline{W}\| - 4 > 0$, which means that $\underline{W} \notin S_2$.

Applying (9.14) and (9.17) we project the $\underline{W}$ matrix to set S, which is the intersection of S_1 and S_2. Matrix W finally concludes to matrix W_4 of Table 9.1. Figure 9.4, shows the iterations at which a projection method was activated. According to the figure, only projection method 2 was activated 15 times before the weight estimates converge to those of W_4.

For the elements of W_4 Eq. (8.22) is now true:

$$\|\underline{W}\| = 3.9988 < 4$$

Table 9.2 W matrices appearing in Examples 9.4 and 9.5

$$W = \begin{bmatrix} 1 & 0.4 & -0.5 & 0.7 & 0 & 0.1 & 0 & 0 \\ 0.1 & 1 & 0 & 0 & 0 & 0.5 & 0 & 0.2 \\ 0.2 & 0 & 1 & 0.9 & 0.7 & 0 & 0.7 & 0 \\ 0.5 & 0 & 0.8 & 1 & 0 & 0 & 0 & 0 \\ 0 & 0 & 0 & 0 & 1 & 0 & 0.4 & 0 \\ 0 & 0 & 0.5 & 0.4 & 0.9 & 1 & 0 & 0.2 \\ 0 & 0 & 0.7 & 0 & 0 & 0 & 1 & 0.1 \\ 0 & 0.4 & 0 & 0 & 0 & 0.7 & -0.1 & 1 \end{bmatrix}$$

$$C_l = \begin{bmatrix} 1 & 1 & 1 & 1 & 1 & 1 & 1 & 1 \end{bmatrix}$$

$$W_l = \begin{bmatrix} 1.0000 & 0.9452 & 0.0760 & 1.0000 & 0 & 0.5454 & 0 & 0 \\ -0.0829 & 1.0000 & 0 & 0 & 0 & 0.3883 & 0 & 0.0498 \\ -0.3736 & 0 & 1.0000 & 0.3264 & 0.2834 & 0 & 0.2955 & 0 \\ 1.0000 & 0 & 1.0000 & 1.0000 & 0 & 0 & 0 & 0 \\ 0 & 0 & 0 & 0 & 1.0000 & 0 & 0.1654 & 0 \\ 0 & 0 & -0.0263 & -0.2667 & 0.4158 & 1.0000 & 0 & -0.3474 \\ 0 & 0 & 0.3382 & 0 & 0 & 0 & 1.0000 & -0.2763 \\ 0 & 0.3244 & 0 & 0 & 0 & 0.6382 & -0.1714 & 1.0000 \end{bmatrix}$$

$$W_b = \begin{bmatrix} 1.0000 & 0.5580 & -0.3331 & 0.9115 & 0 & 0.2291 & 0 & 0 \\ -0.0114 & 1.0000 & 0 & 0 & 0 & 0.4320 & 0 & 0.1085 \\ 0.0093 & 0 & 1.0000 & 0.7093 & 0.5615 & 0 & 0.5655 & 0 \\ 0.6402 & 0 & 0.9107 & 1.0000 & 0 & 0 & 0 & 0 \\ 0 & 0 & 0 & 0 & 1.0000 & 0 & 0.3180 & 0 \\ 0 & 0 & 0.1767 & -0.0095 & 0.6026 & 1.0000 & 0 & -0.1362 \\ 0 & 0 & 0.5145 & 0 & 0 & 0 & 1.0000 & -0.0929 \\ 0 & 0.3690 & 0 & 0 & 0 & 0.6746 & -0.1293 & 1.0000 \end{bmatrix}$$

$$C_b = \begin{bmatrix} 1.4054 & 0.8656 & 0.4996 & 1.3138 & 0.8861 & 0.3186 & 0.7202 & 0.9399 \end{bmatrix}$$

Example 9.4 **LPM and FCN size insufficiency**

Let the 8 nodes FCN with the initial matrix of weights W and vector C_l containing all the c_{l_i} parameters of FCN given in Table 9.2. Except from the diagonal elements, which have the value 1 and the zero elements all other weights were randomly selected. With these weights and sigmoid parameters the FCN reaches an equilibrium point given by the following vector A.

$$A = \begin{bmatrix} 0.7960 & 0.8233 & 0.9531 & 0.8855 & 0.7461 & 0.9311 & 0.8295 & 0.8520 \end{bmatrix}^T$$

Suppose that for the 8 nodes FCN the desired state is equal to:

$$A^{\text{des}} = \begin{bmatrix} 0.95 & 0.71 & 0.75 & 0.95 & 0.69 & 0.58 & 0.67 & 0.78 \end{bmatrix}^T$$

By repetitively applying (9.6) and (9.7) the W matrix concludes to W_l given in Table 9.2. However, with this weight matrix, the FCN node values vector do not converge to the desired one but to:

$$A = \begin{bmatrix} 0.9492 & 0.71 & 0.75 & 0.9328 & 0.69 & 0.58 & 0.67 & 0.78 \end{bmatrix}^T$$

where nodes C_1 and C_4 fail to converge to the desired values. This happens because there are not enough nodes affecting nodes C_1 and C_4. From equation $A_4{}^{\text{des}}(k) = f(w_4 * A^{\text{des}}(k-1))$ one can see that in order to have $A_4{}^{\text{des}}(k) = 0.95$ then

$$A_4{}^{\text{des}} = f(w_{41} * A_1{}^{\text{des}} + w_{43} * A_3{}^{\text{des}} + w_{44} * A_4{}^{\text{des}}) = 0.95$$

where $A_i{}^{\text{des}}$ are the desired values of nodes C_i and f is the sigmoid function with $c_{l_i} = 1$. Solving the above equation using the maximum allowed values for w_{ii} parameters, which is $w_{ii} = 1$, we conclude to:

$$A_4^{\text{des}} = f(1 * A_1{}^d es + 1 * A_{3d}es + 1 * A_4{}^{\text{des}}) = 0.95.$$

From the vector A^{des}, we have that, $A_1{}^{\text{des}} = 0.95$, $A_3{}^{\text{des}} = 0.75$ and $A_4{}^{\text{des}} = 0.95$, so the equation conlcudes to:

$$A_4{}^{\text{des}} = f(1 * 0.95 + 1 * 0.75 + 1 * 0.95) \Rightarrow A_4{}^{\text{des}} = f(2.65) \Rightarrow$$

$$\Rightarrow A_4{}^{\text{des}} = \frac{1}{1 + e^{-2.65}} \Rightarrow A_4{}^{\text{des}} = 0.9340$$

which is not the desired value. If there were more nodes affecting node C_4 then it could be possible to achieve the desired state. In the next example, it is shown that the requirement for larger FCN graphs is alleviated when the bilinear approach is used.

9.6.2 M: Comparisons Between the Two Models

Example 9.5 **Use of the BPM to solve the size insufficiency problem**
Suppose again that for the 8 nodes FCN the desired state is equal to the A^{des} of example 9.4. Then, by applying the proposed algorithm and repetitively applying (9.18), (9.19) and (9.20), matrix W concludes to W_b and vector C_l concludes to C_b, both given in Table 9.2. With these parameters the FCN converges to

$$A = \begin{bmatrix} 0.95 & 0.71 & 0.75 & 0.95 & 0.69 & 0.58 & 0.67 & 0.78 \end{bmatrix}^T$$

which is equal to the desired one. Therefore, by including the option of inclinations' adjustment the need for a larger FCN structure has been removed. It has to be noted here that, under different initial W ans C_l values, the algorithm might conclude to W_b and C_b that are different from those of Table 9.2. Nevertheless, this different set of weights and inclination parameters will serve the purpose of driving the FCM to the same desired equilibrium point.

Example 9.6 **Successful use of the bilinear algorithm without projection**
We consider again the 8 nodes FCN. The initial matrix of weights W and vector C_l containing all the c_{l_i} parameters of FCN are shown in Table 9.2. Except from the diagonal elements, which have the value 1 and the zero elements all other weights were randomly selected. With these weights and sigmoid parameters the FCN reaches an equilibrium point given by the following vector A.

$$A = \begin{bmatrix} 0.7960 & 0.8233 & 0.9531 & 0.8855 & 0.7461 & 0.9311 & 0.8295 & 0.8520 \end{bmatrix}^T$$

This FCN equilibrium point fulfills Eq. (8.16) because

$$\left(\sum_{i=1}^{8} \ell_i^2 c_{l_i}^2 \|w_i\|^2 \right)^{1/2} = 0.4586 < 1$$

Moreover, Eq. (8.16) is also fulfilled at the origin, e.g., when ℓ_i is computed with the arguments of its exponential functions being zero. In this case

$$\left(\sum_{i=1}^{8} \ell_i^2 c_{l_i}^2 \|w_i\|^2 \right)^{1/2} = 0.9523 < 1$$

Therefore, as explained in Chap. 8 (Remark 8.3), the FCN can reach the above equilibrium state regardless its initial condition and its path to the equilibrium.

Suppose that for the 8 nodes FCN the desired state is equal to:

$$A^{\text{des}} = \begin{bmatrix} 0.76 & 0.71 & 0.75 & 0.67 & 0.69 & 0.58 & 0.67 & 0.78 \end{bmatrix}^T$$

By repetitively applying (9.18), (9.19) and (9.20) and allowing changes in the diagonal elements, matrix W concludes to W_1 and vector C_l concludes to C_{l_1}, both given in Table 9.3.

For the matrices W_1 and C_{l_1} Eq. (8.16) at A^{des} is true: $\left(\sum_{i=1}^{8} \ell_i^2 c_{l_i}^2 \|w_i\|^2 \right)^{1/2} = 0.4323 < 1$ The same holds when Eq. (8.16) is computed at the origin, where: $\left(\sum_{i=1}^{8} \ell_i^2 c_{l_i}^2 \|w_i\|^2 \right)^{1/2} = 0.5633 < 1$

Example 9.7 **Activation of the projection methods in the bilinear algorithm**
Suppose that the initial C_l vector is now changed and is chosen to be the C_l' in Table 9.3. Suppose also that the desired state is:

Table 9.3 W and *inclination* parameters of Examples 9.6 and 9.7

$$W_1 = \begin{bmatrix} 0.9914 & 0.3919 & -0.5085 & 0.6924 & 0 & 0.0934 & 0 & 0 \\ 0.0194 & 0.9247 & 0 & 0 & 0 & 0.4385 & 0 & 0.1173 \\ 0.0500 & 0 & 0.8519 & 0.7677 & 0.5638 & 0 & 0.5677 & 0 \\ 0.2833 & 0 & 0.5861 & 0.8089 & 0 & 0 & 0 & 0 \\ 0 & 0 & 0 & 0 & 0.9369 & 0 & 0.3387 & 0 \\ 0 & 0 & 0.1701 & 0.1053 & 0.5965 & 0.7449 & 0 & -0.1431 \\ 0 & 0 & 0.5398 & 0 & 0 & 0 & 0.8569 & -0.0666 \\ 0 & 0.3738 & 0 & 0 & 0 & 0.6786 & -0.1247 & 0.9712 \end{bmatrix}$$

$$C_{l_1} = \begin{bmatrix} 0.9866 & 0.8803 & 0.5603 & 0.5917 & 0.9161 & 0.3470 & 0.7639 & 0.9495 \end{bmatrix}$$

$$C_l' = \begin{bmatrix} 5 & 5 & 5 & 5 & 5 & 5 & 5 & 5 \end{bmatrix}$$

$$W_2 = \begin{bmatrix} 0.7234 & 0.5335 & 0.5723 & 0.6285 & 0 & 0.5464 & 0 & 0 \\ 0.3008 & 0.3927 & 0 & 0 & 0 & 0.3387 & 0 & 0.3234 \\ 0.5539 & 0 & 0.7532 & 0.6394 & 0.6844 & 0 & 0.6370 & 0 \\ 0.4527 & 0 & 0.5444 & 0.5186 & 0 & 0 & 0 & 0 \\ 0 & 0 & 0 & 0 & 1.0000 & 0 & 1.0000 & 0 \\ 0 & 0 & 0.2746 & 0.2250 & 0.3579 & 0.3551 & 0 & 0.1991 \\ 0 & 0 & 0.3703 & 0 & 0 & 0 & 0.4807 & 0.1248 \\ 0 & 0.1890 & 0 & 0 & 0 & 0.3470 & -0.0840 & 0.5042 \end{bmatrix}$$

$$C_{l_2} = \begin{bmatrix} 0.7665 & 0.8170 & 1.0222 & 1.0032 & 1.4532 & 1.0779 & 1.9903 & 2.6611 \end{bmatrix}$$

$$A^{\text{des}} = \begin{bmatrix} 0.87 & 0.71 & 0.95 & 0.79 & 0.93 & 0.79 & 0.85 & 0.89 \end{bmatrix}^T$$

Applying (9.18), (9.20), (9.19) is not sufficient any more, because condition (8.16) is not fulfilled at every step of the algorithm and therefore the convergence of the FCN to the desired equilibrium point cannot be always guaranteed. Instead, using the modified with projection updating laws (9.28), (9.29), (9.31), (9.32), (9.33), matrix W concludes to W_2 and vector c_l to C_{l_2}, both given in Table 9.3. Figure 9.5 shows the activation (if any) of projection methods at each iteration of the updating algorithm. In this example, during the updating procedure, only projection methods 1.B and 2.B are activated. Projection method 2.B is activated for iterations $k = 3, 4, 5$, while projection method 1.B for $k > 22$ until FCNs parameters finally conclude to the ones corresponding to the desired equilibrium point A^{des}.

For the matrices W_2 and C_{l_2} Eq. (8.16) computed at A^{des} is true:

$$\left(\sum_{i=1}^{8} \ell_i^2 c_{l_i}^2 \, \|w_i\|^2 \right)^{1/2} = 0.6289 < 1$$

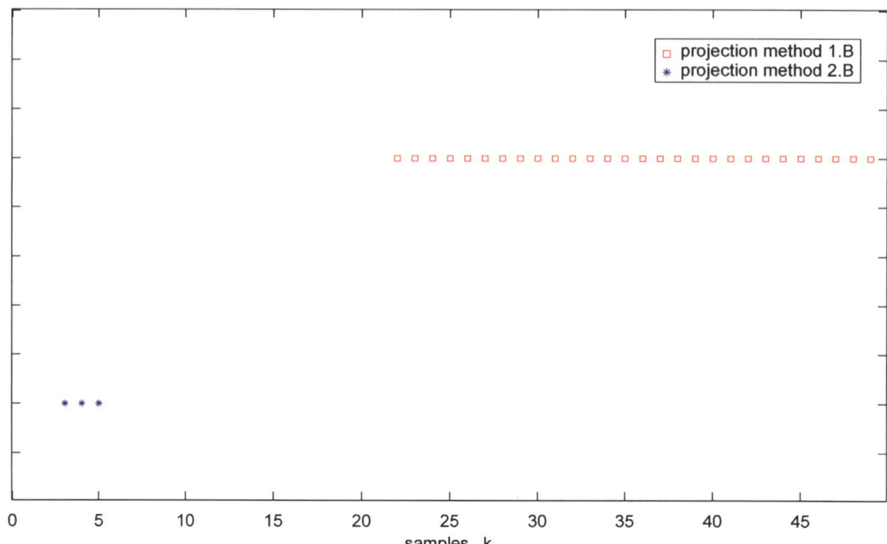

Fig. 9.5 Activated projection methods at updating iteration k

Fig. 9.6 The FCN graph for
Example 9.8

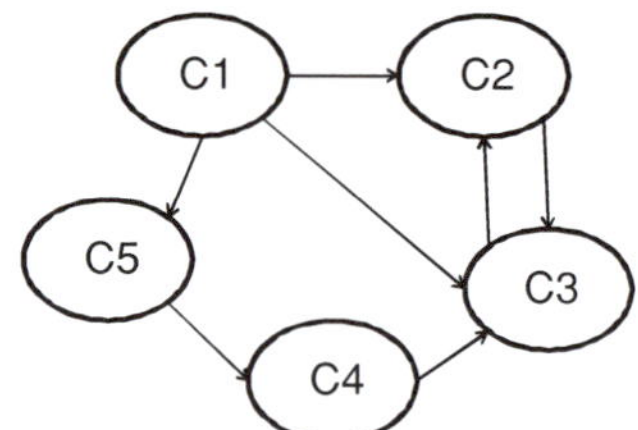

Example 9.8 **Comparison of the two algorithms in a state tracking problem**
Let the FCN shown in Fig. 9.6. The node $C1$ is an input node with its values arised
from equation $\frac{1}{2}sin^2(x)$. The initial values of the nodes $C3$, $C4$, and $C5$ are equal
to 0.6. For each x we want the value of node $C2$ track the value $\cos^2(x)$. In this
example, we are not concerned with the values of the other nodes.

The initial values of W and C_l parameters are shown in Table 9.4. Since one
parameter set is not sufficient to capture the $C1 \rightarrow C2$ associations for every x,
more than one parameter sets have to be estimated. We are using here the exhaustive
case, where for each $C1 \rightarrow C2$ association we estimate a different set of parameters.
To get the estimates we use the two different approaches "linear" versus "bi-linear"
and examine if the estimates produced by each approach are actually appropriate for
"capturing" the association. We do that for two characteristic values of x as well as
for a series of consecutive x values.

We choose first the two desired characteristic values of node vector A^{des} to be:

$$A^{des1} = \begin{bmatrix} 0.25\ 0.5\ 0.6\ 0.6\ 0.6 \end{bmatrix}^T , \, for\ x = pi/4$$

Table 9.4 W and *inclination* matrices of Example 9.8

$$W = \begin{bmatrix} 0 & 0 & 0 & 0 & 0 \\ 0.2000 & 1.0000 & 0.8000 & 0 & 0 \\ 0.3000 & 0.6500 & 1.0000 & 0.5500 & 0 \\ 0 & 0 & 0 & 1.0000 & 0.6700 \\ 0.7800 & 0 & 0 & 0 & 1.0000 \end{bmatrix}$$

$$C_l = \begin{bmatrix} 1 & 1 & 1 & 1 & 1 \end{bmatrix}$$

$$W_{sb1} = \begin{bmatrix} 0 & 0 & 0 & 0 & 0 \\ -1.0000 & 1.0000 & -1.0000 & 0 & 0 \\ 0.0838 & -0.4911 & 1.0000 & -0.3941 & 0 \\ 0 & 0 & 0 & 1.0000 & -0.3242 \\ -0.3891 & 0 & 0 & 0 & 1.0000 \end{bmatrix}$$

$$C_{sb1} = \begin{bmatrix} 1 & 12.9736 & 1 & 1 & 1 \end{bmatrix}$$

$$W_{sb2} = \begin{bmatrix} 0 & 0 & 0 & 0 & 0 \\ 0.4617 & 1.0000 & -0.9966 & 0 & 0 \\ 0.2610 & -0.3777 & 1.0000 & -0.1135 & 0 \\ 0 & 0 & 0 & 1.0000 & -0.3242 \\ -0.7910 & 0 & 0 & 0 & 1.0000 \end{bmatrix}$$

$$C_{sb2} = \begin{bmatrix} 1 & 0.9997 & 0.9997 & 1 & 1.0027 \end{bmatrix}$$

$$W_{sl1} = \begin{bmatrix} 0 & 0 & 0 & 0 & 0 \\ 0.4644 & 1.0000 & -0.9978 & 0 & 0 \\ 0.2610 & -0.3778 & 1.0000 & -0.1135 & 0 \\ 0 & 0 & 0 & 1.0000 & -0.3242 \\ -0.7866 & 0 & 0 & 0 & 1.0000 \end{bmatrix}$$

$$A_{\text{system1}} = \begin{bmatrix} 0.25 & 0.5 & 0.6 & 0.6 & 0.6 \end{bmatrix}$$

$$W_{sl2} = \begin{bmatrix} 0 & 0 & 0 & 0 & 0 \\ -1.0000 & 1.0000 & -1.0000 & 0 & 0 \\ 0.0839 & -0.4912 & 1.0000 & -0.3941 & 0 \\ 0 & 0 & 0 & 1.0000 & -0.3242 \\ -0.3891 & 0 & 0 & 0 & 1.0000 \end{bmatrix}$$

$$A_{\text{system2}} = \begin{bmatrix} 1 & 0.3272 & 0.6 & 0.6 & 0.6 \end{bmatrix}$$

$$A^{des2} = \begin{bmatrix} 1 & 0 & 0.6 & 0.6 & 0.6 \end{bmatrix}^T, \; for \; x = pi/2$$

The Bilinear approach

For the two distinct desired node vectors given above, we repetitively apply (9.18), (9.19),(9.20), (9.28), (9.29), (9.31), (9.32) and (9.33) to get the corresponding weights

Fig. 9.7 Tracking of the desired value for node 2 using the two alternative approaches

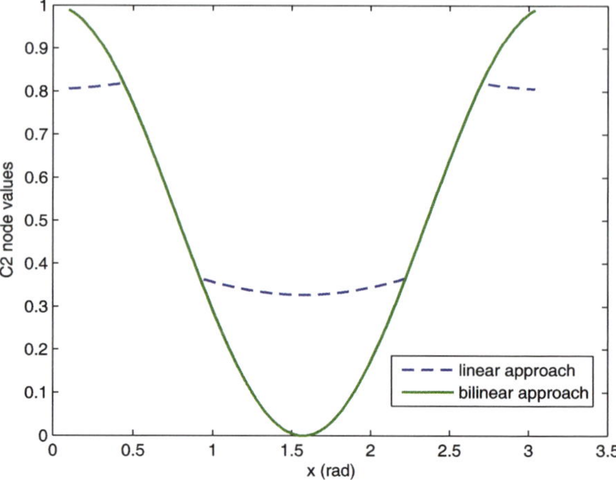

Fig. 9.8 Evolution of an inclination parameter estimate in respect to x

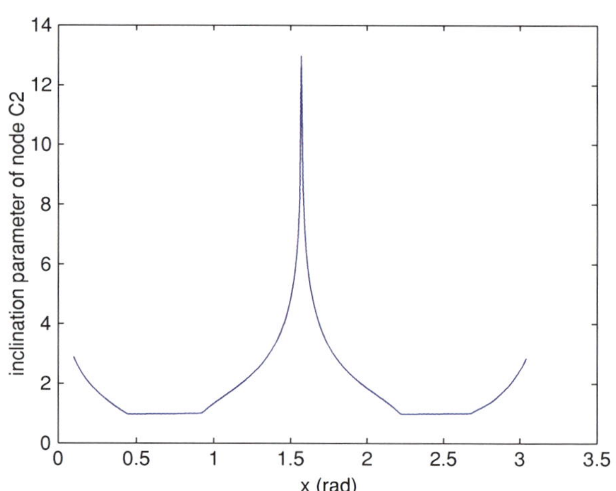

matrices W_{sb1} and W_{sb2} as well as the corresponding inclinations vectors C_{sb1} and C_{sb2}. All values are given in Table 9.4. With these estimated parameters and the corresponding input value $C1$ we get exactly the desired node vectors. We repeat the procedure for the x values between 0 and pi. By using the estimated parameters and the corresponding values of $C1$ for each x the FCN concludes to the values of node $C2$ shown in Fig. 9.7, which are the desired values corresponding to equation $\cos^2(x)$. The inclination values of c_{l2} in respect to x are also displayed in Fig. 9.8. It is apparent that the procedure gives the appropriate estimates for achieving the goal.

The Linear approach

We repeat the same experiment using the "linear" approach. For the two distinct desired node vectors, (9.6) and (9.7) are repetitively applied providing the corresponding weight matrices W_{sl1} and W_{sl2}, both given in Table 9.4. We recall that in the linear approach the inclination parameters are kept always 1. With these weights the FCN concludes to node vectors A_{system1} and A_{system2} respectively, shown in Table 9.4. One may observe that , when $x = pi/2$, using the linear approach the state for node $C2$ concludes to the value of 0.3272 instead of 0. By repeating the procedure for all x values between 0 and pi the corresponding $C2$ values are shown to the plot of Fig. 9.7 which, clearly, is not always the plot $\cos^2(x)$. This is an immediate consequence of the linear's approach inability to give always the appropriate parameter estimates, provided that the FCN graph is kept constant.

9.7 Summary

Based on the conclusions drawn from Chap. 8, new algorithms were proposed for the adaptive estimation of the FCN parameters that fulfill the FCN equilibrium equation. Two formulations were followed. In the first one, we used a linear parametric model to describe the FCN equilibrium equation, taking into account only the interconnection weights of the FCN graph. In the second formulation, a bilinear model was used, taking into account both the weights and the inclination parameters of the nodes' sigmoid functions. The FCN parameters that solve the equilibrium equation are obtained repetitively by using appropriate adaptive parameter estimation algorithms; one for the linear and one for the bilinear case. Moreover, appropriate modifications of these algorithms were derived in the form of parameter *projection* methods, which prevent the estimated parameters from drifting to values that violate the existence and uniqueness of equilibrium points conditions. Simulations carried out on selected numerical examples demonstrate the effectiveness of both approaches and highlight their differences. In general, the bilinear approach has the advantage that allow the use of smaller FCN structures, but it requires the estimation of the inclination parameters of the sigmoid functions of the FCN nodes.

References

Boutalis, Y., Kottas, T., & Christodoulou, M. (2009). Adaptive estimation of fuzzy cognitive maps with proven stability and parameter convergence. *IEEE Transactions on Fuzzy Systems, 17*, 874–889.

Ioannou, P., & Fidan, B. (2006). *Adaptive control tutorial*. Philadelphia: SIAM.

Kottas, T., Boutalis, Y., & Christodoulou, M. (2012). Bi-linear adaptive estimation of fuzzy cognitive networks. *Applied Soft Computing, 21*. doi:10.1016/j.asoc.2012.01.025

Chapter 10
Framework of Operation and Selected Applications

10.1 Framework of Operation of FCNs

As shown in Chap. 9 the concepts values of the FCN with a specified matrix W and an inclination parameters vector C_l have a unique solution as far as (8.16) or (8.22) (when $c_{l_i} = 1, \forall i$) is fulfilled. The perspective of transforming FCNs into a modeling and control alternative requires, first to update its weight matrix W so that the FCN can capture different mappings of the real system and second to store these different kind of mappings. The operational framework of FCN was first introduced in Kottas et al. (2007) and Boutalis et al. (2009). It has been proposed as an operational extension framework of traditional FCM, which updates its weights and reaches new equilibrium points based on the continuous interaction with the system it describes. Moreover, as it has already been shown in Chap. 8, for each equilibrium point a meta-rule of the form "If inputs and weights then fixed point" can be provided. Therefore, a fuzzy rule-based storage mechanism can be devised that connects the inputs to the FCN weights, which together with the inputs will produce the desired equilibrium point. The introduction of the storage mechanism facilitates and speeds up the FCN operation. The components of FCN are briefly presented below.

10.1.1 Training Using Close Interaction with the Real System

The operation of the FCN in close cooperation with the real system it describes might require continuous changes in the weight interconnections, depending on the input received from the real system. Figure 10.1 presents the interactive operation of the FCN with the physical system it describes. The weight updating procedure is analyzed below.

Y. Boutalis et al., *System Identification and Adaptive Control*,
Advances in Industrial Control, DOI: 10.1007/978-3-319-06364-5_10,
© Springer International Publishing Switzerland 2014

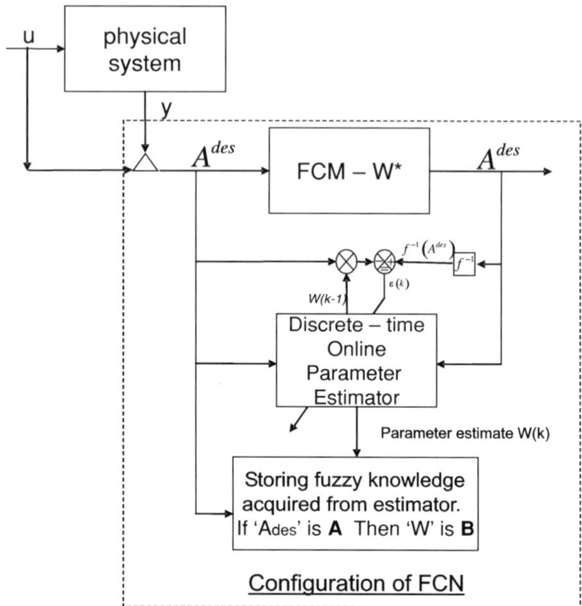

Fig. 10.1 Interactive operation of the FCN with the physical system

10.1.2 Weight Updating Procedure

The updating method takes into account input u and output y from the real system. These u and y values combine to give the A^{des} vector, which is equal to:

$$A^{des} = \begin{bmatrix} U_1 & \cdots & U_m & A_{m+1} & \cdots & A_{m+n} \end{bmatrix}^T.$$

Using the repetitive procedure of the updating algorithm described in Chap. 9 one may estimate W^* matrix for which (9.3) is true. It is pointed out that the inverse function of (9.3) is easily calculated by (9.4). Once the weight estimates converge to their new values, the association between the weights and the desired node values (including the input ones) can be stored using fuzzy rules.

10.2 Storage Mechanism Based on Fuzzy Rule Databases

The procedure described in the previous section modifies FCN's knowledge about the system by continuously modifying the weight interconnections and consequently the node values. During the repetitive updating operation, the procedure uses input from the system variables, producing a new weight matrix for each new equilibrium state. It is desirable to device a storage mechanism for keeping these weights for probable future use in a decision making or control application. It would be also preferable this

Fig. 10.2 An FCN with two
input nodes

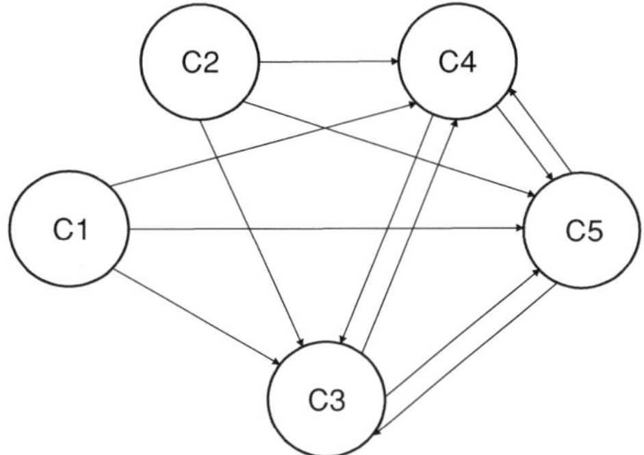

storage to allow weight retrieval even in situations, where the equilibrium conditions
had not been exactly met during the training phase. To this end, we propose storing
the previous acquired operational situations in a fuzzy if-then rule database, which
associates in a fuzzy manner the various weights with the corresponding equilibrium
node values. The procedure is explained as follows:

Suppose for example that the FCN of Fig. 10.2 with $c_{l_i} = 1, \forall i$ has a unique
equilibrium point

$$A = \begin{bmatrix} U1 & U2 & A1 & A2 & A3 \end{bmatrix}^T$$

which is connected with the weight matrix W_α with the following values:

$$W_\alpha = \begin{bmatrix} \alpha_{11} & 0 & 0 & 0 & 0 \\ 0 & \alpha_{22} & 0 & 0 & 0 \\ \alpha_{31} & \alpha_{32} & \alpha_{33} & \alpha_{34} & \alpha_{35} \\ \alpha_{41} & \alpha_{42} & \alpha_{43} & \alpha_{44} & \alpha_{45} \\ \alpha_{51} & \alpha_{52} & \alpha_{53} & \alpha_{54} & \alpha_{55} \end{bmatrix}.$$

It should be pointed out that according to Fig. 10.2 all diagonal elements of matrix W_α
are zero, because there is not any node self-feedback. In the general case, however,
$\alpha_{ii} \neq 0$. In order that A is a unique solution of (9.3) weight matrix W has to be
such that inequality (8.22) is fulfilled. For weight matrix W_α inequality (8.22) takes
the form:

$$\sum_{i=1}^{5} \sum_{j=1}^{5} \alpha_{ij}^2 < 16$$

where $n = 5$ is the number of concepts of the FCN.

Suppose also that the FCN in another operation point is related to the following
weight matrix W_β, which also fulfils (8.22):

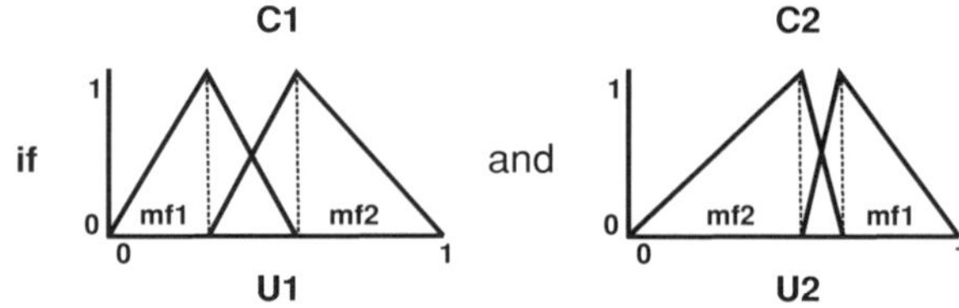

Fig. 10.3 Left-hand side (if-part) membership functions derived from 2 equilibrium points

$$
W_\beta = \begin{bmatrix} \beta_{11} & 0 & 0 & 0 & 0 \\ 0 & \beta_{22} & 0 & 0 & 0 \\ \beta_{31} & \beta_{32} & \beta_{33} & \beta_{34} & \beta_{35} \\ \beta_{41} & \beta_{42} & \beta_{43} & \beta_{44} & \beta_{45} \\ \beta_{51} & \beta_{52} & \beta_{53} & \beta_{54} & \beta_{55} \end{bmatrix}
$$

with the unique equilibrium point being:

$$
A = \begin{bmatrix} U'1 & U'2 & B1 & B2 & B3 \end{bmatrix}^T .
$$

Inequality (8.22) for the weight matrix W_β has now the form:

$$
\sum_{i=1}^{5} \sum_{j=1}^{5} \beta_{ij}^2 < 16.
$$

The fuzzy rule database, which is obtained using the information from the two previous equilibrium points, is depicted in Figs. 10.3 and 10.4 and is resolved as follows:

There are two rules for the description of the above two different equilibrium situations:

Rule 1
if node $C1$ is mf1 *and* node $C2$ is mf1
then w_{31} is mf1 *and* w_{41} is mf1 *and* w_{51} is mf1 *and* w_{32} is mf1 *and* w_{42} is mf1 *and* w_{52} is mf1 *and* w_{43} is mf1 *and* w_{53} is mf1 *and* w_{34} is mf1 *and* w_{54} is mf1 *and* w_{35} is mf1 *and* w_{45} is mf1.

Rule 2
if node $C1$ is mf2 *and* node $C2$ is mf2
then w_{31} is mf2 *and* w_{41} is mf2 *and* w_{51} is mf2 *and* w_{32} is mf2 *and* w_{42} is mf2 *and* w_{52} is mf2 *and* w_{43} is mf2 *and* w_{53} is mf2 *and* w_{34} is mf2 *and* w_{54} is mf2 *and* w_{35} is mf2 *and* w_{45} is mf2.

The number and shape of the fuzzy membership functions of the variables of both sides of the rules are gradually modified as new desired equilibrium points appear to the system during its training. To add a new triangular membership function in the fuzzy description of a variable, the new value of the variable must differ from one already encountered value more than a specified threshold. Figure 10.5 depicts this

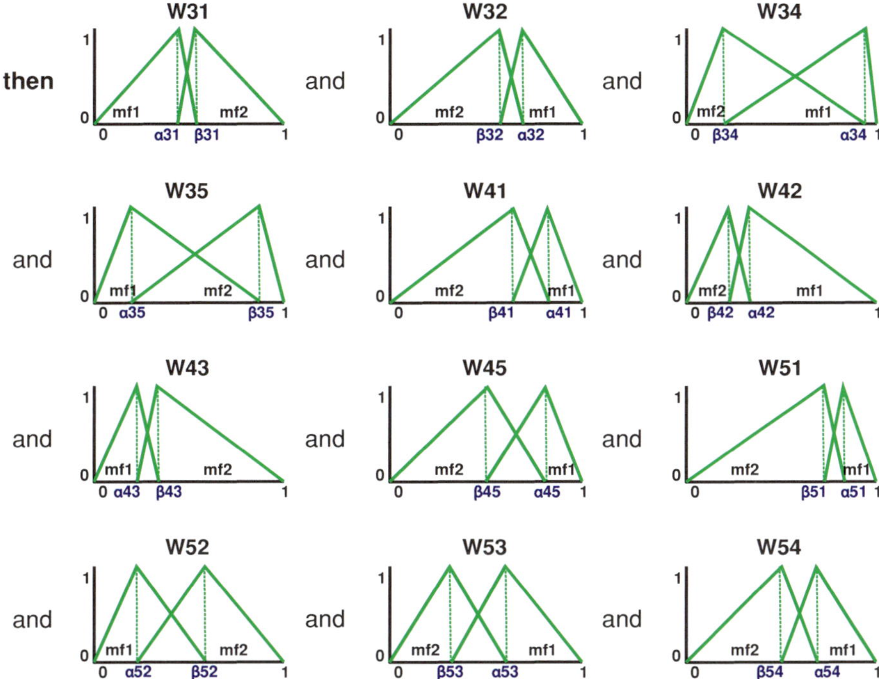

Fig. 10.4 Right-hand side (then-part) membership functions derived from 2 equilibrium points

procedure. The initial value, α, determines the value of the variable that corresponds to the pick of the triangular membership function (mf1) for the fuzzy description of the value. When a new value, β, of the variable appears in a new equilibrium condition, it is compared with the previously encountered value α. If $|\alpha - \beta|$ exceeds a specified threshold, c, then a new triangular membership function (mf2) is created and the initial triangular function (mf1) is modified as shown in Fig. 10.5b. If $|\alpha - \beta|$ does not exceed the threshold the initial fuzzy partition of the variable remains unchanged (Fig. 10.5c). The threshold comes usually as a compromise between the maximum number of allowable rules and the detail in fuzzy representation of each variable. Instead of triangular membership functions, one could also use other types of membership functions, like Gaussian or trapezoidal, to create the fuzzy rule database.

One interesting perspective of the procedure described above is that, the whole scheme can be used as an automatic fuzzy rule generator engine, which is based on system's sampled data and re-encodes the acquired knowledge in a form that embodies the fuzzy values of the FCN parameters. This possibility is based solely on the requirement that the FCN works repetitively until it reaches an equilibrium point and that this equilibrium is associated with a particular set of weights and input values.

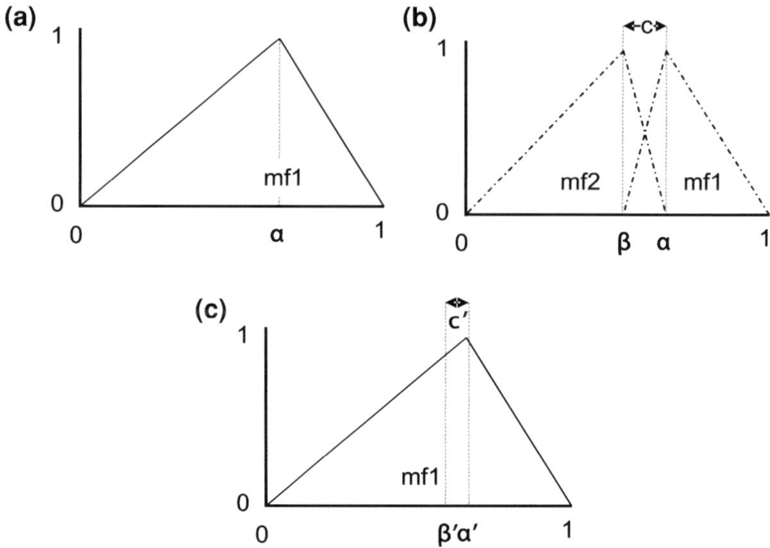

Fig. 10.5 Conditions for creating new triangular membership function. **a** Present triangular with pick corresponding to the value a of the variable, **b** $c >$ *specified threshold* (creation of new triangular) and **c** $c <$ *specified threshold* (noncreation of new triangular)

Fig. 10.6 An estimator of the real system state using the trained FCN

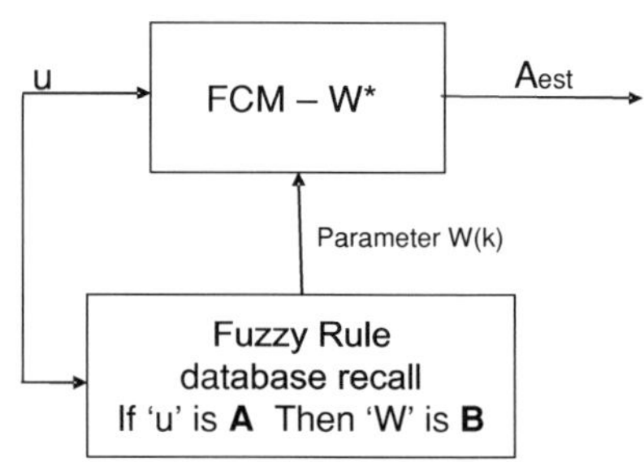

10.2.1 Using the Acquired Knowledge

After the FCN has been trained and the knowledge acquired has been stored in the Fuzzy Rule Database, it can be used in place of the unknown system to predict its behavior in changing condition determined by u. This kind of usage is illustrated in Fig. 10.6, where the FCN estimates the behavior of the system after it recalls the relative weight values using the Fuzzy Rule database and conventional Mamdani-type inference procedure. Such a kind of operation has been first employed in Kottas et al. (2007), where however the weight estimation was not performed using the algorithms presented in Chap. 9 . FCNs equipped with the proposed storage mechanism can also be used as a part of an adaptive control scheme of unknown nonlinear plant predicting the required system inputs, which will drive the plant to a desired performance.

This way of operation has been demonstrated in Kottas et al. (2006a) where the method was used to control a pilot anaerobic wastewater treatment unit. However, the weight updating in Kottas et al. (2006a) is using a conventional delta rule, which does not guarantee that the weights reach always values fulfilling the convergence conditions.

10.3 Applications

The framework of operation described in the previous sections has found a number of applications, which have been presented in the previous years, as the framework was evolved. A noticeably successful application concerned its use for the maximum power point tracking (MPPT) of photo voltaic (PV) arrays based on real meteorological data. The framework was actually used as a fuzzy rule depository, which associates meteorological data (irradiation and temperature) with current and voltage values that will drive the PV array in its maximum performance. An early version of the application appeared in Karlis et al. (2007), where the framework operated using only off-line training. A better approach which combines off-line and on-line training with the use of a conventional fuzzy controller appeared in Kottas et al. (2006b) and later in Kottas et al. (2010), demonstrating optimal performance and adaptation abilities when the PV array characteristics change due to hardware decay or aging. The control of a real nonlinear wastewater treatment bioprocess using the FCN framework has been presented in Kottas et al. (2006a), where the control inputs are determined using the fuzzy rule depository and an inverse procedure using a gradient technique. In all these applications, the FCN parameters estimation was performed using an early approach, appearing in Kottas et al. (2007), where the existence of equilibrium points is not guaranteed according to the theory of Chaps. 8 and 9, but it is rather based on a heuristic procedure.

In the following sections, we present applications that utilize the complete framework of operation, including the parameter estimation algorithms given in Chap. 9. The first application concerns the control of the well-known *inverted pendulum* benchmark. An initial off-line training creates the FCN's fuzzy rule database. Assuming that this depository of fuzzy meta rules is complete, the appropriate control signal is derived using a simple inversion formula, which utilizes the fact that sigmoid functions of the FCN nodes are invertible. Comparisons between linear and bilinear FCN modeling are presented, supporting the theoretical, from Chap. 9, expectation that the bilinear approach outperforms the linear, especially when the system states reach their marginal values. Another variation of the same application assumes that the physical system undergoes parameter changes, rendering the available fuzzy rule depository insufficient. In this case, an adaptive procedure is activated enhancing the fuzzy rule depository based on the new data. This application demonstrates the reliable use of the FCN framework for the inverse control of a plant that it is by design affine in the control and can be approximated by a proper FCN.

The second application is using the FCN framework to drive a hydroelectric plant to its maximum power production, taking into account the dam's water level and

the critical demands in respect to the network's voltage. In this example, the FCN
is initially off-line trained by experts' beliefs regarding its maximum power mode
of operation and is further on-line trained by successful operation decisions of the
human operators in failure situations.

The last application demonstrates the use of the FCN framework to act both at a
local and coordination level. A smart grid of renewable power sources is coordinated
by a FCN so that the available alternative power sources are utilized according to
the grid's power demands and the current climatological data. At the same time, the
various power sources of the grid are optimally regulated by FCN or conventional.

10.3.1 Conventional Benchmark: The Inverted Pendulum

In the inverted pendulum system, shown in Fig. 4.2, a pole is attached to the top of a
cart equipped with a motor that drives it along a horizontal track. Taking into account
only angular position and velocity, a simplified model of the inverted pendulum is
given by:

$$\dot{x}_1 = x_2$$
$$\dot{x}_2 = \frac{mlx_2^2 \sin(x_1) - g(M + m)x_1}{ml \cos(x_1) - l(M + m)} + \frac{1}{ml \cos(x_1) - l(M + m)}F \qquad (10.1)$$

where $x_1 = \vartheta$ is the angle of the pole in respect to its upright position, $x_2 = \dot{\vartheta}$ is
its angular velocity, $g = 9.81\,\mathrm{m/s^2}$, $l = L/2 = 1.0\,\mathrm{m}$ is the half length of the pole,
$m = 1.0\,\mathrm{kg}$ is the mass of the pole, $M = 2.5\,\mathrm{kg}$ is the mass of the cart and F is the
control input.

10.3.1.1 Stability Analysis

A stability analysis around pole's equilibrium point x_e (the vertical position) can
reveal useful conclusions in respect to the allowable margins of the applied control
input forces. We use Lyapunov's direct method for the stability analysis of the inverted
pendulum (Passino and Yurkovich 1998). Equation (10.1) can be rewritten as:

$$\dot{x}_1 = x_2 = f_1(x)$$
$$\dot{x}_2 = \frac{mlx_2^2 \sin(x_1) - g(M + m)x_1}{ml \cos(x_1) - l(M + m)} + \frac{1}{ml \cos(x_1) - l(M + m)}F = f_2(x)$$

F is the force applied in the cart in order to control the pole. Assume that $F_{0,0} = 0$
so that the equilibrium is preserved.

We choose
$$V(x) = \frac{1}{2}x_1^2 + \frac{1}{2}x_2^2$$

so that

$$\nabla V(x) = \left[x_1, \, x_2 \right]^T$$

and

$$\dot{V} = \left[x_1, \, x_2 \right] \left[\begin{array}{c} x_2 \\[4pt] \dfrac{mlx_2^2 \sin(x_1) - g(M+m)x_1}{ml\cos(x_1) - l(M+m)} + \dfrac{1}{ml\cos(x_1) - l(M+m)}F \end{array} \right]$$

and we would like $\dot{V} < 0$ to prove asymptotic stability.

We have

$$x_2 \left(x_1 + \frac{mlx_2^2 \sin(x_1) - g(M+m)x_1}{ml\cos(x_1) - l(M+m)} + \frac{1}{ml\cos(x_1) - l(M+m)}F \right) < -\beta.$$

If for some fixed $\beta > 0$ (note that $x_2 \neq 0$) then:

$$x_1 + \frac{mlx_2^2 \sin(x_1) - g(M+m)x_1}{ml\cos(x_1) - l(M+m)} + \frac{1}{ml\cos(x_1) - l(M+m)}F < \frac{-\beta}{x_2}.$$

Regarding this equation we have that:

$$F < (ml\cos(x_1) - l(M+m)) \left(-\frac{\beta}{x_2} - x_1 \right) - \left(mlx_2^2 \sin(x_1) - g(M+m)x_1 \right)$$

$$(10.2)$$

on $x \in B(h)^1$ for some $h > 0$ and $\beta > 0$. When Eq. (10.2) holds then this control law F ensures that asymptotic stability for the solution x_e holds.

10.3.1.2 FCN Design for the Inverted Pendulum Dynamics

The FCN designed for the inverted pendulum is shown in Fig. 10.7. The description of each node and the range of the values it represents are given in Table 10.1. It is assumed that the nodes receive values in the interval $[0 \ldots 1]$, that correspond to the physical range values of the concepts they represent.

The weight matrix of FCN is depicted below:

$$W = \begin{bmatrix} 0 & 0 & 0 & 0 & 0 \\ 0 & 0 & 0 & 0 & 0 \\ 0 & 0 & 0 & 0 & 0 \\ w_{C(F)C(x_1(t+\Delta T))} & w_{C(x_1(t))C(x_1(t+\Delta T))} & w_{C(x_2(t))C(x_1(t+\Delta T))} & 1 & 0 \\ w_{C(F)C(x_2(t+\Delta T))} & w_{C(x_1(t))C(x_2(t+\Delta T))} & w_{C(x_2(t))C(x_2(t+\Delta T))} & 0 & 1 \end{bmatrix}$$

---

[1] $B(h) = \{x \in \Re^n : |x| < h\}$ is a ball centered at the origin $(0,0)$ with a radius of h and $|\cdot|$ is a norm on $\Re^2$.

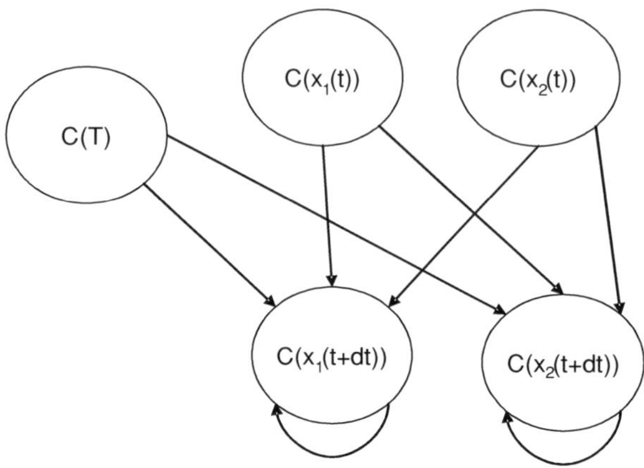

Fig. 10.7 An FCN designed to control an inverted pendulum

Table 10.1 Concept nodes and physical range values for the FCN of the inverted pendulum

Node	Description	Range of values
$C(F)$	The force applied on the cart	$[0\ldots 1] \to [-F_{max}\ldots F_{max}]\,\mathrm{N}$
$C(x_1(t))$	The angle of the pole at time t	$[0\ldots 1] \to [-\pi/2\,(\mathrm{rad})\ldots \pi/2\,(\mathrm{rad})]$
$C(x_2(t))$	The angular velocity of the pole at time t	$[0\ldots 1] \to [-\pi/2\,(\mathrm{rad/s})\ldots \pi/2\,(\mathrm{rad/s})]$
$C(x_1(t+\Delta T))$	The angle of the pole at time $t+\Delta T$	$[0\ldots 1] \to [-\pi/2\,(\mathrm{rad})\ldots \pi/2\,(\mathrm{rad})]$
$C(x_2(t+\Delta T))$	The angular velocity of the pole at time $t+\Delta T$	$[0\ldots 1] \to [-\pi/2\,(\mathrm{rad/s})\ldots \pi/2\,(\mathrm{rad/s})]$

Since nodes $C(T)$, $C((x_1(T))$ and $C((x_1(T+\Delta T)))$ influence but are not influenced by the other nodes the respective rows of the matrix are zero. Having built and defined the FCN structure, we may now train it to capture the system dynamics by using an off-line training procedure and following the storage mechanism explained in Sects. 10.1 and 10.2.

10.3.1.3 Off-Line Training of the FCN

Collecting Training Data for the FCN
From the dynamic equations given in Eq. (10.1), we collect the training data using the following procedure:

for $x_1(t) = -x_{max} : \pi/8 : x_{max}$
  for $x_2(t) = -x_{max} : \pi/8 : x_{max}$
    for $F = -F_{max} : 7.5 : F_{max}$
      calculate $x_1(t+\Delta T)$ and $x_2(t+\Delta T)$ using Eq. (10.1) with $\Delta T = 0.001$.

In the above procedure, for each sample $x_1(t)$, $x_2(t)$ the selected F can be appropriately modified such that Eq. (10.2) is fulfilled. In Eq. (10.2) we select

Table 10.2 Conversion formulas of physical variable values to FCN node values

Node	Conversion formula
$C(F)$	$C(F) = \dfrac{F * 0.5 + F_{max}/2}{F_{max}}$
$C(x_1(t))$	$C(x_1(t)) = \dfrac{x_1(t) * 0.5 + x_{max}/2}{x_{max}}$
$C(x_2(t))$	$C(x_2(t)) = \dfrac{x_2(t) * 0.5 + x_{max}/2}{x_{max}}$
$C(x_1(t + \Delta T))$	$C(x_1(t + \Delta T)) = \dfrac{x_1(t + \Delta T) * 0.5 + x_{max}/2}{x_{max}}$
$C(x_2(t + \Delta T))$	$C(x_2(t + \Delta T)) = \dfrac{x_2(t + \Delta T) * 0.5 + x_{max}/2}{x_{max}}$

parameter $\beta = 0.1$. A common value for x_{max} is $\pi/2$, while F_{max} varies depending on construction characteristics of the pendulum.

As an example, for $F_{max} = 30\,\text{N}$, if $x_1 = -\pi/4 = -0.7854$, $x_2 = \pi/8 = 0.3927$ and $F = 0$ then $x_1(t+\Delta T) = -0.7820$ and $x_2(t+\Delta T) = 0.3831$. Indeed, since the angular velocity is positive, it is expected the negative angle to become less negative and at the same time, due to zero force, the angular velocity is reduced.

Using this procedure we collect 729 training data samples.

Convert Training Data into FCN Vectors

Since by design FCN node values are in the range $[0 \dots 1]$ we have to convert the training data into FCN node values. Table 10.2 gives the conversion formulas that translate each physical variable to its corresponding node value.

As a result, using the above conversion coefficients with $F_{max} = 30$ and $x_{max} = \pi/2$ a training sample vector containing the values: $x_1 = -\pi/4 = -0.7854$, $x_2 = \pi/8 = 0.3927$ and $F = 0$ then $x_1(t + \Delta T) = -0.7820$ and $x_2(t + \Delta T) = 0.3831$. is translated to the following FCN nodes vector:

$$A^{\text{des}} = \begin{bmatrix} 0.5 & 0.25 & 0.6250 & 0.2511 & 0.6219 \end{bmatrix}^T .$$

Another training sample vector containing the values: $x_1 = \pi/8$, $x_2 = -\pi/4$ and $F = 15$ then $x_1(t + \Delta T) = 0.3919$ and $x_2(t + \Delta T) = -0.7861$, is translated to the following FCN nodes vector:

$$A^{\text{des}} = \begin{bmatrix} 0.7500 & 0.6250 & 0.2500 & 0.6247 & 0.2498 \end{bmatrix}^T .$$

In a more general form, the node vector for the FCN of the Inverted Pendulum can be written as:

$$A^{\text{des}} = \begin{bmatrix} C(T) & C(x_1(t)) & C(x_2(t)) & C((x_1(t + \Delta T))) & C((x_2(t + \Delta T))) \end{bmatrix}^T .$$

This training data constitutes the input to the FCN, where according to Fig. 10.1 the weights W of the FCN are updated according to Eqs. 9.7, 9.14, 9.15 and stored as fuzzy knowledge into a fuzzy rule database. For these weights W Eq. (9.1) is also true so that one can ensure that the FCN converges to a unique equilibrium point.

Storing Knowledge From Previous Operating Conditions

The following procedure modifies FCN's knowledge about the system by continuously modifying the weight interconnections and consequently the node values. During the repetitive updating operation, the procedure uses input from the systems training data, producing a new weight matrix for each new equilibrium state. It is desirable to devise a storage mechanism for keeping these weights for probable future use to control the Pendulum. It would be also preferable this storage to allow weight retrieval even in situations, where the equilibrium conditions were not been exactly met during the training phase. To this end, we use the storing mechanism of Sect. 10.2, where the previous acquired operational situations are stored in a fuzzy if-then rule database, which associates in a fuzzy manner the various weights with the corresponding equilibrium node values. The procedure is explained as follows. Suppose for example that the FCN of Fig. 10.7 has a unique equilibrium point:

$$A_1 = [0.5\ 0.25\ 0.6250\ 0.2511\ 0.6219]^T$$

which is connected with the weight matrix W:

$$W_1 = \begin{bmatrix} 0 & 0 & 0 & 0 & 0 \\ 0 & 0 & 0 & 0 & 0 \\ 0 & 0 & 0 & 0 & 0 \\ -1.0000 & -0.8754 & -1.0000 & 1.0000 & 0 \\ -0.1197 & 0.0851 & -0.1371 & 0 & 1.0000 \end{bmatrix}$$

in order that A_1 is a unique solution of (9.1) weight matrix W has to be such that inequality (8.22) is fulfilled. Indeed, for weight matrix W_1 inequality (8.22) takes the form:

$$\left(\sum_{i=1}^{n} \|w_i\|^2 \right)^{1/2} = 2.1924 < 4$$

where $n = 5$ is the number of concepts of the FCN.

Suppose also that the FCM in another operation point A_2 is related to weight matrix W_2 both given below.

$$A_2 = [0.7500\ 0.6250\ 0.2500\ 0.6247\ 0.2498]^T$$

$$W_2 = \begin{bmatrix} 0 & 0 & 0 & 0 & 0 \\ 0 & 0 & 0 & 0 & 0 \\ 0 & 0 & 0 & 0 & 0 \\ -0.2486 & -0.1243 & 0.0643 & 1.0000 & 0 \\ -1.0000 & -0.8979 & -1.0000 & 0 & 1.0000 \end{bmatrix}$$

matrix W_2 also fulfils (8.22):

$$\left(\sum_{i=1}^{5} \|w_i\|^2 \right)^{1/2} = 2.2108 < 4.$$

The fuzzy rule database, which is obtained using the information from the two previous equilibrium points, is resolved as follows:

There are two rules for the description of the above two different equilibrium situations:

Rule 1

if node $C(x_1)$ is mf1 *and* node $C(x_2)$ is mf1 *and* node $C(x_1(t + \Delta T))$ is mf1 *and* node $C(x_2(t + \Delta T))$ is mf1

then w_{41} is mf1 *and* w_{42} is mf1 *and* w_{43} is mf1 *and* w_{51} is mf1 *and* w_{52} is mf1 *and* w_{53} is mf1

Rule 2

if node $C(x_1)$ is mf2 *and* node $C(x_2)$ is mf2 *and* node $C(x_1(t + \Delta T))$ is mf2 *and* node $C(x_2(t + \Delta T))$ is mf2

then w_{41} is mf2 *and* w_{42} is mf2 *and* w_{43} is mf2 *and* w_{51} is mf2 *and* w_{52} is mf2 *and* w_{53} is mf2

where each membership function is deriving from the two equilibrium points acquired from matrices A_1, W_1, A_2 and W_2 and presented bellow:

$$A_1 = \begin{bmatrix} 0.5 & 0.25_{\text{mf1}} & 0.6250_{\text{mf1}} & 0.2511_{\text{mf1}} & 0.6219_{\text{mf1}} \end{bmatrix}^T$$

$$W_1 = \begin{bmatrix} 0 & 0 & 0 & 0 & 0 \\ 0 & 0 & 0 & 0 & 0 \\ 0 & 0 & 0 & 0 & 0 \\ -1.0000_{\text{mf1}} & -0.8754_{\text{mf1}} & -1.0000_{\text{mf1}} & 1.0000 & 0 \\ -0.1197_{\text{mf1}} & 0.0851_{\text{mf1}} & -0.1371_{\text{mf1}} & 0 & 1.0000 \end{bmatrix}$$

$$A_2 = \begin{bmatrix} 0.7500 & 0.6250_{\text{mf2}} & 0.2500_{\text{mf2}} & 0.6247_{\text{mf2}} & 0.2498_{\text{mf2}} \end{bmatrix}^T$$

$$W_2 = \begin{bmatrix} 0 & 0 & 0 & 0 & 0 \\ 0 & 0 & 0 & 0 & 0 \\ 0 & 0 & 0 & 0 & 0 \\ -0.2486_{\text{mf2}} & -0.1243_{\text{mf2}} & 0.0643_{\text{mf2}} & 1.0000 & 0 \\ -1.0000_{\text{mf2}} & -0.8979_{\text{mf2}} & -1.0000_{\text{mf2}} & 0 & 1.0000 \end{bmatrix}.$$

It is pointed out that the node $C(F)$ has been deliberately kept out of the rules, because in the control scheme that will be described in the next section, it will be estimated using an inverse procedure.

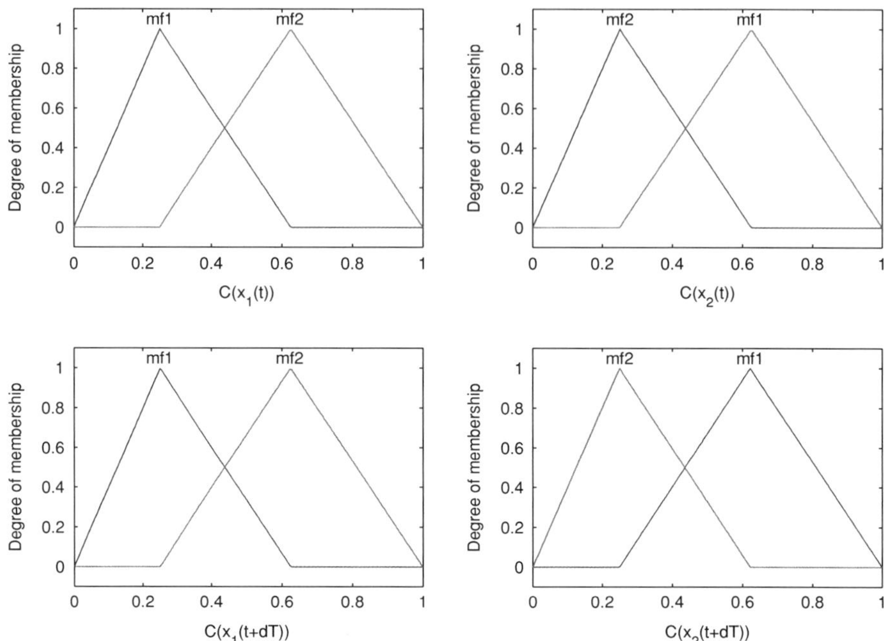

Fig. 10.8 Left-hand side (if-part)

The number and shape of the fuzzy membership functions of the variables of both sides of the rules are gradually modified, according to the procedure described in Sect. 10.2 and Fig. 10.5, as new desired equilibrium points appear to the system during its training. If, for example, the coefficient c in creating new triangular membership functions for nodes $C(x_1(t))$, $C(x_2(t))$, $C((x_1(t + \Delta T)))$, $C((x_2(t + \Delta T)))$ and for the six (6) weights in Inverted Pendulum example is set to $c = 0.01$, then the maximum number of triangular membership functions that can appear during the off-line training are 100 for each node and 200 for each weight interconnection. In practice, however, ever for that value of c the number of required membership functions is far below the maximum number.

The created fuzzy membership functions of the fuzzy rule databases *Rule1* and *Rule2* are depicted in Figs. 10.8 and 10.9.

For implementation reasons, one should consider the fact that, as it was explained in Sect. 8.3.2 and Example 9.4, small FCNs that use the linear FCN parametric model (only weights are estimated) can not always converge to node values that are close to their margins (0 or 1). In this case the FCN does not offer a reliable approximation of the real system. This can be overcome by two ways. First, one has to enlarge the margins of the physical variables considered, so that the conversion through the formulas of Table 10.2 provide node values that are not very close to 0 or 1. For example, one could use wider margins for the values of force, so that the conversion of a really encounter force value is converted to a node value far from 0 or 1. The same can be applied on the state variables. The second way is to use

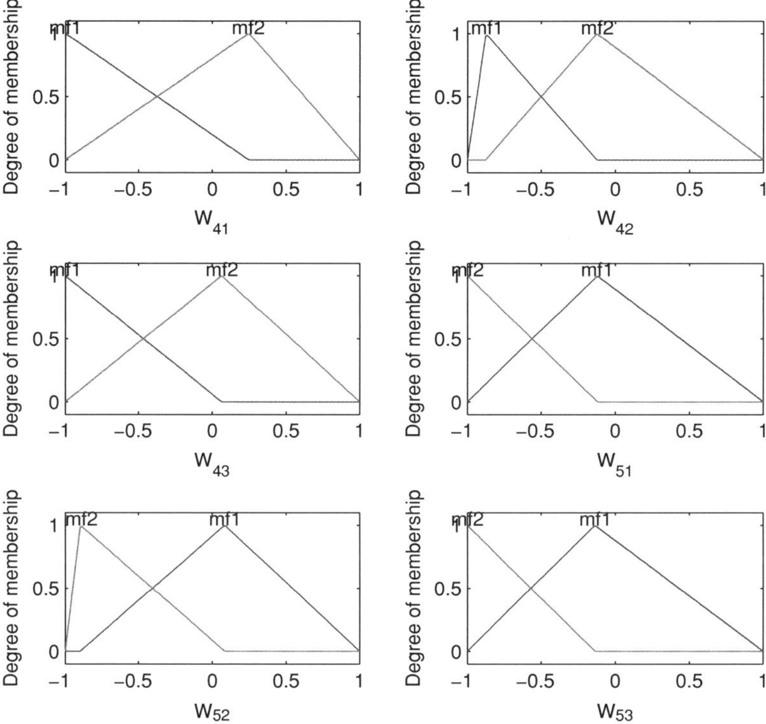

Fig. 10.9 Right-hand side (then-part)

a FCN with bilinear parametric modeling (both weights and inclination parameters are estimated), because as explained in Example 9.5, marginal values can be better approached by increasing some inclination parameters. The combination of the two ways may also be applied.

10.3.1.4 Control of the Inverted Pendulum Using the Linear Parametric Model

We assume that the linear parametric model is used for the estimation of the FCN parameters. After the FCN has been trained and the knowledge acquired has been stored in the Fuzzy Rule Database, it can be used to control the behavior of the Inverted Pendulum. The proposed approach is based on an inverse procedure, which uses the fact that the FCN sigmoid functions are invertible and the FCN structure permits to derive the control input values from the desired states. The FCN can be used either to drive the pole of the pendulum to track a desired state or simply to balance the pole around the vertical position. The control of the Pendulum is based only on the initial off-line training described above, where we used $F_{\max} = 120$. The control scheme is explained below.

Suppose that the desired state of the FCN is:

$$A = \left[\, A(F)\ A(x_1)\ A(x_2)\ A^{\text{des}}(x_1(t + \Delta T))\ A^{\text{des}}(x_2(t + \Delta T))\,\right]^T .$$

From the Fuzzy Rule database which is of the form:

if node $C(x_1)$ is $A(x_1)$ *and* node $C(x_2)$ is $A(x_2)$ *and* node $C(x_1(t + \Delta T))$ is $A^{\text{des}}(x_1(t + \Delta T))$ *and* node $C(x_2(t + \Delta T))$ is $A^{\text{des}}(x_2(t + \Delta T))$

then w_{41} is $w1$ *and* w_{42} is $w2$ *and* w_{43} is $w3$ *and* w_{51} is $w4$ *and* w_{52} is $w5$ *and* w_{53} is $w6$

one obtains the values of the weights using conventional fuzzy inference and defuzzification procedure.

Since $C(F)$, $C(x_1)$ and $C(x_2)$ are steady nodes, then for the FCN nodes we have that:

$$A(F) = A(F)$$

$$A(x_1) = A(x_1)$$

$$A(x_2) = A(x_2)$$

$$A^{\text{des}}(x_1(t + \Delta T)) = f\left(w1 * A(F) + w2 * A(x_1) + w3 * A(x_2) + A^{\text{des}}(x_1(t + \Delta T))\right)$$

$$A^{\text{des}}(x_2(t + \Delta T)) = f\left(w4 * A(F) + w5 * A(x_1) + w6 * A(x_2) + A^{\text{des}}(x_2(t + \Delta T))\right)$$

which can be rewritten:

$$f^{-1}\left(A^{\text{des}}(x_1(t + \Delta T))\right) = w1 * A(F) + w2 * A(x_1) + w3 * A(x_2) + A^{\text{des}}(x_1(t + \Delta T))$$
$$f^{-1}\left(A^{\text{des}}(x_2(t + \Delta T))\right) = w4 * A(F) + w5 * A(x_1) + w6 * A(x_2) + A^{\text{des}}(x_2(t + \Delta T))$$

where f is the sigmoid function $f(x) = \frac{1}{1 + \exp^{-x}}$, $A(x_1)$ is the node value of state x_1 at time t, $A(x_2)$ is the node value of state x_2 at time t, $w1, w2, w3, w4, w5, w6$ are the estimated FCN weight values, $A^{\text{des}}(x_1(t + \Delta T))$, $A^{\text{des}}(x_2(t + \Delta T))$ are node values of the desired states of the pole of the Inverted Pendulum at time $(t + \Delta T)$ and $A(F)$ is the node value of the unknown force to be applied on the cart.

We can now estimate the force F using the following formula:

$$
\begin{aligned}
A(F) = &\left(f^{-1}\left(A^{\text{des}}(x_1(t + \Delta T))\right) + f^{-1}\left(A^{\text{des}}(x_2(t + \Delta T))\right)\right. \\
&- w2 * A(x_1) - w3 * A(x_2) - A^{\text{des}}(x_1(t + \Delta T)) \\
&\left. - w5 * A(x_1) - w6 * A(x_2) - A^{\text{des}}(x_2(t + \Delta T))\right) / (w1 + w4) . \quad (10.3)
\end{aligned}
$$

From Eq. 10.3 all parameters are known, so $A(F)$ is estimated in the interval $[0\ \ 1]$ and after it is converted to the interval $[-120\ \ 120]$ gives F that is applied as input to the cart. It should be mentioned that Eq. (10.3) does not guarantee that the estimated F always fulfills (10.2). However, during the simulations carried out there was not

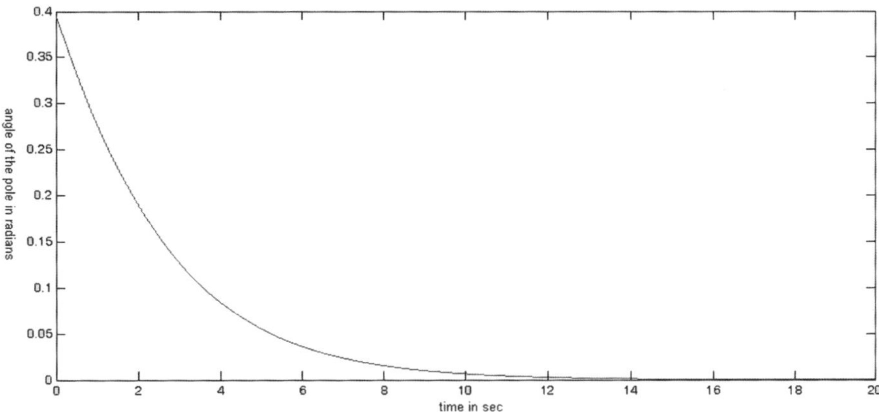

Fig. 10.10 State $x_1(t)$ for initial condition $x_1(0) = \pi/8$, $x_2(0) = \pi/8$

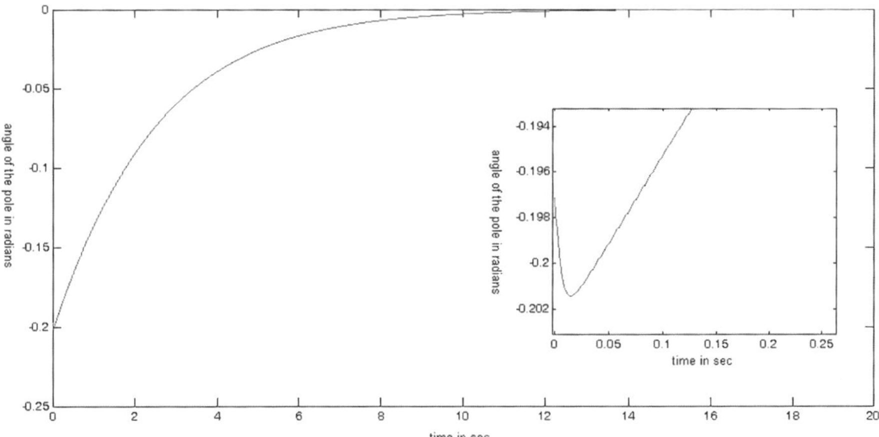

Fig. 10.11 State $x_1(t)$ for initial condition $x_1(0) = -\pi/16$, $x_2(0) = -\pi/4$

any violation of (10.2). In case this happens one could possibly take some corrective action and modify F accordingly.

Figures 10.10 and 10.11 show the balancing of the pole around the vertical position when the initial state values are $x_1(0) = \pi/8$, $x_2(0) = \pi/8$ and $x_1(0) = -\pi/16$, $x_2(0) = -\pi/4$ respectively. The ideal smooth convergence(without oscillations) of the state $x_1(t)$ to zero is due to the fact that in our control paradigm no restrictions are imposed regarding the position of the cart. Moreover due to simulation ideal force values can be applied at each sampling time, something that cannot be always met in a real life experiment.

It can be observed that the proposed technique performs very well in keeping the pole vertical. Tracking performance will be demonstrated below and will be compared with the corresponding performance of a controller designed using the bilinear parametric model of the FCN.

It worth mentioning in this point that, the proposed inverse control scheme can be equally applied on non-Brunovsky, general systems with any number of input and state variables, provided that it assumes, by nature or design, an affine in the control form.

10.3.1.5　Control of the Inverted Pendulum Using the Bilinear Parametric Model

Assume that the bilinear parametric model is used for the estimation of the FCN parameters. In this case, both weights and sigmoid inclination parameters are estimated. As it has already been stressed, we expect better approximation performance of the FCN, especially when its nodes are required to reach lower or upper limit values.

We assume again that in the training procedure we use $F_{max} = 120$. After the FCN has been trained and the knowledge acquired has been stored in the Fuzzy Rule Database, it can be used to control the behavior of the Inverted Pendulum. The proposed approach is based on a similar inverse procedure as with the linear parametric model, which uses the fact that the FCN sigmoid functions are invertible and the FCN structure permits to derive the control input values from the desired states. The control scheme is explained below.

Suppose that the desired state of the FCN is:

$$A = \left[A(F)\ A(x_1)\ A(x_2)\ A^{des}(x_1(t + \Delta T))\ A^{des}(x_2(t + \Delta T)) \right]^T .$$

From the Fuzzy Rule database which is of the form:

if node $C(x_1)$ is $A(x_1)$ *and* node $C(x_2)$ is $A(x_2)$ *and* node $C(x_1(t + \Delta T))$ is $A^{des}(x_1(t + \Delta T))$ *and* node $C(x_2(t + \Delta T))$ is $A^{des}(x_2(t + \Delta T))$

then w_{41} is $w1$ *and* w_{42} is $w2$ *and* w_{43} is $w3$ *and* w_{51} is $w4$ *and* w_{52} is $w5$ *and* w_{53} is $w6$ *and* $c_{C(x_1(t+\Delta T))}$ is $c4$ *and* $c_{C(x_2(t+\Delta T))}$ is $c5$

one obtains the values of the weights and the values of inclination parameters of the sigmoid functions of nodes 4 $(C(x_1(t + \Delta T)))$ and 5 $(C(x_2(t + \Delta T))$ using conventional fuzzy inference and defuzzification procedure.

Since $C(F)$, $C(x_1)$ and $C(x_2)$ are steady nodes, then for the FCN nodes we have that:

$$A(F) = A(F)$$

$$A(x_1) = A(x_1)$$

$$A(x_2) = A(x_2)$$

$$A^{des}(x_1(t + \Delta T)) = f\Big(c4 * w1 * A(F) + c4 * w2 * A(x_1) + c4 * w3 * A(x_2)$$

$$+ c4 * A^{des}(x_1(t + \Delta T))\Big)$$

$$A^{\text{des}}(x_2(t + \Delta T)) = f\Big(c5 * w4 * A(F) + c5 * w5 * A(x_1) + c5 * w6 * A(x_2)$$
$$+ c5 * A^{\text{des}}(x_2(t + \Delta T))\Big)$$

which can be rewritten:

$$f^{-1}\Big(A^{\text{des}}(x_1(t + \Delta T))\Big) = c4 * w1 * A(F) + c4 * w2 * A(x_1)$$
$$+ c4 * w3 * A(x_2) + c4 * A^{\text{des}}(x_1(t + \Delta T))$$
$$f^{-1}\Big(A^{\text{des}}(x_2(t + \Delta T))\Big) = c5 * w4 * A(F) + c5 * w5 * A(x_1)$$
$$+ c5 * w6 * A(x_2) + c5 * A^{\text{des}}(x_2(t + \Delta T))$$

where f is the sigmoid funtion $f(x) = \big(1/(1 + \exp^{-c*x})\big)$. $A(x_1)$, $A(x_2)$ are the node values of the respective states at time t, $w1, w2, w3, w4, w5, w6$ are the estimated FCN weight values, $c4$ and $c5$ are the estimated inclination parameter values of the sigmoid function f for nodes 4 and 5 and $A^{\text{des}}(x_1(t + \Delta T))$, $A^{\text{des}}(x_2(t + \Delta T))$ are the node values of the respective states at time $(t + \Delta T)$. Finally, $A(F)$ is the node value of the unknown force to be applied on the cart.

We can now estimate the force F using the following formula:

$$A(F) = (f^{-1}\Big(A^{\text{des}}(x_1(t + \Delta T))\Big) + f^{-1}\Big(A^{\text{des}}(x_2(t + \Delta T))\Big)$$
$$- c4 * w2 * A(x_1) - c4 * w3 * A(x_2) - c4 * A^{\text{des}}(x_1(t + \Delta T))$$
$$- c5 * w5 * A(x_1) - c5 * w6 * A(x_2)$$
$$- c5 * A^{\text{des}}(x_2(t + \Delta T)))/ (c4 * w1 + c5 * w4) \tag{10.4}$$

from Eq. (10.4) all parameters are known, so the force F is estimated in the interval $[0 \quad 1]$ and after it is converted in the interval $[-120 \quad 120]$ it is applied as input to the cart. It should be mentioned that, as in the linear approach, Eq. (10.4) does not guarantee that the estimated F always fulfills (10.2).

Next, we proceed to the presentation of simulation results comparing the performance of both linear and bilinear approaches in state tracking paradigms.

Figures 10.12 and 10.13 show the tracking of $x_1^{\text{des}}(t) = (1/4)\cos(2 * t)$ and its velocity $x_2^{\text{des}}(t) = -(1/2)\sin(2 * t)$ by using both approaches. The pole, starting from $x_1(0) = -\pi/4$, $x_2(0) = \pi/8$, is controlled to follow the state trajectory. It can be seen that by using the bilinear approach the pole is reaching faster its goal. Figure 10.14 shows the tracking errors, while Fig. 10.15 shows the forces F applied on the cart. In the figures, only the first seconds are displayed where the difference between the two approaches is noticeable. As time passes, both trajectories converge to the desired ones.

Figures 10.16 and 10.17 shows the difference between the two approaches, when the controller must drive the pole in the trajectory produced by $x_1^{\text{des}}(t) = (2/5)\cos(2t)$, having wider margins than the trajectory produced by $x_1^{\text{des}}(t) =$

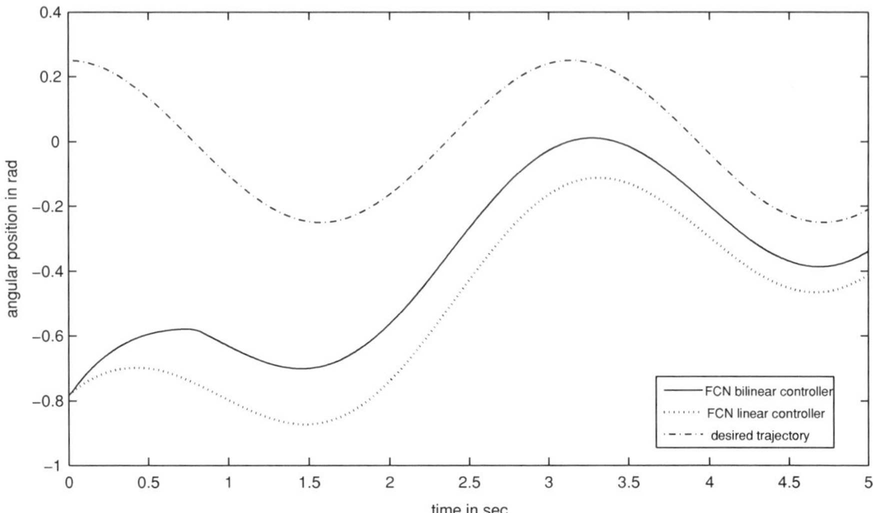

Fig. 10.12 State $x_1(t)$ and its desired value $\frac{\cos(2*t)}{4}$ for initial condition $x_1(0) = -\pi/4$, $x_2(0) = \pi/8$, where *dotted line* represents the desired trajectory, *dashed line* the trajectory of angular position for linear approach and *solid line* the trajectory of angular position for bilinear approach

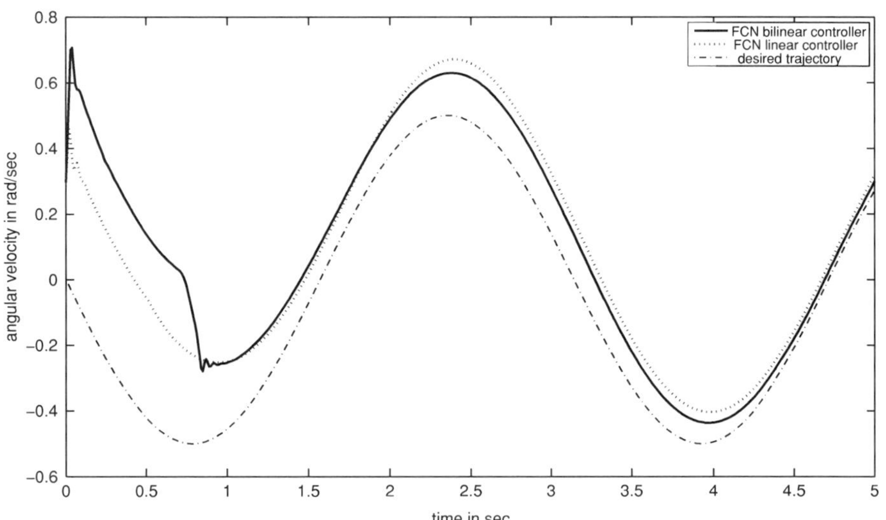

Fig. 10.13 State $x_2(t)$ and its desired value $-\frac{\sin(2*t)}{2}$ for initial condition $x_1(0) = -\pi/4$, $x_2(0) = \pi/8$, where *dotted line* represents the desired trajectory, *dashed line* the trajectory of angular velocity for linear approach and *solid line* the trajectory of angular velocity for bilinear approach

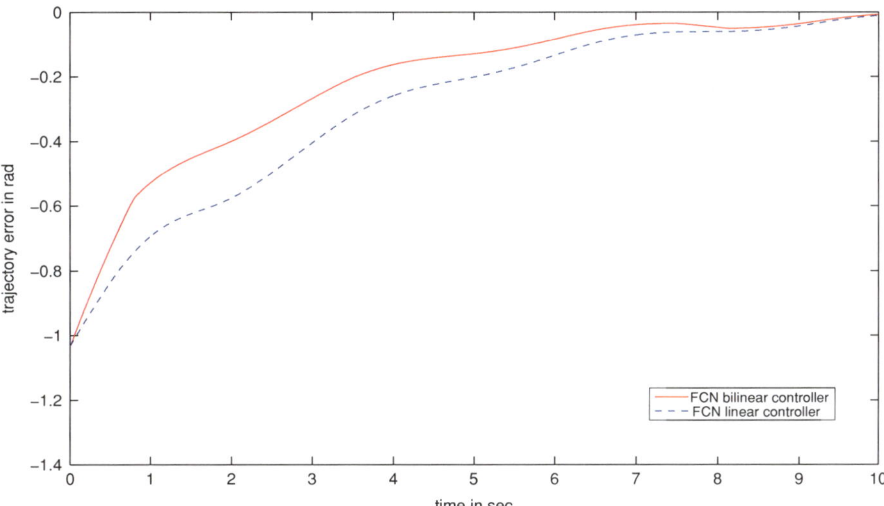

Fig. 10.14 Trajectory error of linear and bilinear approach in respect to the desired value $\frac{\cos(2*t)}{4}$ (*dotted line*) for initial condition $x_1(0) = -\pi/4$, $x_2(0) = \pi/8$, where *solid line* represents the bilinear controller and *dashed line* represents the linear controller

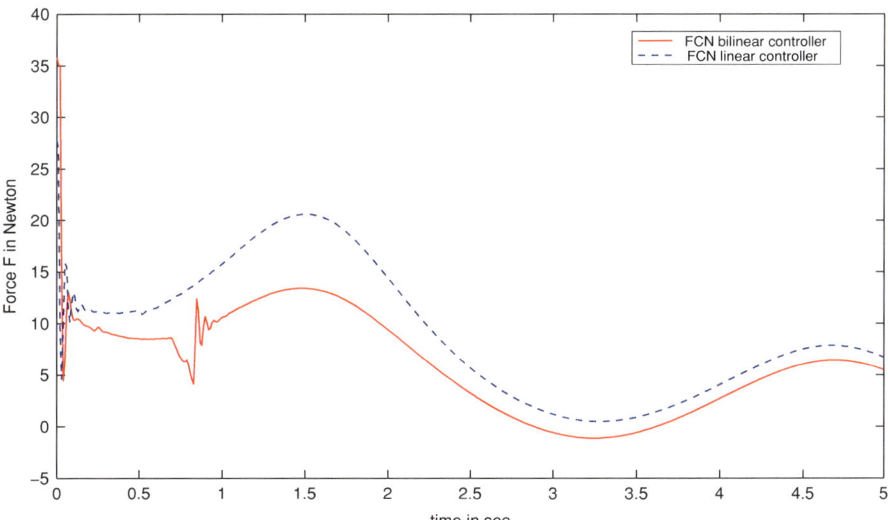

Fig. 10.15 Force F of linear and bilinear approach applied on the cart for tracking state $x_1(t)$ in the desired value $\frac{\cos(2*t)}{4}$ for initial condition $x_1(0) = -\pi/4$, $x_2(0) = \pi/8$. *Solid line* represents the bilinear controller while *dashed line* represents the linear controller

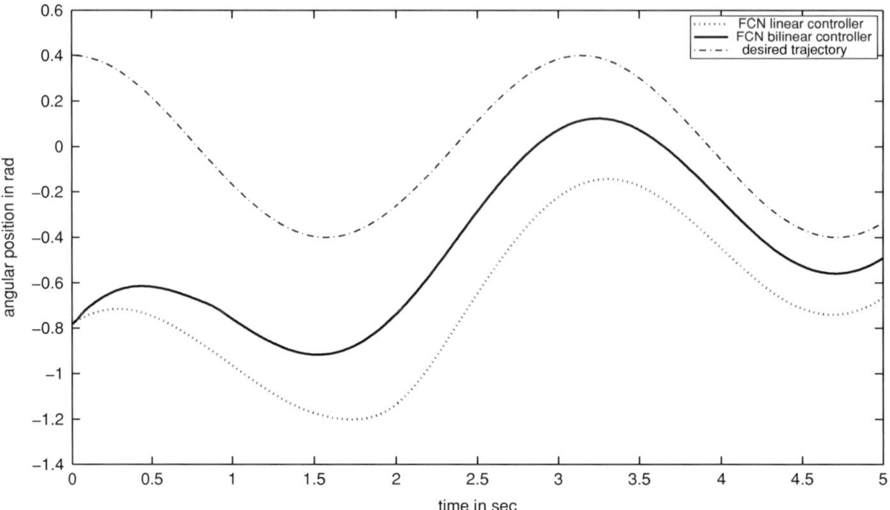

Fig. 10.16 State $x_1(t)$ of linear approach and its desired value $\frac{2\cos(2t)}{5}$ for initial condition $x_1(0) = -\pi/4$, $x_2(0) = \pi/8$, where *dotted line* represents the desired trajectory, *dashed line* the trajectory of angular velocity for linear approach and *solid line* the trajectory of angular velocity for bilinear approach

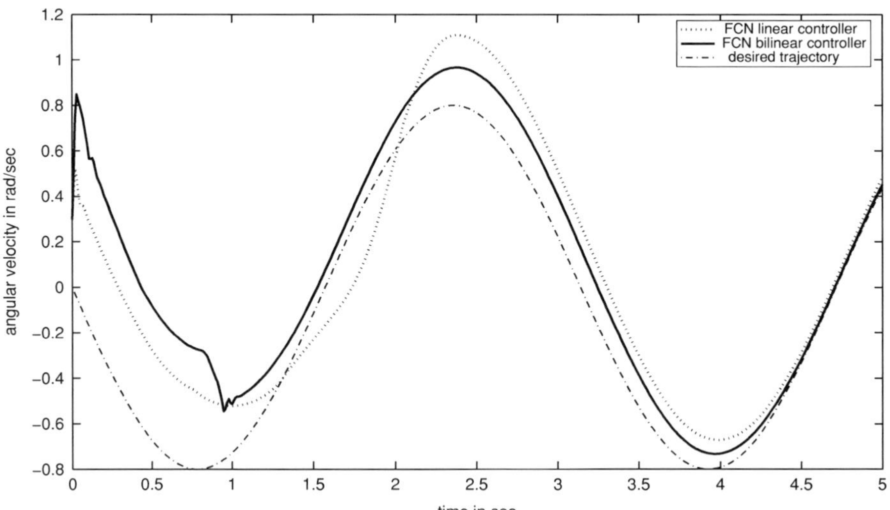

Fig. 10.17 State $x_2(t)$ of linear approach and its desired value $-\frac{4\sin(2t)}{5}$ for initial condition $x_1(0) = -\pi/4$, $x_2(0) = \pi/8$, where *dotted line* represents the desired trajectory, *dashed line* the trajectory of angular position for linear approach and *solid line* the trajectory of angular position for bilinear approach

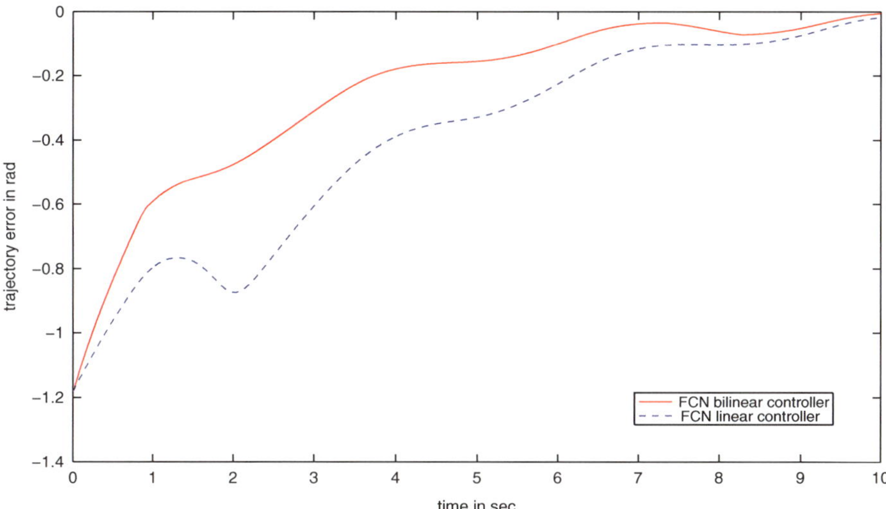

Fig. 10.18 Trajectory error of linear approach in the desired value $\frac{2\cos(2t)}{5}$ for initial condition $x_1(0) = -\pi/4$, $x_2(0) = \pi/8$, where *solid line* represents the bilinear controller while *dashed line* represents the linear controller

$(1/4)\cos(2t)$. It can be seen that, using the bilinear approach the angular position approaches the desired trajectory much quicker and smoother than in the linear approach.

A more detailed view can been seen in Fig. 10.18 where the trajectory error of the angular position is demonstrated.

Figure 10.19 shows the force F applied on the cart, where one can see the problem of inaccurate force derived in the linear approach during seconds 1.5–2.

Figures 10.20 and 10.21 present the failure of the linear approach to reach the desired high values of force F in order to drive pendulum in more demanding trajectory. When the desired trajectory is $x_1(t) = \frac{\cos(2t)}{2}$, the controller using the linear approach is not capable of reaching the desired trajectory. In Fig. 10.20 values of $x_1 < -\pi/2$ are only simulation values, they are not feasible in practice because if $x_1 = -\pi/2$ the pole cannot return to its upright position regardless the applied force.

On the contrary, the controller using the bilinear approach reaches the desired trajectory as Fig. 10.22 shows. The force F produced by the bilinear approach is shown in Fig. 10.23. It can bee seen that in the beginning, when a large force is required, the linear approach fails to give the proper force driving inevitably the pole to its final fall. This does not happen in the bilinear approach, where, as explained in the previous sections, the FCN can successfully handle situations that correspond to marginal node values.

Finally, Fig. 10.24 shows the trajectory of the FCN linear controller when force F is modified at each time instant to fulfill (10.2). It can be observed that the trajectory is shorter than the desired one but the pole is following a similar trajectory with the

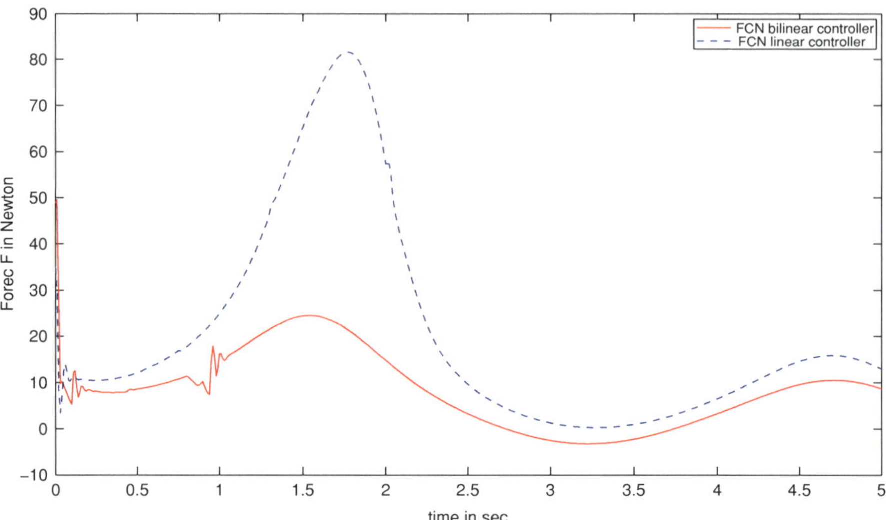

Fig. 10.19 Force T of linear approach applied in the cart for tracking state $x_1(t)$ in the desired value $\frac{2\cos(2t)}{5}$ for initial condition $x_1(0) = -\pi/4$, $x_2(0) = \pi/8$, where *solid line* represents the bilinear controller while *dashed line* represents the linear controller

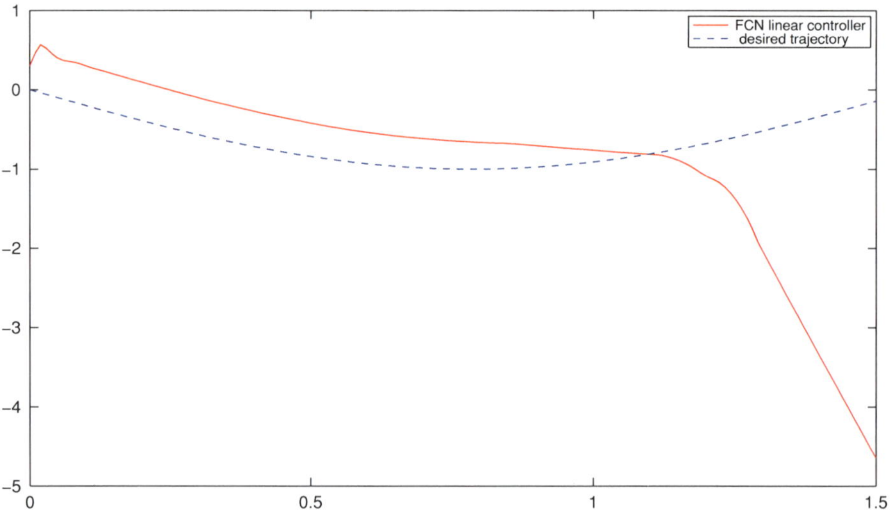

Fig. 10.20 State $x_1(t)$ of linear approach and its desired value $\frac{\cos(2t)}{2}$ for initial condition $x_1(0) = -\pi/4, x_2(0) = \pi/8$, where *solid line* represents the trajectory produced from FCN while the *dashed line* the desired trajectory of the pole

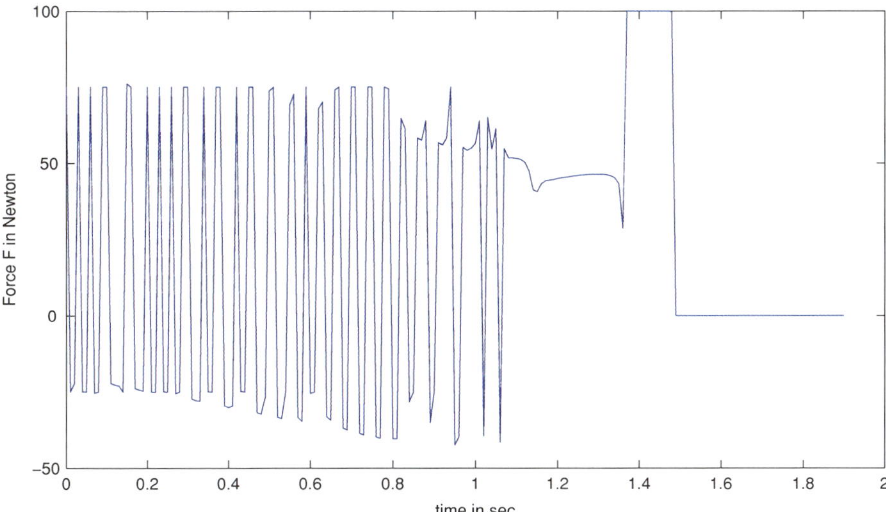

Fig. 10.21 Force F of linear approach applied in the cart for tracking state $x_1(t)$ in the desired value $\frac{\cos(2t)}{2}$ for initial condition $x_1(0) = -\pi/4$, $x_2(0) = \pi/8$

desired one. The modified force F is shown in Fig. 10.25. In the bilinear approach, no need of corrections according to (10.2) was observed.

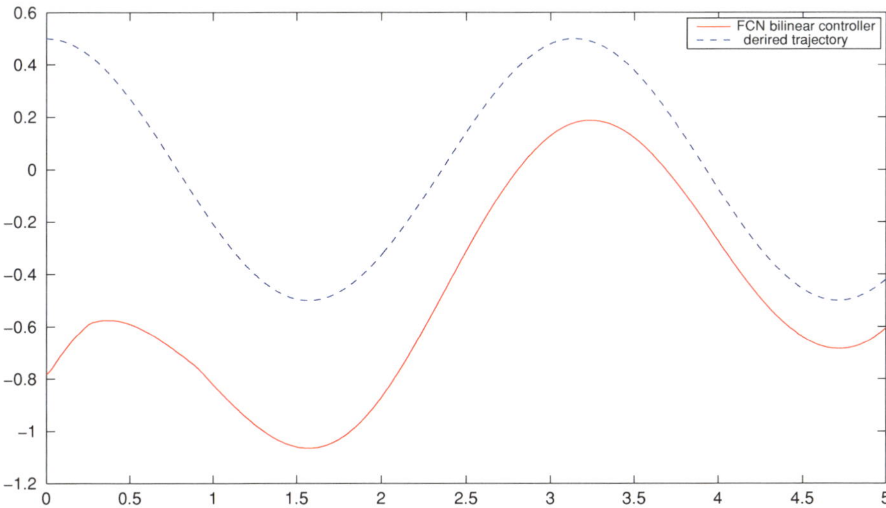

Fig. 10.22 State $x_1(t)$ of bilinear approach and its desired value $\frac{\cos(2t)}{2}$ for initial condition $x_1(0) = -\pi/4$, $x_2(0) = \pi/8$, where *solid line* represents the trajectory produced from FCN while the *dashed line* the desired trajectory of the angular position

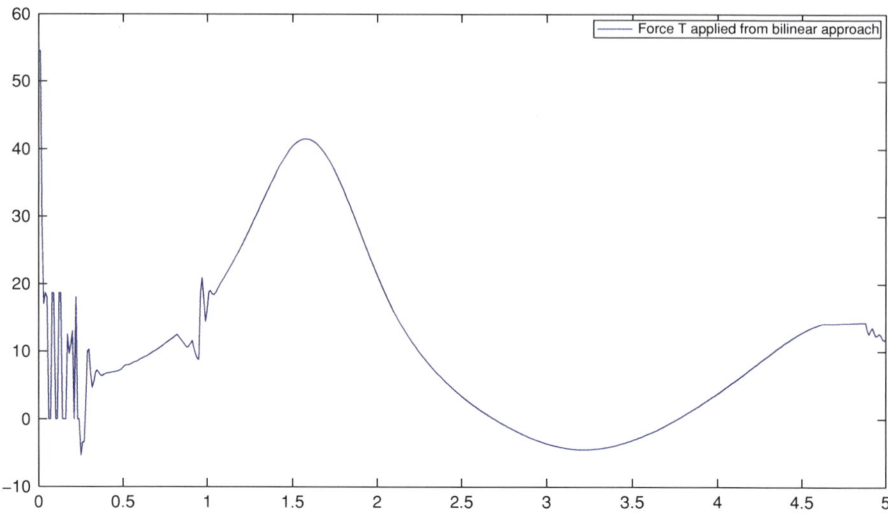

Fig. 10.23 Force F of bilinear approach applied in the cart for tracking state $x_1(t)$ in the desired value $\frac{\cos(2t)}{2}$ for initial condition $x_1(0) = -\pi/4$, $x_2(0) = \pi/8$

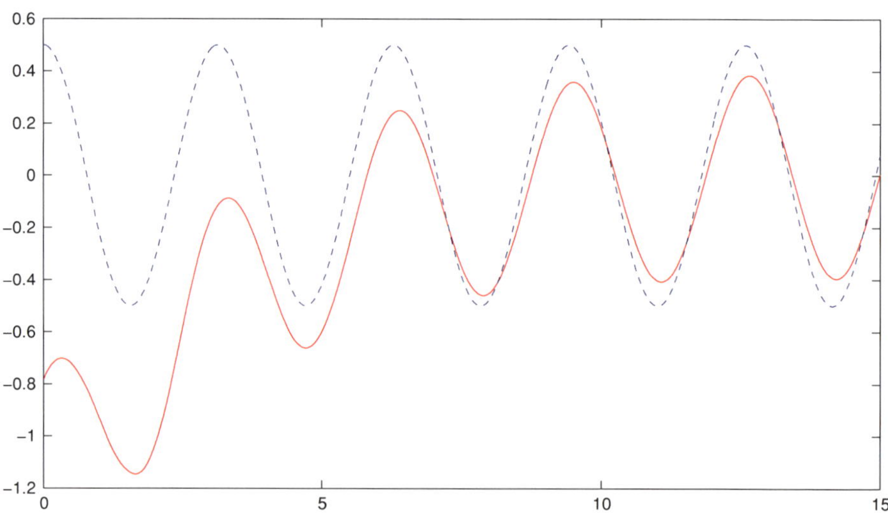

Fig. 10.24 State $x_1(t)$ of linear approach and its desired value $\frac{\cos(2t)}{2}$ for initial condition $x_1(0) = -\pi/4$, $x_2(0) = \pi/8$ when the estimated F fulfills (10.2), where *solid line* represents the trajectory produced from FCN while the *dashed line* the desired trajectory of the angular position

10.3.1.6 Adaptation and Control Under Varying System Parameters

The control procedure described in the previous sections can be extended to cope with system characteristics that has not been encountered during the off-line training of the FCN. In this case, the off-line training and the inverse control scheme will not give accurate results.

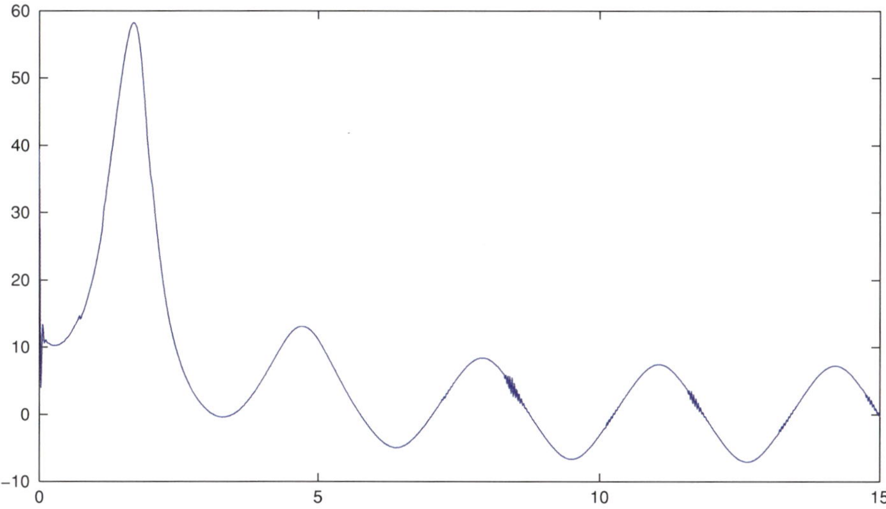

Fig. 10.25 Force F of linear approach applied in the cart for tracking state $x_1(t)$ in the desired value $\frac{\cos(2t)}{2}$ for initial condition $x_1(0) = -\pi/4$, $x_2(0) = \pi/8$ when the estimated F fulfills (10.2)

Suppose for example that the FCN has been trained to model the pendulum with cart mass $M = 2.5\,\text{kg}$. Then, for a starting angular position of the pole at $x_1(0) = -\pi/4$, $x_2(0) = \pi/8$ and a Force $F = 30\,\text{N}$, the next position (after $\Delta T = 0.1\,\text{s}$) will be $x_1(\Delta T) = -\pi/4.5301$, $x_2(\Delta T) = \pi/2.0783$. The FCN node values that correspond to these physical values are given in vector A_1 below. This operation condition has been captured by the FCN during its training and has been connected with the weights and the inclinations parameters W_1 and C_1, respectively.

$$A_1 = \begin{bmatrix} 0.3750 & 0.5625 & 0.25 & 0.3896 & 0.7406 \end{bmatrix}^T$$

$$W_1 = \begin{bmatrix} 0 & 0 & 0 & 0 & 0 \\ 0 & 0 & 0 & 0 & 0 \\ 0 & 0 & 0 & 0 & 0 \\ -0.7577 & -0.8366 & -0.4385 & 1.0000 & 0 \\ 0.1334 & 0.3501 & 0.4889 & 0 & 1.0000 \end{bmatrix}$$

$$C_1 = \begin{bmatrix} 1.0000 & 1.0000 & 1.0000 & 0.9457 & 0.9453 \end{bmatrix}^T.$$

Consider now the case where the real mass of the cart is $M = 7.5\,\text{kg}$. Then, with the same initial state and force values, the next angular position after $\Delta T = 0.1\,\text{s}$ is $x_1(\Delta T) = -\pi/3.9562$, $x_2(\Delta T) = -\pi/5.5417$. In this case, the A vector of the FCN node values is:

$$A_2 = \begin{bmatrix} 0.3750 & 0.5625 & 0.25 & 0.3736 & 0.4098 \end{bmatrix}^T$$

and the corresponding W_2 and C_2 matrices are:

$$W_2 = \begin{bmatrix} 0 & 0 & 0 & 0 & 0 \\ 0 & 0 & 0 & 0 & 0 \\ 0 & 0 & 0 & 0 & 0 \\ -0.7954 & -0.8931 & -0.4636 & 1.0000 & 0 \\ -0.7189 & -0.9283 & -0.0792 & 0 & 1.0000 \end{bmatrix}$$

$$C_2 = \begin{bmatrix} 1.0000 & 1.0000 & 1.0000 & 0.9518 & 0.9080 \end{bmatrix}$$

which are clearly much different than the respective W_1 and C_1. Therefore, it is expected that the inverse control procedure, described in the previous sections, will give erroneous force results failing to drive the real system accurately.

To cope with this situation, we can employ an adaptation mechanism that modifies the FCN framework's acquired knowledge based on the new data. More specifically, when the predicted by the FCN next state is far from the next state of the plant under the current driving force, the FCN knowledge depository is enhanced by incorporating new membership functions and rules that capture the actual association between the current state, the next state and the applied force. The adaptation procedure is applied on each time instant until the FCN framework is sufficiently trained to the actual system parameters.

Figures 10.26 and 10.27 show the performance of the adaptive scheme in tracking the desired trajectories $x_1(t) = \frac{\cos(2t)}{4}$ and $x_2(t) = \frac{-\sin(2t)}{2}$ respectively, starting from $x_1(0) = -\pi/4$, $x_2(0) = \pi/8$. The sharp oscillations are due to the discrepancy between the force values given by the erroneous off-line training and the desired by the actual system force.

The same oscillation is apparent in the force values of Fig. 10.28, which however are gradually taking proper values as the adaptation evolves.

Finally, Fig. 10.29 shows the tracking error that is in accordance with the observations made in the other three figures.

10.3.2 Small Hydro Electric Power Plant

Hydroelectric power is the electric power produced by converting the gravitational power of the falling water. It is one of the most widely used renewable energy sources with a relatively low cost, compared to other forms of renewable sources of electricity production. The size and capacity of hydroelectric facilities varies from the large ones, producing hundreds of Megawatts to more than 10 GW, to the pico ones with power generation less than 5 kW. Small hydro electric power plants with a

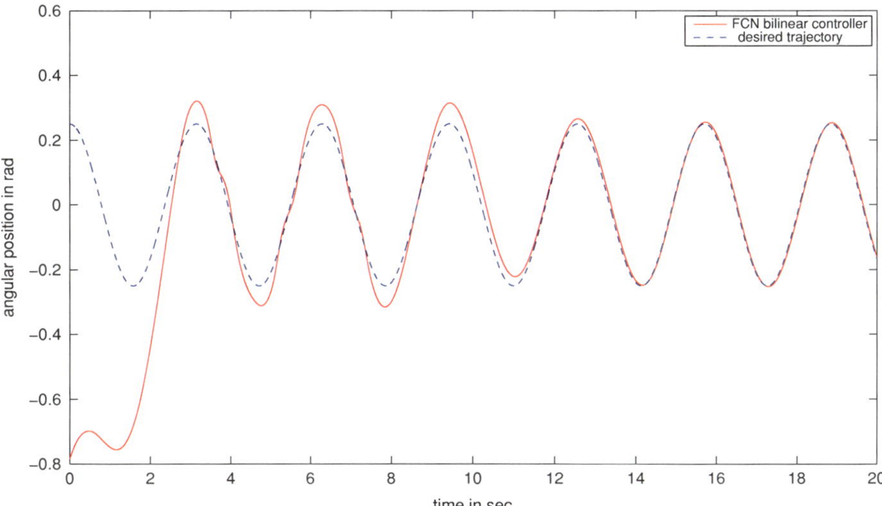

Fig. 10.26 State $x_1(t)$ of adaptive approach and its desired value $\frac{\cos(2t)}{4}$ for initial condition $x_1(0) = -\pi/4, x_2(0) = \pi/8$. The *solid line* represents the trajectory produced by the FCN while the *dashed line* the desired trajectory of the angular position

production under 10 MW are designed in a scale that can serve a small community or an industrial plant. They can be also connected to a conventional electrical distribution network (grid) offering a low-cost renewable energy part of the overall network's consumption. Under proper damps' construction they can also be used by the water management authorities as a local scale water management facility.

The study presented in this section concerns a small hydro electric power plant located in the area of Pindos, West Macedonia village Krania, Grevena, Greece, and is operated by a local company. The topology of this plant is shown in Fig. 10.30. The plant is composed of two hydro tanks in cascade (hydro tank 1 and hydro tank 2). Tank 1 receives water from a supplying river with constant water inflow rate, determined by the water management authority. Note that control valve 1 is usually open, it will close only for maintenance purposes. The inflow from tank 1 to tank 2 is controlled by the control valve 2, while the water inflow to the two power generators is controlled by valves 3 and 4 respectively. Tank 1 has an overflow channel returning overflown water to the river. Tank 2 does not have any overflow channel, therefor its water height should not exceed tank's maximum height. The power produced by the two generators is delivered to the electric grid and the desired voltage is 20 kV. However, the actual voltage delivered to the grid fluctuates, depending on the produced power and the load of the grid that is time varying. In case the voltage exceeds 21.6 kV the plant is cut off from the grid and the power generation is suspended. This cut off and suspension are not desirable because the generators require a certain amount of time ($\sim$30 min) before they reach the desired power and the procedure causes mechanical distress to the plant.

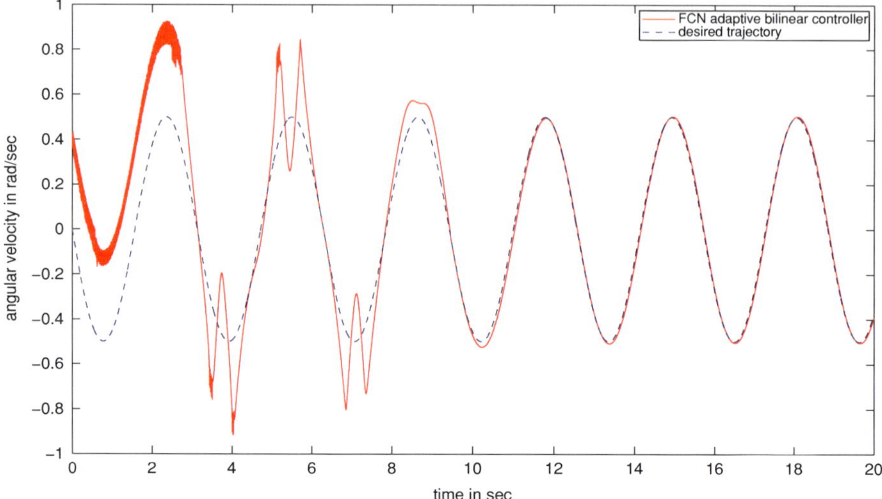

Fig. 10.27 State $x_2(t)$ of adaptive approach and its desired value $\frac{-\sin(2t)}{2}$ for initial condition $x_1(0) = -\pi/4$, $x_2(0) = \pi/8$. The *solid line* represents the trajectory produced by the FCN while the *dashed line* the desired trajectory of the angular velocity

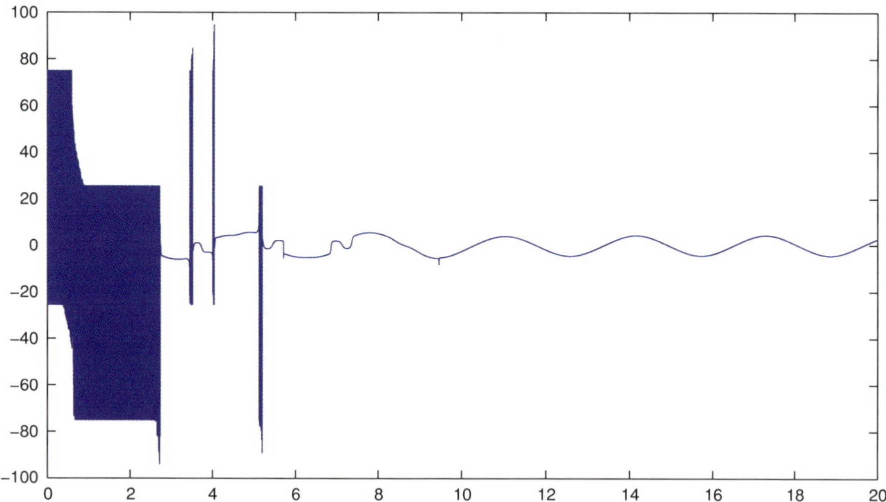

Fig. 10.28 Force F of adaptive approach applied on the cart for tracking state $x_1(t)$ in the desired value $\frac{\cos(2t)}{4}$ starting from initial condition $x_1(0) = -\pi/4$, $x_2(0) = \pi/8$

Given the water inflow rate from the river, the aim of the plant's operator is to produce the maximum amount of power, taking care that the voltage never exceeds 21.6 KV. The overall production procedure is directed using solely the empirical expertise of the engineers of the plant operator and does not rely on any control

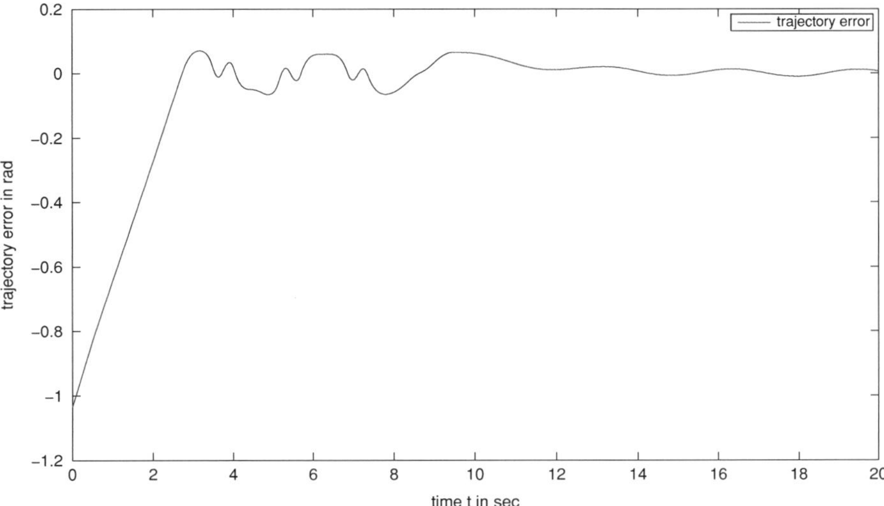

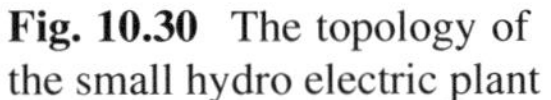

Fig. 10.29 Trajectory tracking error of adaptive approach in respect to the desired value $\frac{\cos(2t)}{4}$ for initial condition $x_1(0) = -\pi/4$, $x_2(0) = \pi/8$

Fig. 10.30 The topology of the small hydro electric plant

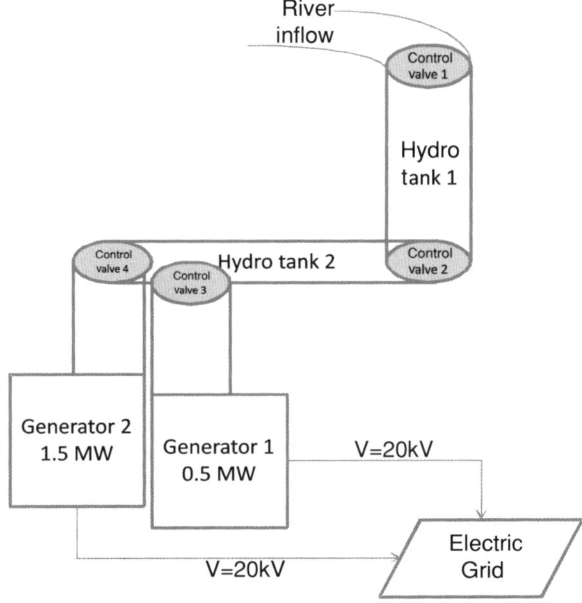

scheme derived from a strict mathematical model. Alternatively, the plant works in an "automatic" mode, where the plant tries at each time instant to increase its power production to the maximum, provided that tank 1 contains enough water. With this procedure the goals are usually met, there are however not seldom instances where the voltage specifications cannot be met, especially at night hours where the

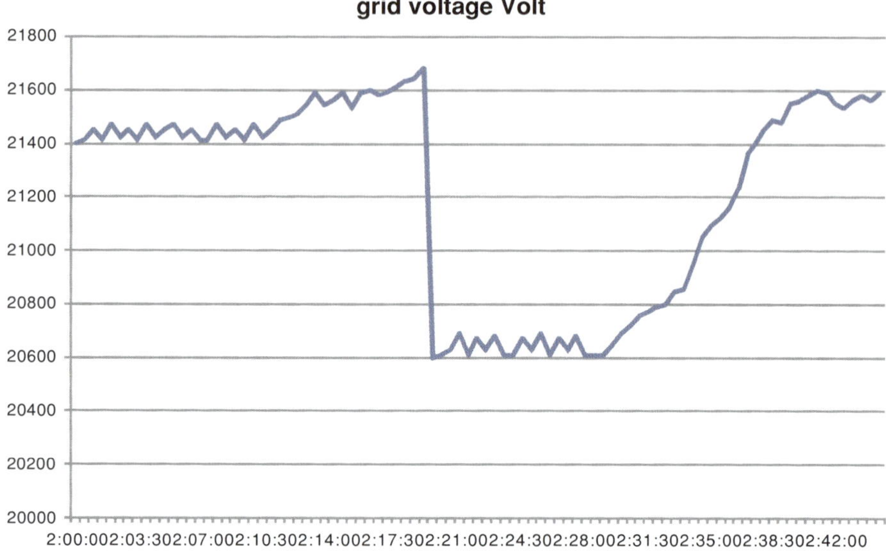

Fig. 10.31   Voltage of the grid of power plant

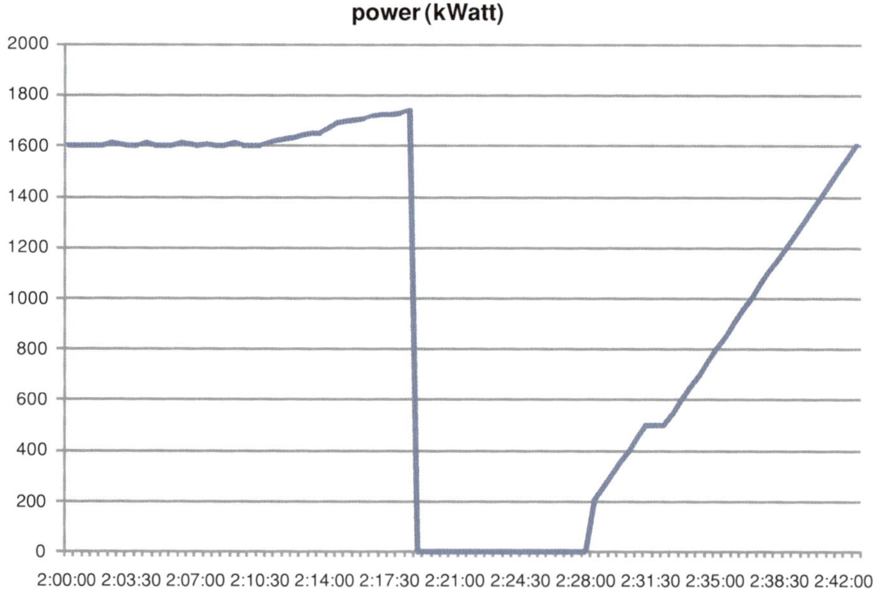

Fig. 10.32   Power of power plant

grids load is reduced. Figures 10.31 and 10.32 demonstrate such a failure instance. When the hydro plant was working at 1.6 MW there was no problem because the voltage was at about 21.5 kV. When the operator tried to achieve a power production

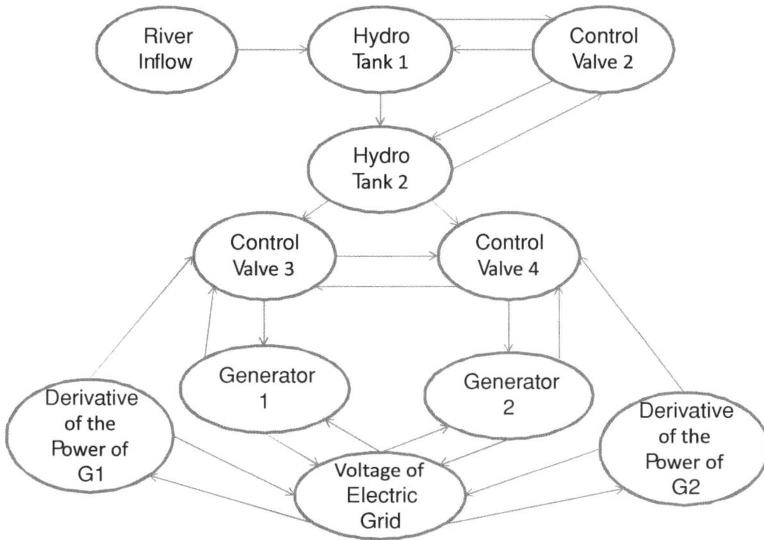

Fig. 10.33 The FCN for hydro electric power plant

of 1.8 MW then the voltage of the grid was raised to 21.7 kV with the result of cutting off the power in the hydro electric power plant.

In the following, we will design a FCN for the hydroelectric Power Plant and train it using the experts' knowledge and data from its operation. The aim is to use the trained FCN for the coordination of the power production in a way that the plant delivers its maximum power without violating the voltage specifications. This way, the FCN will manage to increase the energy delivered to the Grid, especially in the days where the Grid had problems with the value of the voltage.

10.3.2.1 FCN Design for the Hydro Electric Power Plant

The FCN designed to approximate and control the small hydro is shown in Fig. 10.33.

In this figure, the nodes of the FCN are associated with the following physical quantities:

node $C1$ is a steady state node (it can not be influenced by the other nodes) and is the value of the inflow rate coming from the river. It is associated with the water level of the River and is determined by the water management authority (Ministry of Environment, Energy and Climate Change of Greece).

node $C2$ is the water volume in Hydro Tank 1

node $C3$ is the opening of Control Valve 2

node $C4$ is the water volume in Hydro Tank 2

node $C5$ is the opening of Control Valve 3

node $C6$ is the opening of Control Valve 4

Table 10.3 Concept nodes and physical range values for the FCN

Node	Description	Range of values
$C(1)$	River inflow rate	$[0\dots1] \rightarrow [0\,\mathrm{m^3/s}\dots5\,\mathrm{m^3/s}]$
$C(2)$	Water volume in hydro tank 1	$[0\dots1] \rightarrow [0\,\mathrm{m^3}\dots2{,}000\,\mathrm{m^3}]$
$C(3)$	Control valve 2	$[0\dots1] \rightarrow [0\,\mathrm{m^3/s}\dots3\,\mathrm{m^3/s}]$
$C(4)$	Water volume in hydro tank 2	$[0\dots1] \rightarrow [0\,\mathrm{m^3}\dots5{,}000\,\mathrm{m^3}]$
$C(5)$	Control valve 3	$[0\dots1] \rightarrow [0\,\mathrm{m^3/s}\dots1\,\mathrm{m^3/s}]$
$C(6)$	Control valve 4	$[0\dots1] \rightarrow [0\,\mathrm{m^3/s}\dots2\,\mathrm{m^3/s}]$
$C(7)$	Produced power of hydro generator 1	$[0\dots1] \rightarrow [0\,\mathrm{kW}\dots500\,\mathrm{kW}]$
$C(8)$	Produced power of hydro generator 2	$[0\dots1] \rightarrow [0\,\mathrm{kW}\dots1{,}500\,\mathrm{kW}]$
$C(9)$	Power derivative of hydro generator 1	$[0\dots1] \rightarrow [-10\,\mathrm{kW/min}\dots50\,\mathrm{kW/min}]$
$C(10)$	Power derivative of hydro generator 2	$[0\dots1] \rightarrow [-30\,\mathrm{kW/min}\dots150\,\mathrm{kW/min}]$
$C(11)$	Voltage of the electric grid	$[0\dots1] \rightarrow [19.8\,\mathrm{kV}\dots21.8\,\mathrm{kV}]$

node $C7$ is the Produced Power of Hydro Generator 1
node $C8$ is the Produced Power of Hydro Generator 2
node $C9$ is the Derivative of the power of Hydro Generator 1. It captures the rate of increase (or decrease) in power production in consecutive time instants. The sampling time in this plant is taken to be 7 min
node $C10$ is the Derivative of the power of Hydro Generator 2
node $C11$ is the critical node, the Voltage of the Electric Grid, which must always be over 20 kV and less then 21.6 kV.

The description of each node and the range of the values it represents is given in Table 10.3. Note that the water volumes appearing in this table are the active quantities, i.e., the volumes that can be used. The actual volumes of the tanks are much larger. It is assumed that the nodes receive values in the interval $[0\dots1]$, that correspond to the physical range values of the concepts they represent.

The FCN weight matrix assumes the form:

$$
W =
\begin{bmatrix}
1 & 0 & 0 & 0 & 0 & 0 & 0 & 0 & 0 & 0 & 0 \\
w_{21} & 1 & w_{23} & 0 & 0 & 0 & 0 & 0 & 0 & 0 & 0 \\
0 & w_{32} & 1 & w_{34} & 0 & 0 & 0 & 0 & 0 & 0 & 0 \\
0 & w_{42} & w_{43} & 1 & 0 & 0 & 0 & 0 & 0 & 0 & 0 \\
0 & 0 & 0 & w_{54} & 1 & w_{56} & w_{57} & 0 & w_{59} & 0 & 0 \\
0 & 0 & 0 & w_{64} & w_{65} & 1 & 0 & w_{68} & 0 & w_{6,10} & 0 \\
0 & 0 & 0 & 0 & w_{75} & 0 & 1 & 0 & 0 & 0 & w_{7,11} \\
0 & 0 & 0 & 0 & 0 & w_{86} & 0 & 1 & 0 & 0 & w_{8,11} \\
0 & 0 & 0 & 0 & 0 & 0 & 0 & 0 & 1 & 0 & w_{9,11} \\
0 & 0 & 0 & 0 & 0 & 0 & 0 & 0 & 0 & 1 & w_{10,11} \\
0 & 0 & 0 & 0 & 0 & 0 & w_{11,7} & w_{11,8} & w_{11,9} & w_{11,10} & 1
\end{bmatrix}
$$

where its diagonal elements are assumed all constant and equal to 1, reflecting the fact that, in the repetitive procedure that leads to an equilibrium, the value of each node depends from its value at the previous iteration.

10.3.2.2 Training of the FCN

The initial off-line training of the FCN of Fig. 10.33 was based on the experts' opinion regarding the proper operation of the small hydro. The experts expressed their knowledge in the form of the following fuzzy linguistic rules, which are used in the everyday operation of the plant:

1. When the volume of the hydro tank 1 is high then valve 2 should be fully open
2. When the volume of the hydro tank 1 is medium–high then valve 2 should be 75 % open
3. When the volume of the hydro tank 1 is medium then valve 2 should be 50 % open
4. When the volume of the hydro tank 1 is low–medium then valve 2 should be 25 % open
5. When the volume of the hydro tank 1 is low then valve 2 should be fully closed
6. When the volume of the hydro tank 1 is zero then obviously then valve 2 should be fully closed.

The triangular membership functions that are used in these rules appear in Fig. 10.34. It can be observed that the rules are quite simple, they rely mostly on the adjustment of valve 2 and do not survey the way by which the grid corresponds to the derivative of power produced by the two generators, which however is taken into account in the design of the FCN nodes.

In order to perform the initial off-line training of the FCN, we have first to translate into respective linguistic rules involving all of the FCN nodes. In the new rules, which also have the consensus of the experts of the plant, the inflow of the river is taken into account and more emphasis is given to the adjustment of control valves 3 and 4 (nodes $C5$ and $C6$), which have smaller size than valve 1 and at the same time are mechanically adjusted. The translated rules are the following:

Rule 1:   IF node $C1$ is 1.00 THEN node $C2$ is 1.00 and node $C3$ is 1.00 and node $C4$ is 1.00 and node $C5$ is 1.00 and node $C6$ is 1.00 and node $C7$ is 1.00 and node $C8$ is 1.00 and node $C9$ is 1.00 and node $C10$ is 1.00

Rule 2:   IF node $C1$ is 0.75 THEN node $C2$ is 0.75 and node $C3$ is 1.00 and node $C4$ is 1.00 and node $C5$ is 1.00 and node $C6$ is 0.75 and node $C7$ is 1.00 and node $C8$ is 0.75 and node $C9$ is 1.00 and node $C10$ is 1.00

Rule 3:   IF node $C1$ is 0.50 THEN node $C2$ is 0.50 and node $C3$ is 1.00 and node $C4$ is 1.00 and node $C5$ is 1.00 and node $C6$ is 0.50 and node $C7$ is 1.00 and node $C8$ is 0.50 and node $C9$ is 1.00 and node $C10$ is 1.00

Rule 4:   IF node $C1$ is 0.25 THEN node $C2$ is 0.25 and node $C3$ is 1.00 and node $C4$ is 1.00 and node $C5$ is 1.00 and node $C6$ is 0.25 and node $C7$ is 1.00 and node $C8$ is 0.25 and node $C9$ is 1.00 and node $C10$ is 1.00

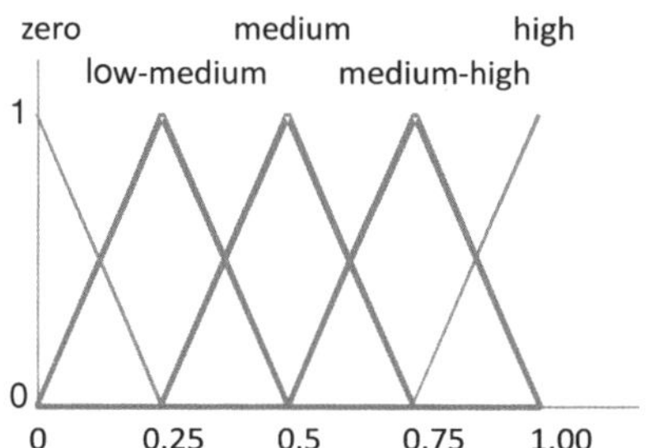

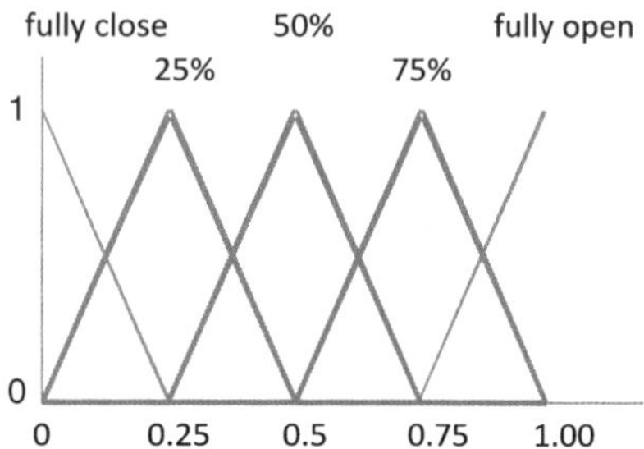

Fig. 10.34 Membership functions for nodes C1–C8

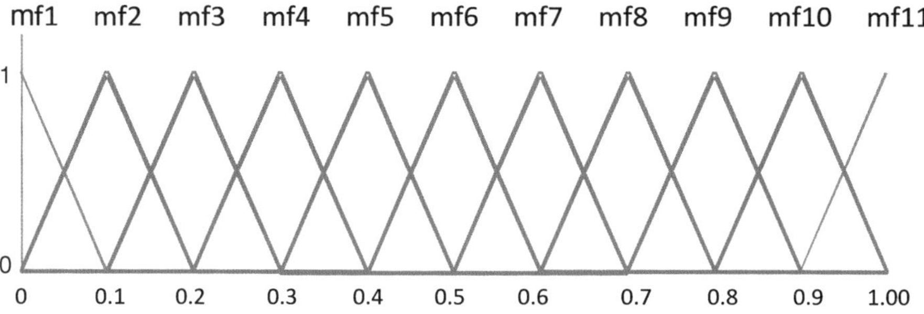

Fig. 10.35 Membership functions for nodes C9–11

Rule 5:   IF node $C1$ is 0.25 THEN node $C2$ is 0.25 and node $C3$ is 1.00 and node
$C4$ is 1.00 and node $C5$ is 1.00 and node $C6$ is 0.25 and node $C7$ is 1.00 and node
$C8$ is 0.25 and node $C9$ is 1.00 and node $C10$ is 1.00

Rule 6:   IF node $C1$ is 0.00 THEN node $C2$ is 0.00 and node $C3$ is 1.00 and node
$C4$ is 1.00 and node $C5$ is 0.00 and node $C6$ is 0.00 and node $C7$ is 0.00 and node
$C8$ is 0.00 and node $C9$ is 1.00 and node $C10$ is 1.00.

The five membership functions presented in Fig. 10.34 were found sufficient to
capture the necessary information of the values of the physical system corresponding
to the nodes 1–8 of the FCN. Nodes 9–11 require more resolution. For these nodes,
we designed 11 membership functions, as Fig. 10.35 shows.

Let's see now how these six rules are transformed into initial FCN training actions.
Take for example Rule 2:

Rule 2:   IF node $C1$ is 75 % open (0.75) and node $C2$ is medium–high (0.75) and
node $C3$ is fully open (1.00) and node $C4$ is high (1.00) and node $C5$ is fully
open (1.00) and node $C6$ is 75 % open (0.75) and node $C7$ is high power (1.00)
and node $C8$ is medium–high power (0.75) and node $C9$ is high (1.00) and node
$C10$ is high (1.00) **THEN node $C11$ is high (1.00)**,

where the value of $C11$ is given by the FCN designer as a plausible conclusion of the fact that $C9$ and $C10$ are both 1.

This FCN node values vector that correspond to this rule Rule is:

$$A = \begin{bmatrix} 0.75 & 0.75 & 1.00 & 1.00 & 1.00 & 0.75 & 1.00 & 0.75 & 1.00 & 1.00 & 1.00 \end{bmatrix}.$$

Using this vector as the desired node values, we will proceed to train the FCN, using the bilinear modeling and the respective parameter estimation algorithm of Chap. 9 to obtain the following weight matrix W and inclination parameters vector C:

$$W = \begin{bmatrix} 1.0000 & 0 & 0 & 0 & 0 & 0 & 0 & 0 & 0 & 0 & 0 \\ 0.1342 & 1.0000 & 0.2451 & 0 & 0 & 0 & 0 & 0 & 0 & 0 & 0 \\ 0 & 1.0000 & 1.0000 & 1.0000 & 0 & 0 & 0 & 0 & 0 & 0 & 0 \\ 0 & 1.0000 & 1.0000 & 1.0000 & 0 & 0 & 0 & 0 & 0 & 0 & 0 \\ 0 & 0 & 0 & 0.8078 & 1.0000 & 0.7332 & 0.9078 & 0 & 0.9078 & 0 & 0 \\ 0 & 0 & 0 & -0.0492 & 0.1548 & 1.0000 & 0 & 0.2627 & 0 & 0.1548 & 0 \\ 0 & 0 & 0 & 0 & 1.0000 & 0 & 1.0000 & 0 & 0 & 0 & 1.0000 \\ 0 & 0 & 0 & 0 & 0 & 0.2457 & 0 & 1.0000 & 0 & 0 & 0.1963 \\ 0 & 0 & 0 & 0 & 0 & 0 & 0 & 0 & 1.0000 & 0 & 1.0000 \\ 0 & 0 & 0 & 0 & 0 & 0 & 0 & 0 & 0 & 1.0000 & 1.0000 \\ 0 & 0 & 0 & 0 & 0 & 0 & 0.8699 & 0.8045 & 0.8699 & 0.8699 & 1.0000 \end{bmatrix}$$

$$C = \begin{bmatrix} 1.4637 & 1.0048 & 1.6832 & 1.6832 & 1.1107 & 0.9088 & 1.5472 & 0.9735 & 2.3208 & 2.3208 & 1.1001 \end{bmatrix}.$$

Using the fuzzy rule storage mechanism described in Sect. 10.2 the triplet A, W and C is connected with the following FCN fuzzy training rule:

IF: node $C1$ is 75 % open (0.75) and node $C2$ is medium–high (0.75) and node $C3$ is fully open (1.00) and node $C4$ is high (1.00) and node $C5$ is fully open (1.00) and node $C6$ is 75 % open (0.75) and node $C7$ is high power (1.00) and node $C8$ is medium–high power (0.75) and node $C9$ is high (1.00) and node $C10$ is high (1.00).

THEN: weight w_{21} is mf12 and w_{23} is mf13 and w_{32} is mf21 and w_{34} is mf21 and w_{42} is mf21 and w_{43} is mf21 and w_{54} is mf19 and w_{56} is mf18 and w_{57} is mf20 and w_{59} is mf20 and w_{64} is mf11 and w_{65} is mf12 and w_{68} is mf13 and $w_{6,10}$ is mf13 and w_{75} is mf21 and $w_{7,11}$ is mf21 and w_{86} is mf13 and $w_{8,11}$ is mf13 and $w_{9,11}$ is mf21 and $w_{10,11}$ is mf21 and $w_{11,7}$ is mf20 and $w_{11,8}$ is mf19 and $w_{11,9}$ is mf20 and $w_{11,10}$ is mf19.

AND: c_1 is mf3 and c_2 is mf2 and c_3 is mf3 and c_4 is mf3 and c_5 is mf2 and c_6 is mf2 and c_7 is mf3 and c_8 is mf2 and c_9 is mf5 and c_{10} is mf5 and c_{11} is mf2.

Note that the diagonal elements of the weight matrix do not participate in the rule because they considered, by design, fixed and equal to 1. The maximum allowable number of membership functions and their type for the weights and inclination parameters are shown in Figs. 10.36 and 10.37. This maximum number per weight or inclination parameter is not necessarily met in practice, it merely determines the maximum allowable resolution.

Fig. 10.36 Membership functions for weights w

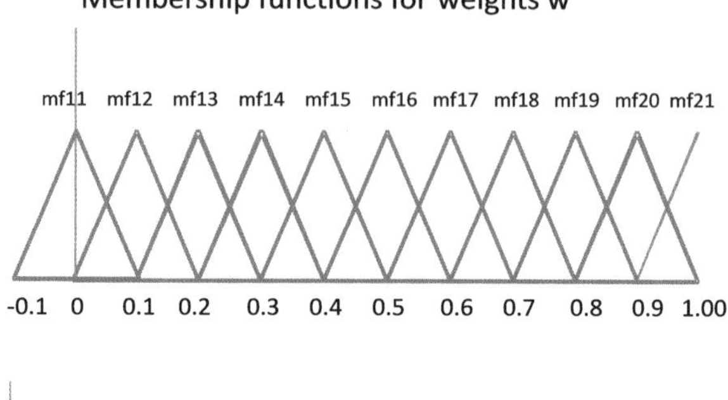

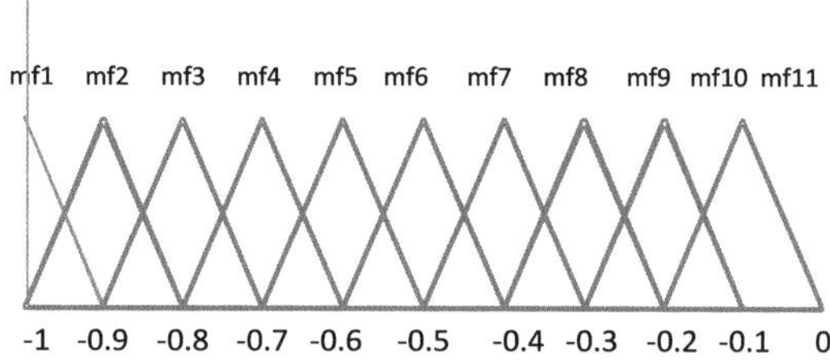

Fig. 10.37 Membership functions for coefficient c

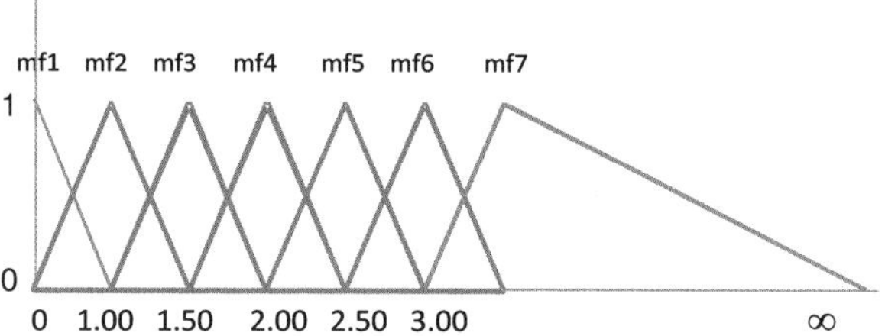

10.3.2.3 Further Training Using Operator's Intuition and Verification

Apart from the initial off-line training described above, the FCN undergoes the understanding of the optimal system operation using the human operators' intuition and data from the way the electric grid reacts to the power raising.

As an example, we saw that when the plant is to be coordinated according to Rule 2:

Rule 2:   IF node $C1$ is 75 % open (0.75) and node $C2$ is medium–high (0.75) and node $C3$ is fully open (1.00) and node $C4$ is high (1.00) and node $C5$ is fully open (1.00) and node $C6$ is 75 % open (0.75) and node $C7$ is high power (1.00) and node $C8$ is medium–high power (0.75) and node **$C9$ is high (1.00) and node $C10$ is high (1.00) THEN node $C11$ is high (1.00)**

the result would be very close to the power cutoff situation because node $C11$ will go beyond 21.6 kV.

In this case, the human operator may use his intuition and modify the knowledge by using the following rule as an improvement of previous Rule 2 :

Rule 2:   IF node $C1$ is 75 % open (0.75) and node $C2$ is medium–high (0.75) and node $C3$ is fully open (1.00) and node $C4$ is high (1.00) and node $C5$ is fully open (1.00) and node $C6$ is 75 % open (0.75) and node $C7$ is high power (1.00) and node $C8$ is medium–high power (0.75) and node **$C9$ is 0.8 and node $C10$ is 0.8 THEN node $C11$ is 0.8**

which is acceptable because node $C11$ corresponds to less than 21.5 kV. Note that the desired range values of $C11$ is [0.7 . . . 0.8]. This rule is verified by the plant's data and is accepted for training the FCN according to the following guidelines.

operators guidelines: The operators have been working the plant for over 4 months and every time they make a good prediction about how the grid reacts they press a button in the computer which means that the selected operation was good and can be translated into a rule for the FCN.

This way of operation of the hydro electric power plant and training of the FCN gave over a thousand of stored rules that capture all the historically good decisions.

10.3.2.4 Control of the Plant Using the FCN

The trained FCN is used as a decision-making consultant of the human operator according to the following procedure: First, the values of nodes $C1$–$C8$ are evaluated using real measurements from the power plant. Then, it puts in nodes $C9$, $C10$ the value 0.8 and using the FCN's fuzzy rule database it takes the weights and inclination values which will produce the FCN's equilibrium value. This equilibrium can be accepted, provided that node $C11$ is in the interval [0.7 . . . 0.8]. In this case, the other node values of the equilibrium and especially those that correspond to the opening of the valves are given as advice to the operator.

IF node $C11$ is over 0.8 then it reduces the values of nodes $C5$ and $C6$ with step 0.1 and resolves the FCN again until node $C11$ is under 0.8.

IF node $C11$ is over 0.6 and under 0.7 it increases the values of nodes $C5$ and $C6$ with step 0.1 until node $C11$ is over 0.7 but always under 0.8.

The performance of the FCN is demonstrated next, focusing mainly in its reaction after a power cut off. Figures 10.38 and 10.39 show the power production and the voltage of the plant when a power cut off occurred after midnight at 3:04:30. The solid line shows the actual values of the plant operating in its "automatic" mode, trying always to reach maximum power. It can be seen that with this kind of operation the plant is driven to a new power cut off. The dotted line shows, at each time instant, the values that would be achieved if the FCN decisions were applied. However, in this example the FCN decisions are not actually applied.

Next, the FCN was used to control the real plant acting as an advisor of the human operator. So far, for validation reasons, the FCN is not connected directly to the SCADA system that controls the plant. Instead, the human operator performs manual actions based on the FCN decisions. Figures 10.40 and 10.41 show the

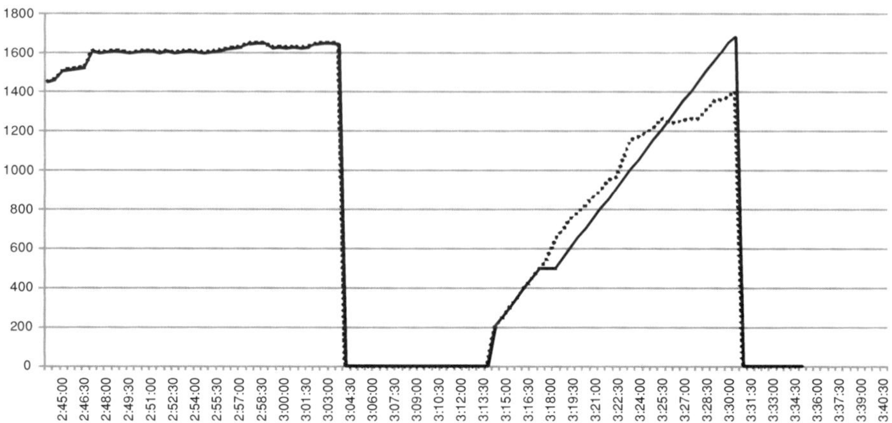

Fig. 10.38 Case study 1: power of the plant. The *straight line* shows the actual power of the plant, while the *dotted line* shows the power that would be produced if the operator used the FCN's advice

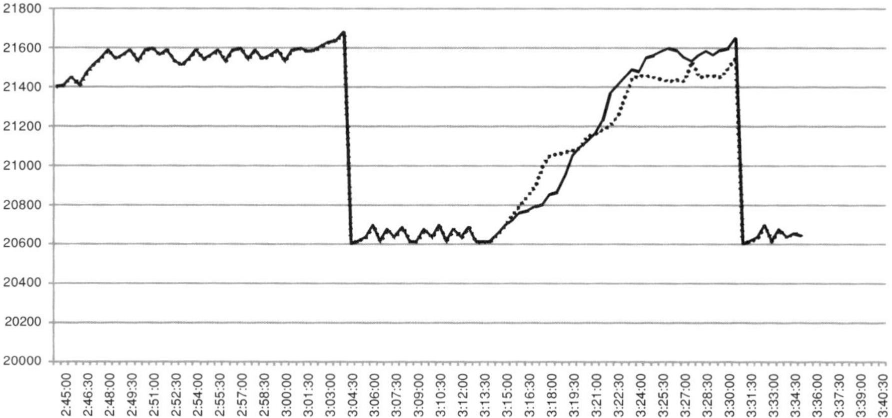

Fig. 10.39 Case study 1: grid voltage. The *straight line* shows the actual voltage, while the *dotted line* shows the voltage that would be achieved if the operator used the FCN's advice

voltage and the power achieved when the operator follows exactly the advice of the FCN. It can be seen that after the power cut off, the FCN decisions push for an immediate rapid rise of the power production, followed by a stabilizing period of a few minutes with mild power fluctuations. The same pattern appears for the next power rise up, until the plant achieves the desired maximum power. With this mode of operation, the voltage does not exceed the limit of 21.6 kV avoiding a next power cut off. It is apparent that by using the FCN's advice there is a significant power gain, especially during the night, because the plant is always on. At the same time, the plant does not face the mechanical distress due to the engagement and disengagement of the machines.

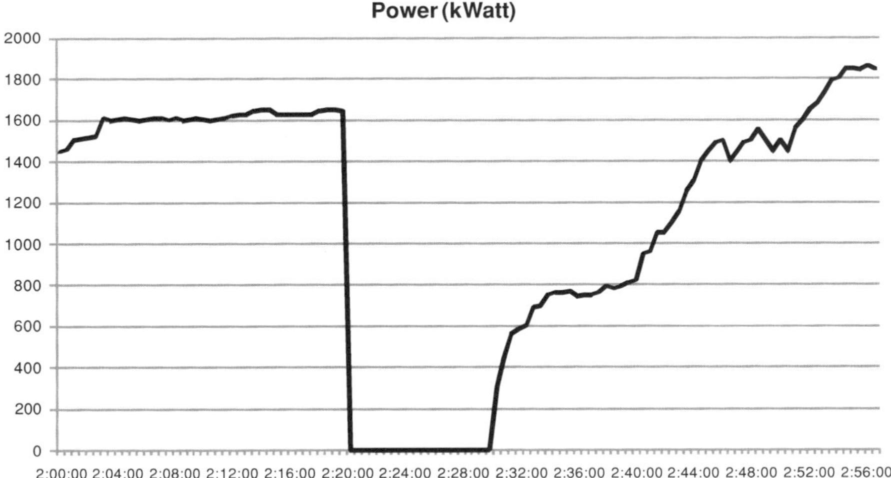

Fig. 10.40 Case study 2: grid voltage of the plant. On the start up the operator handles the plant according to the FCN instructions

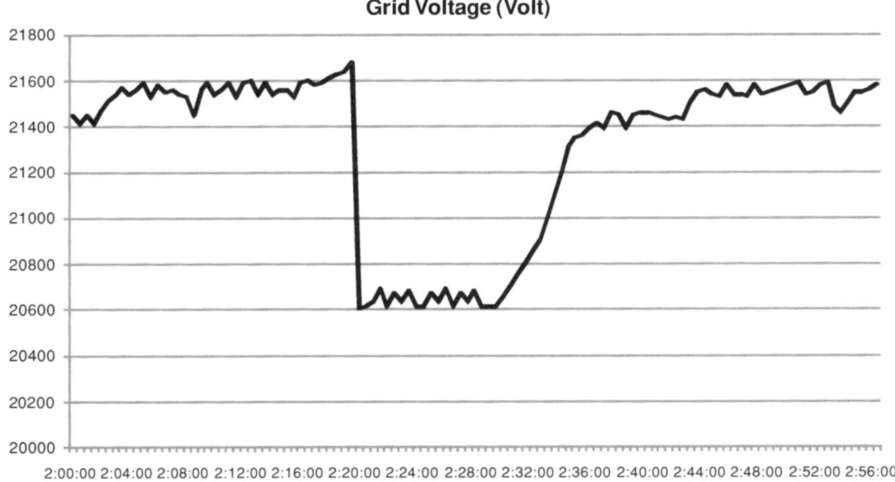

Fig. 10.41 Case study 2: power of the plant. On the start up the operator handles the plant according to the FCN instructions

10.3.3 Coordination of a Smart Grid with Renewable Power Generation Sources

"Smart grid" is the term used to characterize a utility electricity delivery system that has been modernized using computer-based remote control and automation. Such systems are beginning to be used on electricity networks, from the power plants and wind farms all the way to the consumers of electricity. They offer many benefits to utilities and consumers, mostly seen in energy efficiency improvement on the electricity grid and in the users consumed energy. A key feature of the smart grid is automation technology that lets the utility adjust and control each individual device or millions of devices in a centralized or decentralized fashion.

In order to integrate large capacity of renewable energy sources and energy storage technologies into the grid, the respective topology and control infrastructure must be determined. Instead of a topology and infrastructure, most of the reported efforts are focusing on the integration of the grid intelligence on smart noninvasive meters and on the development of Advanced Metering Infrastructure (AMI) in general. From the control point of view, this kind of distribution grid should be divided into smaller, interconnected grids, which operate in parallel. Each smaller grid contains nested subgrids, and so on. Each smaller grid or subgrid should have ability to disconnect from the remaining grid and operate in island mode. This means that it must be self-controlled. Thus, it is the control infrastructure that will transform an existing passive grid into a "smart grid".

In Stimoniaris et al. (2011), two topologies for the smart distribution grid are presented. Based on a small-scale smart grid infrastructure developed at the Technological Institute of Western Macedonia (TEIWM), Greece, the aim of this section is to present an implementation of the second topology in experimental scale and introduce the FCN framework in the control software of the experimental micro-grid, containing various weather-dependent Renewable Energy Sources (PV and wind-generator), battery bank and loads.

The aim of the FCN is to predict and perform the necessary actions for supplying different ancillary services to the grid, such as fast active power compensation, voltage regulation, and back-up supply. In this example, the FCN assumes a distributed form of interconnected FCNs operating either locally to predict the behavior of a special power control unit or at a coordination level that receives inputs from the local controllers and handles their decision making. Based on data from the experimental microgrid, the proposed scheme is simulated in Matlab showing an excellent response of the FCN to different weather and operating conditions. The FCN manages to regulate the PV-inverters power output, the battery bank current and the loads consumption effectively, thus keeping the voltage and frequency of the microgrid in acceptable levels, maintaining at the same time its optimal, in respect to cost, operation.

10.3.3.1 Topology of a Smart Grid

The first distribution system topology is shown in Fig. 10.42 (Stimoniaris et al. 2011). The smart integration module (SIM) will have the following functionalities:

1. Connection to the grid (SD)
2. Ac bus for AC loads and the existing PV systems (Distributed Generators (DG) with inverter)

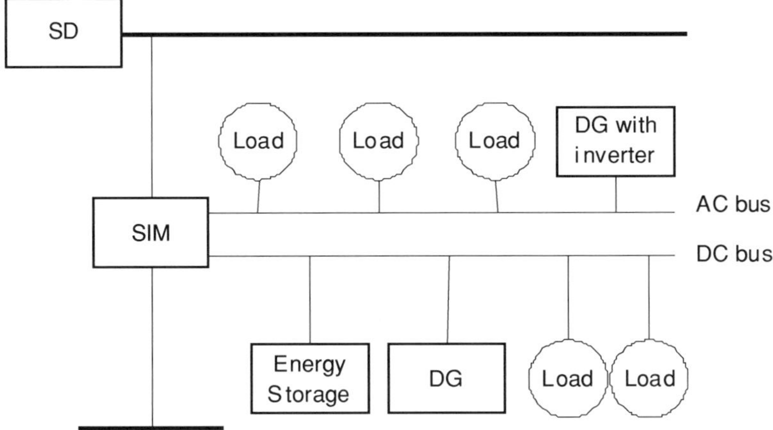

Fig. 10.42 A microgrid-based smart distribution grid topology

3. DC bus for DC loads and connection to energy storage and distributed generation by renewable energy resources (DG)
4. Voltage regulation in steady state and transients
5. Fast real and reactive power compensation
6. Load shedding possibilities.

In this distribution system, autonomous intelligent control will be conducted by all the SIMs of the grid. In order to experimentally investigate this topology, only a SIM and its respective microgrid are needed and should be operated in island or connected to the grid mode. This topology is relatively easy to be implemented in small-scale, since most of the functions of its basic module, SIM, are included in the operation of commercially available island inverters. At the pilot installation of the Technological Institute of Western Macedonia (TEIWM), Greece, the SMA Sunny Island 3324 inverter plays the role of the SIM in the topology of Fig. 10.42. This unit contains DC bus for connection to energy storage, like battery banks or hydrogen fuel cell systems and for connection to distributed generators like PV systems or small wind generators. It also contains AC bus for connection to AC loads and distributed generators, like wind generators and PV systems with their inverters. A second AC bus provides connection to the remaining robust grid. Sunny Island 3324 regulates the voltage of the microgrid, takes care of the real and reactive power compensation and provides a central load shedding possibility, in order to avoid deep discharge of the batteries. The microgrid is shown in Fig. 10.43.

Two Sunny Boy inverters are used to integrate into the system 12 PV panels of total power 2,050 Wp. Islanded operation requires careful matching of available supply, demand and storage, in order to avoid discharging of the batteries in island mode, and subsequent load shedding. A minimum capacity of 100 Ah per 1,000 Wp installed PV power capacity is necessary for stand-alone-islanded systems. Following this rule, the battery bank consists of 24 FLA batteries. Each battery was selected with

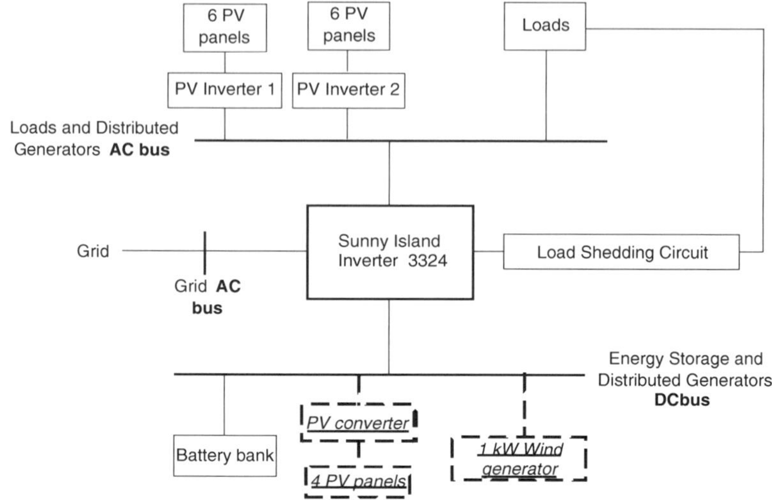

Fig. 10.43 Implementation of the smart distribution grid topology. TEIWM microgrid

300 Ah capacity and 2 V nominal voltage. They are connected in two parallel branches and each of them contains 12 batteries. Thus, a 24 V DC voltage output, which is the voltage of the DC bus of the island inverter and a total capacity of 600 Ah (300 + 300 Ah) are achieved. Considering that the installed PV power capacity is 2,050 Wp, the 600 Ah capacity are three times larger than the minimum required battery capacity for islanded operation. The available AC loads are a 90 W fridge, four compact fluorine light lamps of 144 W in total, 12 incandescent lamps of 1,400 W in total and an electric motor with a load that requires 1,000 W peak. The maximum load is 2,934 W.

10.3.3.2   Control of a Smart Grid

The second envisioned topology proposed does not include a DC-bus. Every DC load, generator or energy storage apparatus should be equipped with its DC–AC inverter and connected to the AC grid via a smart device called special control unit (SCU). SCUs are also necessary for the AC loads (or group of loads) and generators as well. The topology is shown in Fig. 10.44.

The special control units (SCUs) are categorized in load SCUs (L-SCUs), storage SCUs (S-SCUs), generators SCUs (G-SCUs) and interconnection SCUs (I-SCUs), for connecting the small grid of the AC bus to the remaining grid (SD). All SCUs will consist of a simple metering module, a communication module, an activation module (actuator) and a "smart" module which handles the decision making of each SCU and thus determines the type of each SCU (L-, S-, G-, or I-SCU).

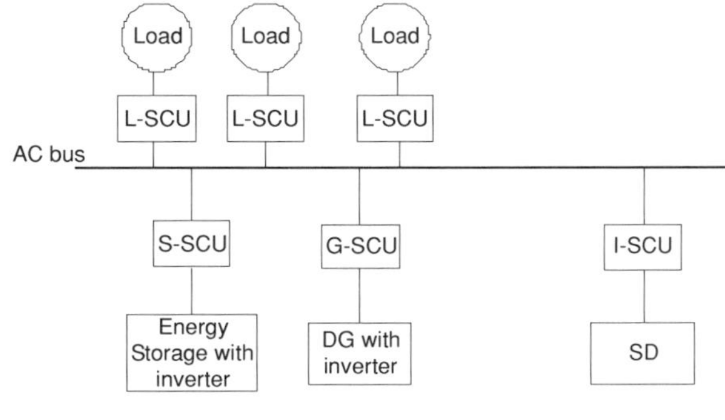

Fig. 10.44 The second smart distribution grid topology

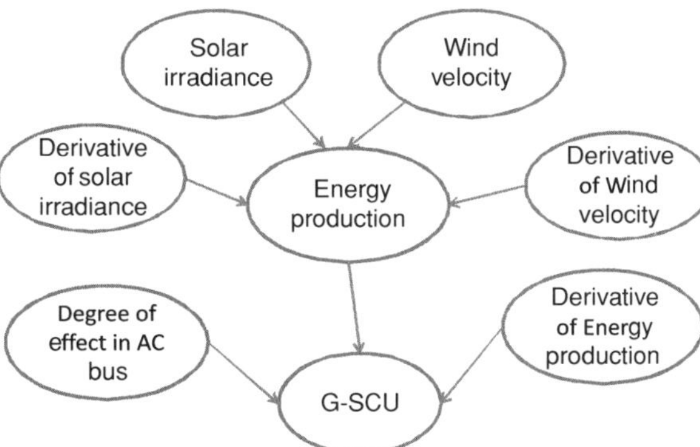

Fig. 10.45 The FCN designed for G-SCUs controls units

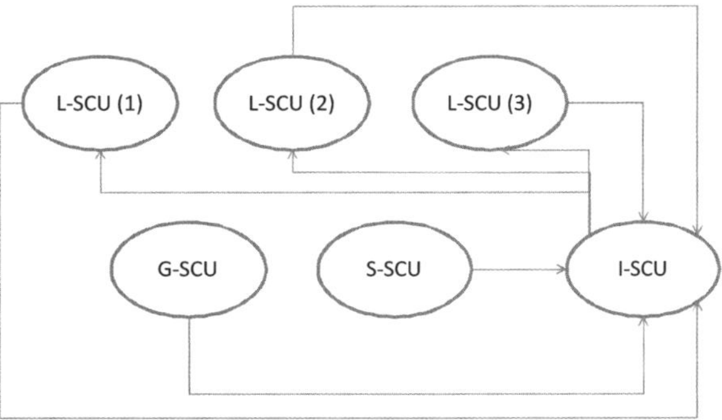

Fig. 10.46 The FCN designed for the I-SCU control unit

10.3.3.3 The FCN Designed to Coordinate the Smart Grid

For the second smart grid topology, given in Sect. 10.3.3.2, a number of interconnected FCNs have been designed according to Figs. 10.45 and 10.46.

A G-SCU FCN is used to predict the power that will be produced in the next 15 min by the renewable power resource (sun or wind) that represents. Its prediction will be based on training data. The FCN of Fig. 10.45 concerns the PV array production and takes into account meteorological data, as well as measurements of the rate of current real energy production of the generator and its contribution to the overall

Table 10.4 Concept nodes and physical range values for the FCN of the G-SCU control unit

Node	Description	Range of values
$C(1)$	The solar irradiance	$[0\ldots1] \rightarrow [0\,\text{W/m}^2 \ldots 1{,}000\,\text{W/m}^2]$
$C(2)$	The wind velocity	$[0\ldots1] \rightarrow [0\,\text{m/s} \ldots 2\,\text{m/s}]$
$C(3)$	The derivative of solar irradiance	$[0\ldots1] \rightarrow [-10\,\text{mW/m}^2/\text{s} \ldots 10\,\text{mW/m}^2/\text{s}]$
$C(4)$	The derivative of wind velocity	$[0\ldots1] \rightarrow [-0.1\,\text{m/s}^2 \ldots 0.1\,\text{m/s}^2]$
$C(5)$	The energy production	$[0\ldots1] \rightarrow [0\,\text{Wh} \ldots 360\,\text{Wh}]$
$C(6)$	The degree of effect in AC bus	$[0\ldots1] \rightarrow [0\ldots1]$
$C(7)$	The derivative of energy production	$[0\ldots1] \rightarrow [-10\,\text{Wh} \ldots 10\,\text{Wh}]$
$C(8)$	The G-SCU	$[0\ldots1] \rightarrow [0\,\text{Wh} \ldots 360\,\text{Wh}]$

production of the smart grid (Degree of effect in AC bus). Its prediction outcome will be used by the I-SCU FCN, that will be described next, which will make the final coordination of the produced energy and its allocation to the loads. The FCN nodes for the control unit G-SCU are:

node $C1$ is a steady state node representing the Solar Irradiance
node $C2$ is a steady state node representing the Wind Velocity
node $C3$ is a steady state node representing the derivative of Solar Irradiance
node $C4$ is a steady state node representing the derivative of Wind Velocity
node $C5$ corresponds to the predicted Energy Production
node $C6$ is a steady state node representing the Degree of effect in AC bus
node $C7$ is a steady state node representing the derivative of Energy Production
node $C8$ is the G-SCU that gives the final prediction of energy production taking into account its degree of effect and the derivative of the actual produced energy.

The description of each node and the range of the values it represents is given in Table 10.4. It is assumed that the nodes receive values in the interval $[0\ldots1]$ that correspond to the physical range values of the concepts they represent. Note that values of solar irradiance or wind velocities that may be above the specified limits are truncated to their maximum values.

The weight matrix for the FCN of the G-SCU control unit is:

$$W_{\text{gscu}} = \begin{bmatrix} 1 & 0 & 0 & 0 & 0 & 0 & 0 & 0 \\ 0 & 1 & 0 & 0 & 0 & 0 & 0 & 0 \\ 0 & 0 & 1 & 0 & 0 & 0 & 0 & 0 \\ 0 & 0 & 0 & 1 & 0 & 0 & 0 & 0 \\ w_{51} & w_{52} & w_{53} & w_{54} & 1 & 0 & 0 & 0 \\ 0 & 0 & 0 & 0 & 0 & 1 & 0 & 0 \\ 0 & 0 & 0 & 0 & 0 & 0 & 1 & 0 \\ 0 & 0 & 0 & 0 & w_{85} & w_{86} & w_{87} & 1 \end{bmatrix}.$$

A I-SCU FCN is used to coordinate the power production and its allocation to the loads, taking into account the predictions of G-SCU FCNs and the demands of

Table 10.5 Concept nodes and physical range values for the FCN of the I-SCU control unit

Node	Description	Range of values
$C(1)$	The L-SCU(1)	$[0\dots1] \rightarrow [0\,W\dots534\,W]$
$C(2)$	The L-SCU(2)	$[0\dots1] \rightarrow [0\,W\dots1,400\,W]$
$C(3)$	The L-SCU(3)	$[0\dots1] \rightarrow [0\,W\dots1,000\,W]$
$C(4)$	The G-SCU	$[0\dots1] \rightarrow [0\,Wh\dots360\,Wh]$
$C(5)$	The S-SCU	$[0\dots1] \rightarrow [0\,Wh\dots14,400\,Wh]$
$C(6)$	The I-SCU	$[0\dots1] \rightarrow [0\,W\dots5,000\,W]$

the loads. The I-SCU handles the power that can be received by the external grid as well as the power coming from the batteries and PV production. Wind generators are not considered in this example, although they could equally participate through their respective G-SCU. The prediction of the G-SCU FCNs is important because the I-SCU will coordinate the use of its resources according to this prediction, the load demands and available successful coordination paradigms taken from human experts. The FCN nodes for the control unit I-SCU are:

node $C1$ represents the Critical loads(1)
node $C2$ represents the Critical loads(2)
node $C3$ represents the Critical loads(3)
node $C4$ represents the G-SCU
node $C5$ represents the S-SCU
node $C6$ represents the I-SCU,

where the term critical load determines the load that, according to the consumer, should be served anyway. It is not always constant, it is time-dependent and may be reduced if the consumer is persuaded to lower his demands. The description of each node and the range of the values it represents is given in Table 10.5. It is assumed that the nodes receive values in the interval $[0\dots1]$, that correspond to the physical range values of the concepts they represent. Note that the three load groups present different critical values because are loads from different consumer groups. For example, home consumers have different critical consumption demands than industrial or public service consumers. Note also that the value of the I-SCU node determines numerically the power that is imported to the microgrid from external resources (the main grid or non renewable sources), which should be kept low for optimal operation. It is supposed that the I-SCU distributes the internal power production to the loads and then seeks for supplementary power from external sources.

The weight matrix for the FCN of the I-SCU control unit is:

$$
W_{\text{iscu}} =
\begin{bmatrix}
w_{11} & 0 & 0 & 0 & 0 & w_{16} \\
0 & w_{22} & 0 & 0 & 0 & w_{26} \\
0 & 0 & w_{33} & 0 & 0 & w_{36} \\
0 & 0 & 0 & 1 & 0 & 0 \\
0 & 0 & 0 & 0 & w_{55} & 0 \\
w_{61} & w_{62} & w_{63} & 0 & 0 & w_{66}
\end{bmatrix}.
$$

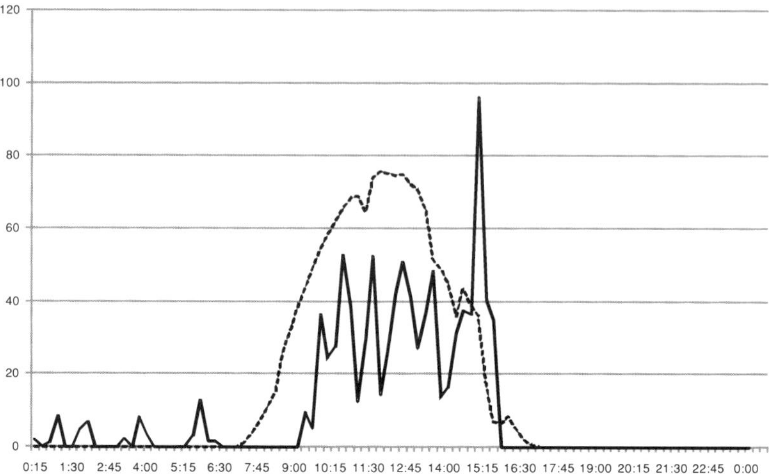

Fig. 10.47 Meteorological data for FCN G-SCU, where *solid line* is the wind speed in m/min and *dotted line* is the solar irradiation in W/(10 m^2)

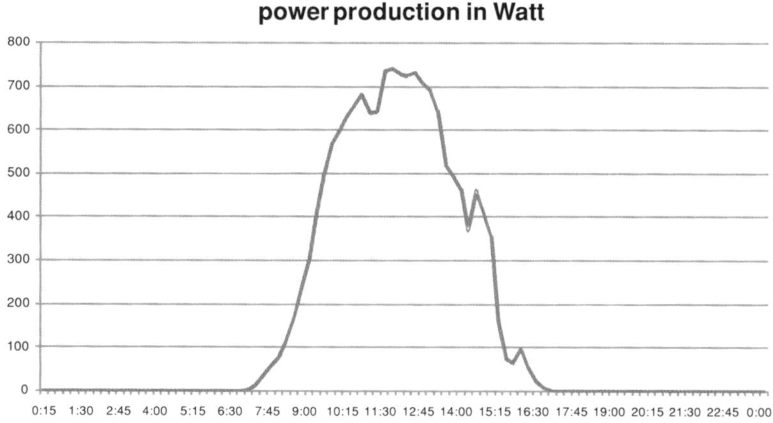

Fig. 10.48 Power production of the PV in Watt

10.3.3.4 Training of the G-SCU FCN

As has already been explained, the G-SCU FCN will be used to provide the prediction of the power production of the PV in the next 15 min. To this end, it has first to be trained, off-line initially and on-line in the sequel during the operation of the overall scheme. Meteorological data and the respective power production of the PV system within the next 15 min are given to the FCN in order to learn the response of the real plant to different meteorological conditions. Figures 10.47 and 10.48 demonstrate 24 h meteorological data and the respective power production. Note that the units of

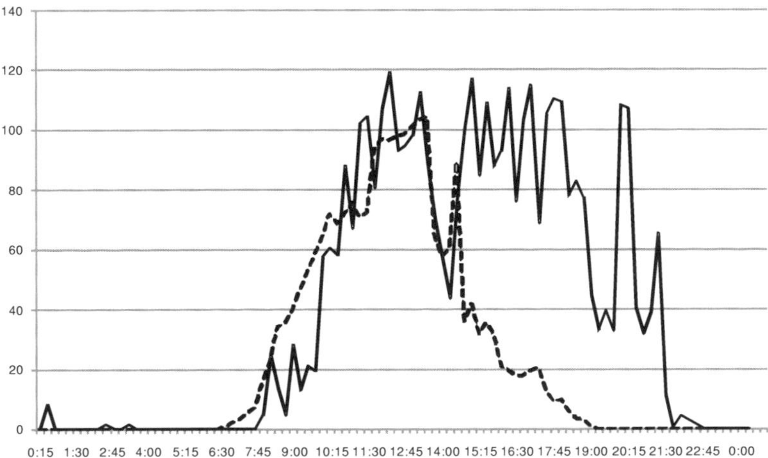

Fig. 10.49 Meteorological data for FCN G-SCU, where *solid line* is the wind speed in m/min and *dotted line* is the solar irradiation in $W/(10\,m^2)$

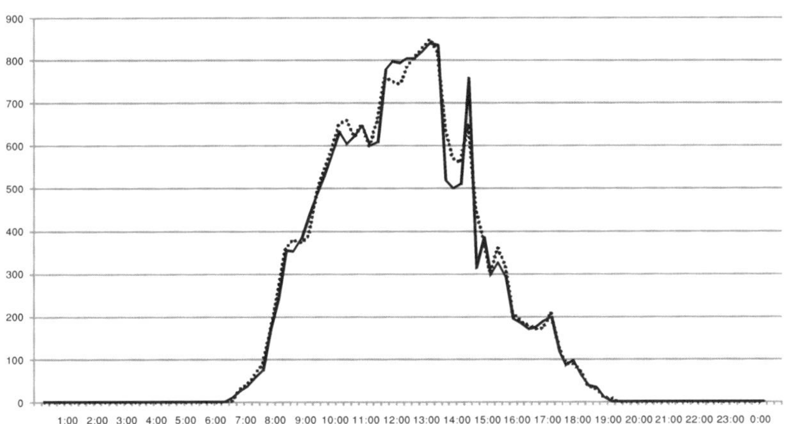

Fig. 10.50 Power prediction given by FCN, where *solid line* is the actual power arrived from the PVs of G-SCU and *dotted line* is the power predicted by FCN in Watt

solar irradiance and wind speed have been slightly modified so that both curves fit to the figure.

Trained from these and many other meteorological data, the FCN provides the predicted power of the PVs. For example, for the example day given in Fig. 10.49, the predicted power which is calculated from the FCN is depicted in Fig. 10.50.

Figure 10.50 demonstrates the predicted power for the next 15 min of the time we calculate. The FCN G-SCU was trained during the operation of the smart grid in TEIWM with training sequences of the form

IF Solar irradiance in time t is mf_a and wind velocity in time t is mf_b and derivative of solar irradiance in time t is mf_c and Derivative of Wind Velocity in time t is mf_d

THEN Energy production in time $(t + Dt)$ is mf_e. Note that Energy production is the energy that the PVs are producing after 15 min since the meteorological data were invoked, that is Dt is equal to 15 min.

For this rule, FCN G-SCU will produce a weight matrix which is:

$$W_{\text{gscu}_1} = \begin{bmatrix} 1 & 0 & 0 & 0 & 0 \\ 0 & 1 & 0 & 0 & 0 \\ 0 & 0 & 1 & 0 & 0 \\ 0 & 0 & 0 & 1 & 0 \\ w_{51} & w_{52} & w_{53} & w_{54} & w_{55} \end{bmatrix}.$$

Note that only the first upper 5×5 submatrix of W_{gscu} is shown. In the next step the FCN further calculates the weights matrices for the next FCN GSCU submatrix according to the next rule.

IF Energy production prediction in time $(t + Dt)$ is mf_e and Degree of effect in time t is mf_A and Derivative of Energy Production is mf_B **THEN** G-SCU in time $(t + Dt)$ is mf_C.

Such a rule will produce the lower right submatrix of W_{gscu} having the form:

$$W_{\text{gscu}_2} = \begin{bmatrix} 1 & 0 & 0 & 0 \\ 0 & 1 & 0 & 0 \\ 0 & 0 & 1 & 0 \\ w_{85} & w_{86} & w_{87} & w_{88} \end{bmatrix}.$$

A more concrete example of the above procedure, employing numerical values, is given as follows:

IF Solar irradiance is equal to 600 W/m^2 (values of node is 0.6—with mf7) and Wind velocity is 28 m/min (value of node is 0.2—with mf3) and derivative of Solar irradiance is equal to -5 mW/m^2/s (value of node is 0.3—with mf4) and Derivative of Wind Velocity is -0.4 m/s^2 (value of node is 0.3—with mf4) **THEN** Energy production is 162 Wh (value of node is 0.6—with mf7).

Also:

IF Energy Production is 162 Wh (value of node is 0.6—with mf7) and Degree of effect in AC bus is 0.5 and Derivative of Energy production is -2 Wh (value of node is 0.4—with mf5) **THEN** G-SCU is 0.3

For this example the following W and C overall FCN matrices are produced.

$$W = \begin{bmatrix} 1.0000 & 0 & 0 & 0 & 0 & 0 & 0 & 0 \\ 0 & 1.0000 & 0 & 0 & 0 & 0 & 0 & 0 \\ 0 & 0 & 1.0000 & 0 & 0 & 0 & 0 & 0 \\ 0 & 0 & 0 & 1.0000 & 0 & 0 & 0 & 0 \\ 0.2298 & 0.2505 & 0.2453 & 0.2453 & 1.0000 & 0 & 0 & 0 \\ 0 & 0 & 0 & 0 & 0 & 1.0000 & 0 & 0 \\ 0 & 0 & 0 & 0 & 0 & 0 & 1.0000 & 0 \\ 0 & 0 & 0 & 0 & -0.5548 & -0.3914 & -0.2280 & -0.0173 \end{bmatrix}$$

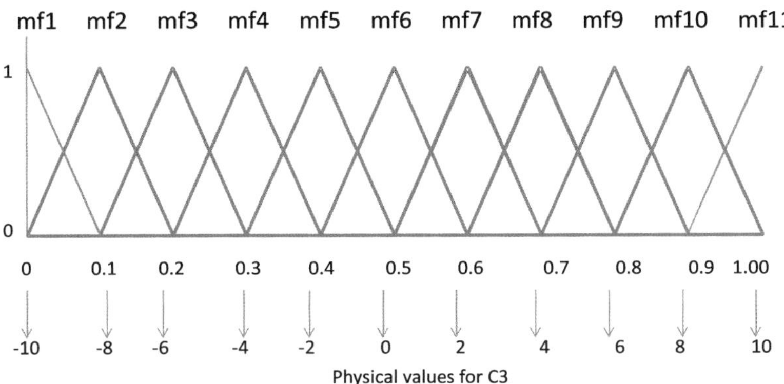

Fig. 10.51 The memberships function for nodes of FCN GSCU. The example of node $C3$ of GSCU is given

$$C = \begin{bmatrix} 1.0000 & 1.0000 & 1.0000 & 1.0000 & 0.8257 & 1.0000 & 1.0000 & 1.3549 \end{bmatrix}.$$

Using the fuzzy rule storage mechanism described in Sect. 10.2 the triplet A, W and C, with A denoting the vector of node values, is connected with the following FCN fuzzy training rules:

IF: node $C1$ is 0.6 and node $C2$ is 0.2 and node $C3$ is 0.3 and node $C4$ is 0.3
THEN: weight w_{51} is 0.2298 (mf13) and w_{52} is 0.2505 (mf14) and w_{53} is 0.2453 (mf13) and w_{54} is 0.2453 (mf13) and w_{55} is 1.0000 (mf21)
AND: c_5 is mf2

Also:

IF: node $C5$ is 0.6 and node $C6$ is 0.5 and node $C7$ is 0.3
THEN: weight w_{85} is -0.5548 (mf4) and w_{86} is -0.3914 (mf7) and w_{87} is -0.2280 (mf9) and w_{88} is -0.0173 (mf11)
AND: c_8 is mf3.

Note also that mf for each value of concept C is calculated according to Fig. 10.51, while for each weight value w_{ij} and for each inclination coefficient c_i is calculated according to Figs. 10.36 and 10.37 respectively.

Following the above training procedure, the FCN was trained in detail for more than 6 months under various meteorological conditions to produce more than 3.000 rules. Figure 10.50 demonstrate the performance of the trained FCN for the meteorological data of Fig. 10.49.

10.3.3.5 Training of I-SCU FCN

The training of the FCN is made by using successful coordination decisions, taken by human operators. With the objective to meet voltage specifications and to achieve optimal, in respect to cost and critical loads satisfaction, operation, some rule of thumbs have been found very useful. The lowering of the consumption of nonrenewable power sources, usually bought from the external grid, is a key objective. One such a rule of thumb is that, one should try to keep the power from non renewable or external sources used from the I-SCU lower than the 25 % of predicted G-SCU, while S-SCU should be over 50 % of the value of G-SCU.

Apart from such rules of thumb, other successful decisions that are taken from expert grid operators, arrived from personal instant intuition, are gathered and used for the training of the FCN. Different operators, working in different shifts, produce successful decisions that complement other operators' decisions. With this training procedure, over 2,550 successful coordination paradigms have been gathered and used to train the I-SCU FCN. In order to facilitate the optimal negotiation of the I-SCU with the consumer demands the paradigms are directed to two different fuzzy rule databases according to the procedure described by the following example:

Given that the G-SCU prediction is 0.4 and taking into account that the rule of thumb requires the S-SCU to be 0.6 (50 % over G-SCU) and the I-SCU to be 0.1 (75 % lower then G-SCU)
THEN for the load nodes L-SCU (1) is 0.7 and L-SCU (2) is 0.4 and L-SCU (3) is 0.8

For this training data the FCN produces the next matrices:

$$A_{\mathrm{iscu}_1} = \begin{bmatrix} 0.7 & 0.4 & 0.8 & 0.4 & 0.6 & 0.1 \end{bmatrix}$$

$$W_{\mathrm{iscu}_1} = \begin{bmatrix} 1.0000 & 0 & 0 & 0 & 0 & 0.9781 \\ 0 & -1.0000 & 0 & 0 & 0 & 0.1583 \\ 0 & 0 & 1.0000 & 0 & 0 & 1.0000 \\ 0 & 0 & 0 & 1.0000 & 0 & 0 \\ 0 & 0 & 0 & 0 & 1.0000 & 0 \\ -1.0000 & -0.8723 & -1.0000 & 0 & 0 & 0.5569 \end{bmatrix}$$

$$C_{\mathrm{iscu}_1} = \begin{bmatrix} 1.0620 & 1.0554 & 1.5403 & 1.0000 & 1.0000 & 1.2253 \end{bmatrix}.$$

Using the fuzzy rule storage mechanism described in Sect. 10.2 the triplet A, W and C is connected with a training rule. However, this rule is stored in two forms, denoted below as $rule_1$ and $rule_2$, carrying different degree of detail in its IF part. In the first database the rule is stored as:

IF $rule_1$:	node $C4$ is 0.4 and $C5$ is 0.6

THEN: weight w_{11} is 1.0000 and w_{16} is 0.9781 and w_{22} is -1.0000 and w_{26} is 0.1583 and w_{33} is 1.0000 and w_{36} is 1.0000 and w_{61} is -1.0000 and w_{62} is -0.8723 and w_{63} is -1.0000 and w_{66} is 0.5569

AND: c_1 is 1.0620 and c_2 is 1.0554 c_3 is 1.5403 and c_6 is 1.2253

while in the second database it is stored in its detailed form:

IF $rule_2$: node $C1$ is 0.7 and node $C2$ is 0.4 and node $C3$ is 0.8 and node $C4$ is 0.4 and $C5$ is 0.6

THEN: weight w_{11} is 1.0000 and w_{16} is 0.9781 and w_{22} is -1.0000 and w_{26} is 0.1583 and w_{33} is 1.0000 and w_{36} is 1.0000 and w_{61} is -1.0000 and w_{62} is -0.8723 and w_{63} is -1.0000 and w_{66} is 0.5569

AND: c_1 is 1.0620 and c_2 is 1.0554 c_3 is 1.5403 and c_6 is 1.2253

The way the 2 databases are used is explained in the following two case examples.

Case 1:

Assume that the load demands correspond to the following node values: L-SCU (1) $= 0.9$, L-SCU (2) $= 0.6$ and L-SCU (3) $= 0.9$. Given that the G-SCU prediction is 0.4 and the required by the rule of thumb S-SCU is 0.6 (50 % over G-SCU), the execution of the rule from the first database (IF $rule_1$) derives the matrices W_{iscu_1} and C_{iscu_1}. With these matrices the FCN concludes to the A vector which is the A_{iscu_1} above. The three first elements of A_{iscu_1} are actually the proposals of the FCN to the three loads. In case the loads accept the proposal the interconnection (I-SCU) and storage (S-SCU) control units will be instructed to run according to the FCN proposal for nodes $C5$ and $C6$, which are 0.1 and 0.6 respectively.

If the loads do not accept the proposal, the FCN will try to find a better solution digging into the second database where the experts decisions have been stored in a more detailed form. Taking into account that the second database (IF $rule_2$ storage form) has been enriched by many successful decisions, then we start inquiring the database by using modified values (with a 0.1 step) of S-SCU until we find a recorded successful operation that can serve the loads. For example the following IF part:

IF L-SCU (1) is 0.9 and L-SCU (2) is 0.6 and L-SCU (3) is 0.85 and G-SCU is 0.4 and S-SCU is 0.6 will result to the following matrices W_{iscu_2} and C_{iscu_2}

$$
W_{\text{iscu}_2} = \begin{bmatrix}
1.0000 & 0 & 0 & 0 & 0 & 1.0000 \\
0 & 0.4725 & 0 & 0 & 0 & 0.7242 \\
0 & 0 & 1.0000 & 0 & 0 & 1.0000 \\
0 & 0 & 0 & 1.0000 & 0 & 0 \\
0 & 0 & 0 & 0 & 1.0000 & 0 \\
-0.8967 & -0.2978 & -0.7968 & 0 & 0 & 0.6007
\end{bmatrix}
$$

$$
C_{\text{iscu}_2} = \begin{bmatrix} 1.9975 & 0.9466 & 1.6520 & 1.0000 & 1.0000 & 0.8985 \end{bmatrix}.
$$

Using these weight matrices, the FCN will conclude to the following A vector:

$$
A_{\text{iscu}_2} = \begin{bmatrix} 0.9 & 0.6 & 0.85 & 0.4 & 0.8 & 0.2 \end{bmatrix}.
$$

It can be seen that the load demands are now satisfied by deviating from the rule of thumb and permitting the S-SCU to be 0.8 and the I-SCU to be 0.2. That is, the grid will accept more external power (the cost is increased) and the batteries will jeopardize a less safe but still acceptable, from previous experience, operation.

Case 2:

Suppose that the consumers ask for L-SCU $(1) = 0.9$, L-SCU $(2) = 0.7$ and L-SCU $(3) = 0.9$. Assume also that the FCN G-SCU predicts a value for G-SCU node that is 0.2 and the S-SCU node is 0.4. Inquiring the first database by the following IF $rule_1$:

IF $rule_1$: node $C4$ is 0.2 and node $C5$ is 0.4

We receive the FCN matrices W_{case2} and C_{case2}

$$
W_{\text{case2}} =
\begin{bmatrix}
0.9967 & 0 & 0 & 0 & 0 & 0.3696 \\
0 & -0.9967 & 0 & 0 & 0 & -0.9967 \\
0 & 0 & -0.9975 & 0 & 0 & -0.6515 \\
0 & 0 & 0 & 1.0000 & 0 & 0 \\
0 & 0 & 0 & 0 & 1.0000 & 0 \\
-0.5755 & -0.9908 & -0.8524 & 0 & 0 & 0.4196
\end{bmatrix}
$$

$$
C_{\text{case2}} = \begin{bmatrix} 0.4557 & 1.0252 & 0.5624 & 1.0000 & 1.0000 & 3.0592 \end{bmatrix}.
$$

With these matrices the FCN will conclude to the following A vector

$$
A_{\text{case2}} = \begin{bmatrix} 0.6 & 0.3 & 0.4 & 0.2 & 0.4 & 0.05 \end{bmatrix},
$$

which is exact in respect to G-SCU and S-SCU but far from the load demands.

Let us assume that L-SCU(3) is compromised with the above proposal, but loads L-SCU (1) and L-SCU (2) are considered very critical and should be equal to their initial demand with values 0.9 and 0.7, respectively. Then, by repetitively modifying (with a 0.1 step) the S-SCU (C5) we are digging into the second database to find a previously encountered successful operation that would match the load demands. Another restriction imposed in this example is that the batteries are not allowed to give more than 50 % of their energy. The following IF $rule_2$ part

IF $rule_2$: L-SCU (1) is 0.9 and L-SCU (2) is 0.7 and L-SCU (3) is 0.4 and G-SCU is 0.2 and S-SCU is 0.5

delivers the FCN matrices W_{case2_n} and C_{case2_n}.

$$
W_{\text{case2}_n} =
\begin{bmatrix}
0.9987 & 0 & 0 & 0 & 0 & 0.9987 \\
0 & 0.9989 & 0 & 0 & 0 & 0.9989 \\
0 & 0 & -0.9989 & 0 & 0 & -0.9989 \\
0 & 0 & 0 & 1.0000 & 0 & 0 \\
0 & 0 & 0 & 0 & 1.0000 & 0 \\
-0.4360 & 0.0510 & 0.7815 & 0 & 0 & 0.4895
\end{bmatrix}
$$

$$C_{\text{case2}_n} = \begin{bmatrix} 1.3746 & 0.6065 & 0.3610 & 1.0000 & 1.0000 & 1.6240 \end{bmatrix}.$$

With these matrices the FCN concludes to a new vector A which is equal to:

$$A_{\text{case2}_n} = \begin{bmatrix} 0.9 & 0.7 & 0.4 & 0.2 & 0.5 & 0.6 \end{bmatrix}.$$

It can be observed that the new equilibrium condition satisfies the new load demands by requiring an increased power consumption from external sources (C6 = 0.6), thus increasing the cost of the microgrid's operation.

The two test cases, presented above, were based on the trained I-SCU FCN using data from successful operational decisions applied on the experimental microgrid setup of TEIWM. They demonstrate that the FCN can give proper decisions that will marshal the pressing consumers' demands with the optimal, in respect to cost and hardware safety, operation of the microgrid, rendering its behavior really "smart".

10.4 Summary

The complete framework of operation of FCN was introduced in this chapter and a number of potential applications that rely on this framework was presented. The FCN framework operates in close interaction with the system that models; it receives data of operation and uses the algorithms developed in Chap. 9 to estimate the corresponding FCN parameters (weights and inclination coefficients). Moreover, the framework uses a fuzzy rule-based mechanism to store the information acquired from distinct system operation conditions. This way, the entire framework may act as a fuzzy rule database generation engine, that creates rules capturing intrinsic system node relationships as they are recoded through the FCN graph.

Next, a number of selected applications were presented, demonstrating the applicability of the framework in traditional control problems or in decision making that supports the operation of selected plants. More specifically, an inverse control approach was presented that was applied on a traditional Brunovsky type benchmark system. However, the same inverse control approach can be applied on any affine in the control system that has been appropriately modeled by the FCN framework. The next application concerned the optimal operation of a real small hydro electric plant and demonstrated the use of the FCN as a decision-making supportive tool that learns from the successful actions of the human operators. Finally, the last application concerned the coordination of a small-scale smart grid facility powered mostly by renewable sources, where many FCN work in cooperation, either in local subgrid tasks or in a coordination level to predict the future power generation and coordinate the power and loads distribution.

References

Boutalis, Y., Kottas, T., & Christodoulou, M. (2009). Adaptive estimation of fuzzy cognitive maps with proven stability and parameter convergence. *IEEE Transactions on Fuzzy Systems, 17*, 874–889.

Karlis, A., Kottas, T., & Boutalis, Y. (2007). A novel maximum power point tracking method for pv systems using fuzzy cognitive networks (fcn). *Electric Power System Research, 77*, 315–327.

Kottas, T., Boutalis, Y., Diamantis, V., Kosmidou, O., & Aivasidis, A. (2006a). A fuzzy cognitive network based control scheme for an anaerobic digestion process. In: *Proceedings of the 14th Mediterranean Conference on Control and Applications, Session TMS Process Control 1, Ancona, Italy* (p. 7), June 28–30, 2006.

Kottas, T., Boutalis, Y., & Karlis, A. (2006b). A new maximum power point tracker for pv arrays using fuzzy controller in close cooperation with fuzzy cognitive networks. *IEEE Transactions on Energy Conversion, 21*, 793–803.

Kottas, T., Boutalis, Y., & Christodoulou, M. (2007). Fuzzy cognitive networks: A general framework. *Inteligent Desicion Technologies, 1*, 183–196.

Kottas, T. L., Karlis, A. D., & Boutalis, Y. S. (2010). *Fuzzy cognitive networks for maximum power point tracking in photovoltaic arrays* (vol. 247, pp. 231–257). New York: Springer.

Passino, K., & Yurkovich, S. (1998). *Fuzzy control*. Reading: Addison Wesley.

Stimoniaris, D., Tsiamitros, D., Kottas, T., Asimopoulos, N., & Dialynas, E. (2011). Smart grid simulation using small-scale pilot installations-experimental investigation of a centrally-controlled microgrid. In: *IEEE PowerTech'11, Trondheim, Norway*.

Epilogue

Having completed the material of the book, we are now able to summarize and have a useful discussion regarding its contents.

The book has been devoted to the presentation of two different modeling approaches, both aiming at system identification and adaptive control of nonlinear, probably unknown systems. Both approaches are placed among the so-called "intelligent" techniques, which are continuously evolving in the last years, seeking, among other, for the development of new approximation models and control techniques.

Part I of the book presented the recurrent neurofuzzy (NF) model. Based on the introduction of a new recurrent neurofuzzy approximation scheme, the system identification problem was addressed and coped with the use of a number of adaptive parameter estimation algorithms. The proposed approximation scheme has novel characteristics in the way that permits the interweave between concepts from high-order neural networks and fuzzy systems. This results in a faster and better system approximation, when compared to older approaches that use recurrent high-order neural networks or adaptive fuzzy systems alone.

Based on the NF approximation scheme several problems of direct and indirect control have been addressed and their behavior under different uncertainties and errors during the modeling procedure was theoretically investigated to conclude useful guidelines for the selection of the design parameters that are used. The existence of the control signal and the robustness of the closed-loop system has been assured by using a novel method, called *parameter hopping*, which suits better to the NF approximation model than the conventional projection approaches of the literature.

Different system classes that were investigated include SISO Brunovsky type systems, multiple input–multiple output square systems (number of input = number of states) and the class of nonsquare systems that permit the matrix pseudoinversion operation to occur. The control algorithms were designed for *affine in the control* systems and the a priori design information is kept minimum; it requires only the selection of some membership function centers related to the fuzzy part of the model and the selection of the algorithm's adaptation gains that should comply with

Y. Boutalis et al., *System Identification and Adaptive Control*,
Advances in Industrial Control, DOI: 10.1007/978-3-319-06364-5,
© Springer International Publishing Switzerland 2014

the theoretical investigation made. The derived NF controllers have been tested by simulations and experimental application on a number of selected benchmark and real-life problems.

Part II of the book presented the FCN model. Fuzzy Cognitive Networks stem from fuzzy cognitive maps (FCM), initially introduced to model complex socioeconomical systems. They both have a common background, namely the influence of the human expert in the design of the cognitive graph assigning concepts to nodes and the common repetitive way of computing their next state based on previous states. FCN, however, came to "cure" a convergence peculiarity of their ancestors, which fail, in many cases, to converge on equilibrium points that capture clear relationships between cause nodes and equilibrium effects.

The material of Part II started by pinpointing this convergence peculiarity. Then, a theoretical analysis, made for FCN equipped with sigmoid functions, produced conditions guaranteeing the existence of equilibriums. The conditions include the FCN weights and the inclination parameters of the sigmoid functions of the nodes. It was established that, when the conditions are met, the guaranteed equilibrium points depend on the values of the input nodes and have a clear connection with the FCN parameters (weights and inclination parameters).

The next step was the derivation of recursive parameters estimation algorithms providing the FCN parameters that are associated with the desired equilibrium points. Two approaches were presented. In the first one, where only the weights of the FCN are estimated, the derivation of the recursive algorithm is based on an underlying linear parametric FCN model. In the second one, where bot weights and sigmoid inclination parameters are estimated, the algorithm uses a bilinear parametric model. Both approaches use appropriate parameter projection laws that keep the estimated parameters within the bounds required by the conditions for the existence of equilibrium points.

With this development, FCN emerged as an operational extension of FCM which assumes, first, that it always reaches equilibrium points during its operation and second, it is in a close interaction with the system it models. Data derived from the operation of the system are used to estimate FCN parameters associated with different system's operational conditions. In the FCN's framework of operation, the knowledge acquired by this procedure is stored in a specially designed fuzzy rule database. It turned out that the entire framework can act as a fuzzy rule generator engine that can be used in decision making and system control. Selected applications on conventional benchmarks and potential real-life applications demonstrated that the FCN framework can be reliably used in inverse control schemes of affine in the control systems, as well as in decision making based on the knowledge stored in the fuzzy rule depository.

A common thread between the neurofuzzy and the FCN models is fuzzy logic and the influence of the experts. In the first model, fuzzy logic and experts' knowledge are involved in the initial design of underlying fuzzy systems, approximating the various parts of the system model. However, this influence is in the sequel minimized because it is, to a great extent, replaced by the high-order neural networks (HONN) and its learning abilities. In the FCN approach, the expert's influence is kept important, mostly in the initial design of the FCN graph that captures probable relationships between the concept nodes. Fuzzy logic is also present, mainly in the form of the fuzzy rule database created by the FCN framework. However, these fuzzy rules do not rely on experts' definitions but on the intrinsic storage mechanism of the framework that builds the rules based on systems operations data.

Another common point in both modeling approaches is adaptability. The neuro-fuzzy model and the control laws that are derived based on it, are by design adaptive. All the presented algorithms are designed to adapt in system changes, based on sampled data. Likewise, the FCN framework may adapt to system changes by modifying its fuzzy rule database, based on system's operation data and the adapted FCN parameters.

Rigorous mathematical proofs support the development in both approaches. In the neurofuzzy approach, well-established paths of the adaptive estimation and the control literature were followed to derive estimation and control algorithms that guarantee the stability of the closed loop system. Moreover, issues of robustness are theoretically investigated and conditions for the operation of the algorithms and the range of design parameters to be selected are derived. From the theoretical analysis, the applicability of the derived control schemes is determined. They can be used for affine in the control systems assuming either a Brunovsky type form or a general multiple input–multiple state form, where each state is influenced by at least one input variable. The presented applications, mostly in engineering problems, supported the theoretical findings. More general system forms could be treated by extending the presented in this book material and constitute issues of further development. The engineer or researcher who wants to apply the neurofuzzy approach has now a clear understanding of its potential and its applicability.

In the FCN approach, the parameters estimation recursive algorithms were for the first time based on a rigorous stability analysis of the error estimation scheme, guaranteeing error convergence to a global minimum. The derived inverse control scheme, that was for the first time based on the FCN framework, is applicable on a large class of general form, affine in the control systems, in a relatively straightforward manner. This opens the road for applying FCN in clear engineering applications. However, in its current development the control algorithm is not tied up with proofs for the stability of the closed loop system, leaving enough space for future improvement. For safe implementation, the engineer should have an initial guess regarding the limits of the control signal that permit stable system operation. Due to its intrinsic fuzzy rule depository nature, the FCN framework can be alternatively used as a supportive decision-making tool of human operators. The applications presented in this book covered mainly engineering problems, wanting to emphasize that in its current development the FCN framework can reliably support traditional engineering applications. However, the FCM literature is full of nonengineering applications covering a wide variety of scientific areas. These areas constitute a potential field of applications of the FCN framework. In general, for complex systems for which there is the expert's advice for the construction of the cognitive graph, the use of the FCN framework will offer a clear advantage.

Index

Y. Boutalis et al., *System Identification and Adaptive Control,*
Advances in Industrial Control, DOI: 10.1007/978-3-319-06364-5,
© Springer International Publishing Switzerland 2014